Clinical and Radiological Examination of the Foot and Ankle

Siddhartha Sharma • Bedri Karaismailoglu
Soheil Ashkani-Esfahani
Editors

Clinical and Radiological Examination of the Foot and Ankle

The Path to Definitive Diagnosis

FARIL International

Springer

Editors
Siddhartha Sharma
Foot and Ankle, Biomechanics Experimentation & Research (FABER) Laboratory
Postgraduate Institute of Medical Education and Research
Chandigarh, Chandigarh, India

Foot and Ankle Research and Innovation Laboratory
Massachusetts General Hospital
Harvard Medical School
Boston, MA, USA

Soheil Ashkani-Esfahani
Foot & Ankle Research and Innovation Laboratory (FARIL)
Massachusetts General Hospital, Harvard Medical School
Boston, MA, USA

Department of Orthopaedic Surgery
Massachusetts General Hospital, Harvard Medical School
Boston, MA, USA

Bedri Karaismailoglu
Department of Orthopaedics and Traumatology
Istanbul University-Cerrahpasa
Istanbul, Istanbul, Turkey

FARIL - Foot and Ankle Research and Innovation Laboratory
Mass General Hospital - Harvard Medical School
Boston, MA, USA

CAST - Cerrahpasa Research Simulation and Design Laboratory
Istanbul University-Cerrahpasa
Istanbul, Turkey

ISBN 978-981-97-4201-1 ISBN 978-981-97-4202-8 (eBook)
https://doi.org/10.1007/978-981-97-4202-8

This Springer imprint is published by the registered company Springer Nature Singapore Pte Ltd.
The registered company address is: 152 Beach Road, #21-01/04 Gateway East, Singapore 189721, Singapore

Dedicated to our families

Foreword

It is with great pleasure that I provide the foreword to this comprehensive volume titled *Clinical and Radiological Examination of the Foot and Ankle: The Path to Definitive Diagnosis.* This body of work is authored by a distinguished group of international surgeons and scientists associated with the Foot & Ankle Research and Innovation Lab (FARIL) at Massachusetts General Hospital, Harvard Medical School, Boston, MA, USA, as well as the FARIL International Initiative that spans multiple continents.

Having dedicated almost three decades to the field of orthopedic surgery and to the intricate subspecialty of foot and ankle most specifically, I continue to be awed by the complexities involved in accurately diagnosing and managing many of the disorders affecting this part of our anatomy. This book arrives as an essential resource to help bridge the existing knowledge gaps and foster a deeper understanding of the current clinical and radiological perspectives impacting the care of challenging foot and ankle pathology.

The contributors to this volume are friends and colleagues from around the world who have shared their wealth of experience and specific areas of expertise to enrich the pages of this book. Their insights broach many of the disease states affecting the foot and ankle patients we collectively serve, rendering this an invaluable educational care tool and practical guide for musculoskeletal clinicians, radiologists, and researchers alike. Each chapter has been crafted to reflect the latest advancements and methodologies in the examination and treatment of specifically controversial or challenging foot and ankle problems.

This publication also serves as an up-to-date educational scaffold for those who are actively training in the fields of orthopedic or radiological medicine. It underscores the importance of a nuanced approach to patient care and emphasizes diagnostic techniques that are pivotal in developing effective treatment plans.

As we continue to make strides in medical and surgical sciences, the cross-specialty collaboration demonstrated by this book is evidence of the significant progress that can be achieved when working in unison. I am confident that *Clinical and Radiological Examination of the Foot and Ankle: The Path to Definitive Diagnosis* will prove to be a cornerstone text for the current and future generations of healthcare professionals dedicated to improving patient outcomes in musculoskeletal care.

Christopher W. DiGiovanni, MD
Department of Orthopaedic Surgery,
Massachusetts General Hospital,
Harvard Medical School
Boston, MA, USA

Foot & Ankle Research and Innovation
Laboratory (FARIL), Massachusetts General Hospital,
Harvard Medical School
Boston, MA, USA

Preface

The examination of the foot and ankle represents a unique challenge in the field of orthopedic surgery and musculoskeletal medicine. This intricate region, with its delicate balance of bones, tendons, ligaments, and joints, is susceptible to a myriad of pathologies that can manifest as complex disorders affecting mobility and quality of life. Hence, the journey to achieving a definitive diagnosis demands a comprehensive understanding of both clinical assessment and radiological interpretation.

Clinical and Radiological Examination of the Foot and Ankle: The Path to Definitive Diagnosis is a pioneering text designed to bridge the gap in resources dedicated to the examination of foot and ankle pathologies. Within these pages, we explore every facet of clinical and radiological evaluation, from fundamental principles to advanced diagnostic techniques, offering a systematic approach to understanding and diagnosing a wide spectrum of conditions affecting this crucial anatomical region.

This book is meticulously crafted to serve as a comprehensive guide for orthopedic surgeons, podiatrists, fellows, residents, and medical students interested in mastering the intricacies of foot and ankle examination. It integrates the latest evidence-based practices, providing detailed insights into the biomechanics, anatomy, and radiology essentials for accurate diagnosis and appropriate decision-making. Each chapter delves into specific pathologies, offering a blend of clinical wisdom and cutting-edge diagnostic methods.

The key features of this book include:

- A detailed exploration of relevant biomechanics, anatomy, and radiology fundamentals.
- In-depth coverage of physical examination techniques for common foot and ankle pathologies.
- Extensive collection of clinical photographs and radiological images, enhancing understanding and aiding in diagnostic accuracy.
- Examination of emerging technologies shaping the future of foot and ankle diagnostics.

This book has come to life through the collaborative efforts of the FARIL International Initiative affiliated with Foot & Ankle Research and Innovation Lab (FARIL), Massachusetts General Hospital, Harvard Medical School. The contributors to this book are experts in the field from around the world, each sharing valuable insights gained through years of clinical practice and research. By combining their expertise, we have created a resource that not only elucidates the complexities of foot and ankle examination but also provides practical guidance for achieving definitive diagnoses.

We invite you to embark on this journey with us—to delve into the nuances of foot and ankle pathology and equip yourself with the knowledge and skills needed to navigate the path to definitive diagnosis. Whether you are a seasoned practitioner seeking to refine your diagnostic acumen or a trainee eager to explore this captivating specialty, *Clinical and Radiological Examination of the Foot and Ankle: The Path to Definitive Diagnosis* promises to be an indispensable companion on your professional voyage.

Chandigarh, Chandigarh, India — Siddhartha Sharma
Istanbul, Istanbul, Turkey — Bedri Karaismailoglu
Boston, MA, USA — Soheil Ashkani-Esfahani

Contents

About the Editors

Siddhartha Sharma is Additional Professor of Orthopedic Surgery at the Postgraduate Institute of Medical Education and Research, Chandigarh, India. He specializes in foot and ankle surgery and limb lengthening and deformity correction.

He completed his Ph.D. at PGIMER, Chandigarh, and pursued a postdoctoral fellowship at the Foot & Ankle Research and Innovation Lab (FARIL) at Massachusetts General Hospital, Harvard Medical School, USA. Currently, he serves as the FARIL International Ambassador from India. He is faculty for AO Trauma and AO Alliance, teaching on various AO basic, advances, and foot and ankle courses. He also holds the position of Secretary at the North Zone Orthopaedic Association (India) and Joint Editor for ASAMI India. He has received numerous research grants internationally and nationally and has authored over 100 publications. He serves as the Associate Editor of the *Journal of Foot and Ankle Surgery* (Asia Pacific) and reviewer for several journals of international repute.

Soheil Ashkani-Esfahani is Assistant Professor of Orthopedic Surgery and the Director of the Foot & Ankle Research and Innovation Lab (FARIL) at Massachusetts General Hospital, Harvard Medical School, USA.

He is a renowned clinician-scientist who has several highly cited publications in world-renowned journals, patents, techniques, and lectures published and presented at various international platforms and events. As the director of research and innovation, he focuses on developing more accurate, faster, and less invasive diagnostic technologies and treatment methods for orthopedic conditions. His work encompasses new imaging technologies such as portable handheld ultrasound, weight-bearing CT scan, needle arthroscopy, using artificial intelligence (AI) in orthopedics, 3D modeling, and device development, patient-centered decision support, and patient-specific management methods.

He has served as the reviewer for globally distinguished journals such as *Foot and Ankle International*, *The Spine Journal*, and Foot and Ankle Surgery. He has received various grants, awards, and accolades from national and international societies and institutions such as AOFAS, EFAS, AAOS, NSF, Harvard Medical School, MIT, and various orthopedic industries. Moreover, he has trained tens of medical students, fellows, residents, and scientists, some of whom are current leaders in the orthopedic world.

Bedri Karaismailoglu serves as a foot and ankle surgeon and Assistant Professor of Orthopedics and Traumatology at Istanbul University-Cerrahpasa. He is the Director of the Cerrahpasa Research, Design, and Simulation Laboratory (CAST), focusing on integrating 3D technologies into healthcare. Previously, he worked at the Foot & Ankle Research and Innovation Lab (FARIL) at Mass General Hospital—Harvard University, Boston, USA, and currently serves as its International Ambassador in Turkey.

He has authored numerous peer-reviewed publications, delivers lectures globally, and has received prestigious honors and grants from organizations such as EFAS, SICOT, ISAKOS, and AOFAS. Actively engaged in international orthopedic associations, he serves as the Vice-Chair of the Asia Pacific Orthopaedic Association Young Surgeon's Forum and contributes as a reviewer for esteemed orthopedic journals.

General Principles and Basics

1

Ankit Dadra, Mandeep S. Dhillon, and Siddhartha Sharma

1.1 Introduction

The foot and ankle serve as the fundamental support structure of the human body, playing a pivotal role in maintaining our characteristic upright posture, and acting as the central point for transmitting the entire body's weight during walking. Comprising 26 bones and encompassing 33 articulations, the foot is designed to offer flexibility on uneven surfaces and to absorb the shocks generated during movement [1].

The intricate workings of this remarkable mechanism often go unnoticed and unappreciated until a malfunction or issue arises. When faced with a foot and ankle pathology, conducting a thorough yet focused examination becomes imperative for attaining an accurate orthopedic diagnosis. It is crucial to understand that diagnostic tests and imaging techniques should act as supplementary tools, and not as replacements of clinical assessments. The clinician must delve into a detailed patient history and conduct a thorough examination, considering the patient's overall health rather than just concentrating on the affected anatomical site. While various methodologies for foot and ankle examination exist, it is essential for orthopedic surgeons to formulate their own approach, ensuring they do not overlook critical examination points. This chapter outlines the fundamental principles of orthopedic clinical examination and offers a general overview of the examination of the foot and ankle. Specific examinations focused on particular pathologies will be discussed in subsequent chapters.

A. Dadra · M. S. Dhillon
Foot and Ankle Biomechanics, Experimentation & Research (FABER) Laboratory, Department of Orthopedics, Postgraduate Institute of Medical Education and Research, Chandigarh, India

S. Sharma (✉)
Foot and Ankle, Biomechanics, Experimentation & Research (FABER) Laboratory, Postgraduate Institute of Medical Education and Research, Chandigarh, Chandigarh, India

Foot and Ankle Research and Innovation Laboratory, Massachusetts General Hospital, Harvard Medical School, Boston, USA

S. Sharma et al. (eds.), *Clinical and Radiological Examination of the Foot and Ankle*, https://doi.org/10.1007/978-981-97-4202-8_1

We also encourage our readers to develop personalized examination approaches that encompass all essential aspects. Diverse practitioners may approach patients with foot or ankle issues differently, but several key components should be an integral part of a comprehensive examination. These include, but are not limited to, assessment of gait, examination of footwear and any orthotic devices in use, and evaluation of the foot and ankle in both standing and sitting positions, following a systematic "look, feel, and move" schema.

1.2 Prerequisites

1.2.1 Consent, Rapport, and Other Important Aspects

Before commencing an examination, it is imperative for a clinician to obtain the patient's consent and the patient's right to decline the examination must be respected. Maintaining proper hand hygiene is of paramount importance both before and after examining each patient, serving as a crucial measure to prevent the transmission of infections.

Given that certain orthopedic tests may be provocative or uncomfortable, establishing a strong rapport with the patient becomes essential prior to the examination. Building rapport with the patient can be effectively nurtured during the process of gathering the patient's medical history. As a general rule, tests that may provoke discomfort should be conducted toward the latter part of the examination. In the case of pediatric patients or children, having their caregiver or parents present can create a more comfortable and cooperative examination setting.

When female patients are being examined by male examiners, it is essential to have a chaperone or a relative of the patient present during the examination. This practice ensures that a respectful and secure environment is maintained throughout the examination process.

1.2.2 Adequate Exposure

To conduct a thorough examination, it is essential that the patient is appropriately undressed. This involves removing socks and shoes from both feet. If the patient is wearing pants or trousers, they should either be entirely removed, taking due care to cover the genitalia, or rolled up to the mid-thighs to fully expose the knees, legs, ankles, and feet from all angles—front, back, and sides. This comprehensive exposure is particularly critical because certain deformities may be concealed if the area is inadequately exposed. For instance, an equinus contracture at the ankle may not be readily apparent if there is a recurvatum deformity at the knee, and the foot might appear to have a normal plantigrade position even in the presence of a significant deformity if proper exposure has not been achieved (Fig. 1.1).

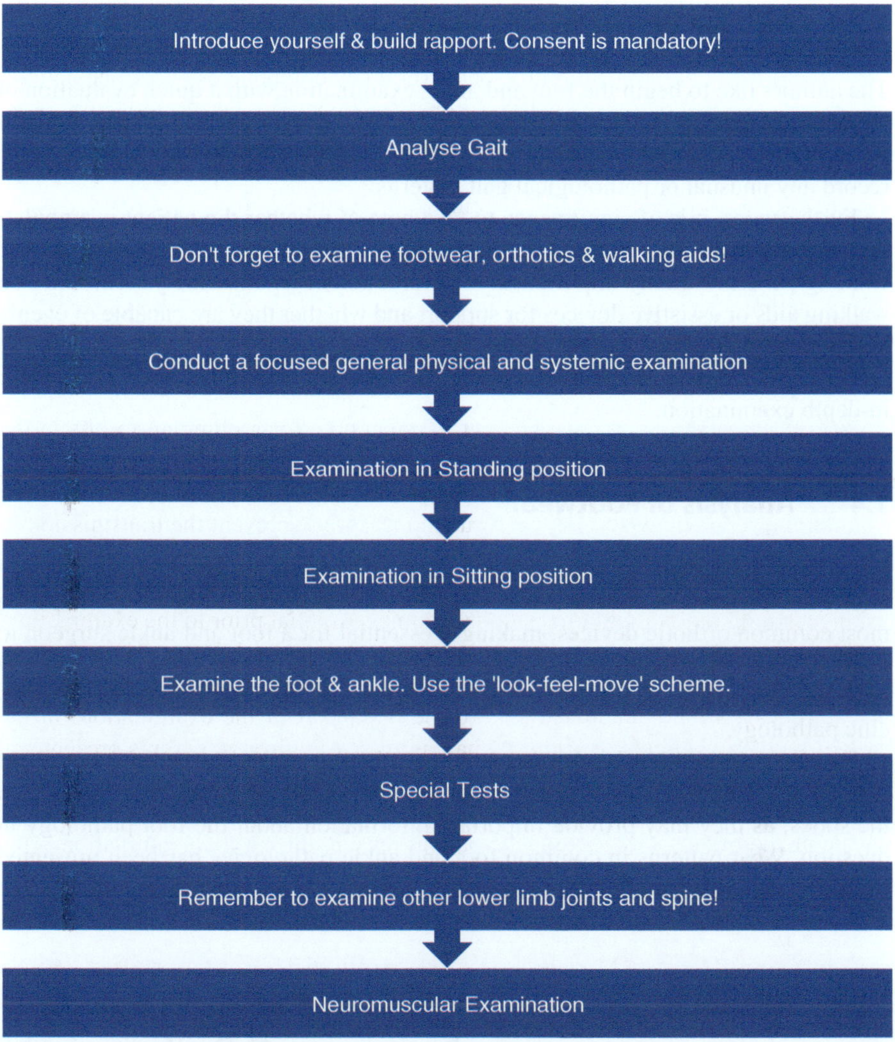

Fig. 1.1 The proposed scheme of foot and ankle examination

1.3 The Foot and Ankle Examination

After having obtained consent, as well as built a good rapport with the patient, we will now look at a general scheme of foot and ankle examination. As pointed out earlier, readers are encouraged to develop their own schemes, while ensuring that no important aspects are left out.

1.3.1 Analysis of Gait and Walking Aids

The authors like to begin the foot and ankle examination with a quick evaluation of the patient's gait. It is not within the scope of this chapter to discuss different gait types in detail. Nevertheless, it is essential to have the patient walk, as well as to record any unusual or pathological gait patterns.

Furthermore, it is of significance to take note of whether the patient is ambulating with a plantigrade foot or if there are any noticeable deformities affecting their walking pattern. Equally important is the evaluation of whether the patient relies on walking aids or assistive devices for support and whether they are capable of evenly bearing weight on both feet during the bipedal gait.

Any deviations or irregularities in the patient's gait should be subject to a more in-depth examination.

1.4 Analysis of Footwear

Examining the patient's choice of footwear is a frequently underestimated aspect of clinical assessment; however, it can yield valuable insights. A shoe is perhaps the most common orthotic devices, making it essential for a foot and ankle surgeon to possess familiarity with its components (Fig. 1.2). This knowledge also helps in recommending any necessary footwear modifications tailored to the patient's specific pathology.

Footwear may either be the source of a patient's complaints or hold important diagnostic clues. It is also imperative to take note of any atypical wear patterns on the shoes, as they may provide important information about the foot pathology in question. Wear patterns in common foot and ankle pathologies has been presented in Table 1.1. Furthermore, it is essential to observe specific shoe modifications, such

Fig. 1.2 Parts of a shoe

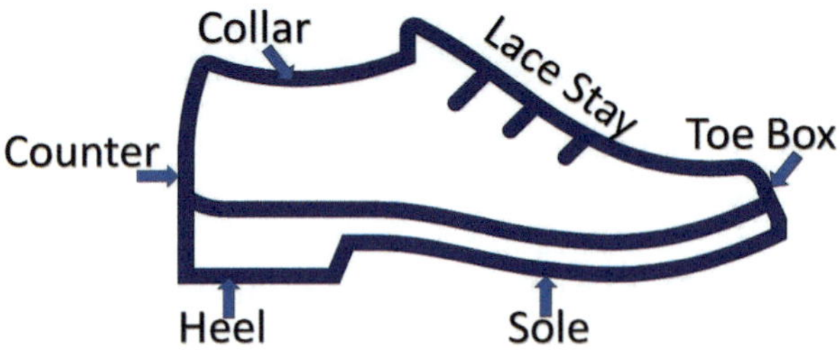

Table 1.1 Shoe wear patterns in different foot and ankle pathologies

Sr no.	Deformity	Wear patterns of shoe
1.	Cavovarus	Lateral border
2.	Flatfoot	Medial border
3.	Equinus	Front (toebox)
4.	Calcaneus	Heel

Table 1.2 Commonly used shoewear modifications

Shoewear modifications		
1.	Medial arch support	Pes planus
2.	Outer border raise	Cavo-varus deformity
3.	Metatarsal pads	Pes cavus/metatarsalgia
4.	Heel raise/sole lift	Limb length discrepancy/fixed equinus
5.	Rocker sole	Great toe arthritis/metatarsalgia, bursitis
6.	SACH (solid ankle cushion heel)	Hindfoot arthritis/ankle fusion, spurs
7.	Metatarsal bar	Metatarsalgia/Morton's neuroma/callosities/ plantar fasciitis

as lifts to address limb length discrepancies or accommodations for foot deformities. The examination should also encompass an evaluation of insoles or sole modifications when applicable (Table 1.2).

When assessing foot and ankle pathologies, it is also crucial to consider the use of any other orthotic devices by the patient. This includes items such as canes, axillary crutches, walkers, ankle–foot orthoses (AFOs), and foot inserts or insoles. These additional orthotic devices can have a significant impact on the patient's gait, stability, and overall orthopedic condition, and they may play a role in the diagnosis and treatment plan. Therefore, a comprehensive examination should encompass the evaluation of all relevant orthotic devices to ensure a thorough understanding of the patient's orthopedic needs.

1.5 General Physical Examination

A quick but focused general physical examination is invaluable and its importance cannot be overemphasized. Loss of eyebrows and typical facies in patient presenting with ulnar neuritis often bilateral and skin lesions is typical of leprosy [2]. "Moon" facies, protuberant abdomen, skin striae, and so on may well reveal chronic steroid intake as the etiology of avascular necrosis of the hip in a patient with hip pain [3]. Similarly, nicotine stains on the teeth and nails and atrophy of the nails indicate that the patient has been a chronic smoker and that his leg pain may be a result of vascular claudication due to Buerger's disease [4]. Bruising over the body, cigarette burn marks, and other "accidental" burn marks in a young child who presents with a history of multiple fractures may hint towards child abuse. Signs that may suggest spinal dysraphism such as the presence of a hairy patch, nevus (birthmark), lipoma (fatty mass), dimple, or sinus in the lumbar region should be looked for [5]. These physical findings might indicate underlying neurological issues that could contribute to foot and ankle problems. Beighton's scoring system should be employed to evaluate hyperlaxity of the joints in various parts of the body, which can provide valuable insights into the patient's condition and help guide the diagnosis and management of their foot and ankle deformities [6]. It is often helpful to revisit the general physical examination after examining the joint, in light of the etiological possibilities.

1.6 Systemic Examination

In addition to the musculoskeletal system, it is crucial to examine all other body systems. A routine and precise evaluation of the nervous, cardiovascular, respiratory, gastrointestinal, and urogenital systems should be developed. Referring to medical textbooks for detailed examination procedures of these systems is encouraged.

1.7 Examination of the Musculoskeletal System

In general, it is advisable to follow the *look-feel-move* scheme of examination for the musculoskeletal system.

1.7.1 Inspection (Look)

Adequate exposure of the area to be examined is essential to avoid missing critical information. It is essential to conduct a thorough inspection from multiple angles, encompassing views from the front, sides, and back in all cases. Specifically, when assessing the lower limbs, examinations should be performed both with the patient in a supine position and, whenever possible, in a standing position. This dual approach is vital as certain abnormalities become apparent only when the patient is bearing weight while standing. This should be standard practice whenever the patient's condition allows it, because weight-bearing assessments offer dynamic insights into joint function and can accentuate deformities, revealing their true extent and magnitude. This is particularly valuable when evaluating issues related to the lower limbs and their weight-bearing capacity (Fig. 1.3). Assessing the position of the heel with respect to the floor is best done from the back, allowing for the observation of any calf asymmetry as well as any swelling in the Achilles tendon or the posteromedial and posterolateral aspects of the ankle. Furthermore, from behind, it is possible to comment on the tibiocalcaneal angle, representing the alignment between the long axis of the tibia and the long axis of the os calcis. This angle is typically around 5° of valgus.

From behind, it is also feasible to evaluate the visibility of the forefoot, which is particularly relevant in cases of posterior tibial tendon disruption. This assessment includes checking for the "too-many-toes sign," where normally, about one and a half toes are visible [7]. If more toes are visible or there is an asymmetry between the two sides, it suggests a dropped arch or a varus/valgus deformity of the foot that requires attention. Detailed inspection of each joint will be covered in respective chapters.

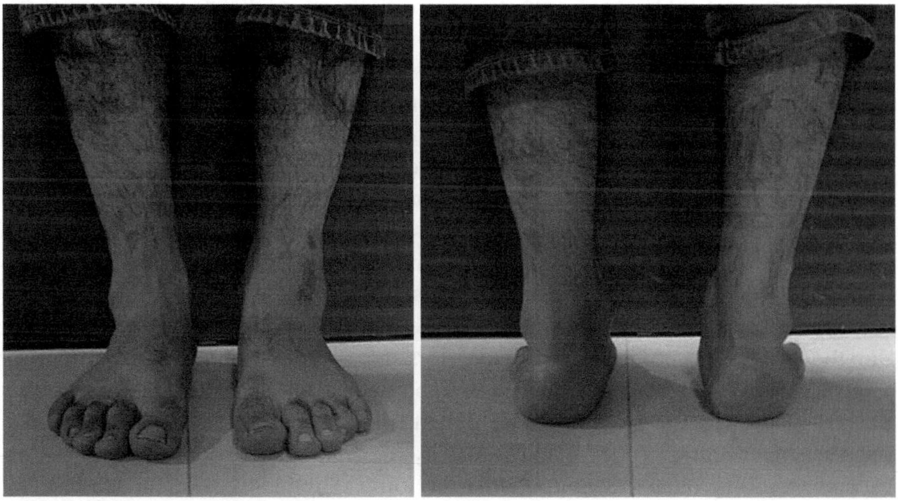

Fig. 1.3 Examination of the foot and ankle from the front and back in a healthy subject. Note is made of the plantigrade position of both feet. The heels are in physiological valgus, as noted from the back

1.7.2 Palpation (Feel)

It is important to ensure that your hands are warm before making contact with the patient, especially in the case of pediatric patients. Palpation begins with assessing joint warmth, followed by tenderness evaluation. Given the complexity of the foot, which comprises 26 bones and numerous articulations, conducting a meticulous examination can indeed be overwhelming. To navigate this complexity effectively, it is valuable to engage the patient in the process by asking them to pinpoint the specific area of pain. This approach not only aids in identifying the exact location of tenderness but also provides valuable diagnostic clues. For example, tenderness in the sinus tarsi can be a significant clinical sign, as it often suggests injury or arthritis affecting the posterior facet of the subtalar joint. Detecting tenderness over the anterior process of the calcaneum is equally important, as fractures in this region can frequently go unnoticed in routine X-ray examinations. Palpation techniques for each joint are discussed in subsequent chapters with relevant pathologies.

1.7.3 Movements and Measurements (Move)

Active and passive movements of the joint under examination should be checked. Comparison with the normal side helps to determine the extent of movement. The "arc of movement" is the sum of movements in any plane, accounting for any deformities. Ankle movement, both active and passive, is assessed with the patient seated, knees flexed and extended, preferably with their legs hanging off the examination table. The typical ankle range of motion comprises around 20° dorsiflexion and 40° plantarflexion [8]. If the ankle cannot reach a neutral position during plantarflexion, it is termed a "calcaneus" position.

When an ankle cannot be dorsiflexed to a neutral position, it is described as being in equinus. In cases where the Achilles tendon complex appears to contribute to this deformity, it becomes crucial to identify whether the gastrocnemius or soleus muscle is responsible for the limited dorsiflexion. This differentiation can be achieved through the Silfverskiold test [9]. It involves assessing ankle dorsiflexion first with the hip and knee extended and then with the hip and knee flexed. If dorsiflexion improves with knee flexion, it suggests a gastrocnemius contracture.

When evaluating passive movement in the foot and ankle, it is important to position the forefoot in supination to rule out dorsiflexion at Chopart's and the midtarsal joints. Additionally, inverting the heel to stabilize the subtalar joint helps ensure that the passive movement assessment is isolated to the ankle joint itself. This approach enhances the accuracy of the examination by focusing on the specific joint of interest and excluding movements from adjacent joints. Subtalar movement can be evaluated with the foot hanging off the edge of the examination table or with the patient lying face down. Specific joint movements are discussed in respective chapters.

1.7.4 Special Tests

Each joint or anatomical area has unique clinical tests for diagnosing various pathologies. The selection of relevant "special" tests depends on the differential diagnoses considered such as the Coleman block test for cavovarus foot, single heel raise test for tibialis posterior dysfunction, and Thompson test for Achilles tendon rupture. Specific tests for each joint are discussed in respective chapters.

1.7.5 Neuromuscular Examination, Vascular Status, and Regional Lymph Nodes

A detailed neuromuscular examination is crucial for assessing foot and ankle conditions. This assessment includes evaluating motor power and sensations using the MRC grading system, checking peripheral pulses, and examining regional lymph nodes. Assessing major muscle groups is critical, with a primary focus on ankle dorsiflexors, plantar flexors, foot evertors, invertors, and toe flexors and extensors. These muscle groups are integral to foot and ankle function, influencing stability, mobility, and balance, and it is essential to check for strength, coordination, and flexibility in these muscles. Any weakness, asymmetry, or limited range of motion may aid in guiding appropriate treatment and rehabilitation strategies for individuals with foot and ankle pathologies. When patients report neurological symptoms in the foot, it is vital to rule out lumbar nerve root tension. Assessing for clonus helps distinguish between upper and lower motor neuron lesions, and sensory evaluation aids in identifying patients at risk for ulceration and differentiating between conditions such as polio and spina bifida, both involving lower motor neuron lesions.

Besides considering the possibility of distant neurological diseases affecting the foot and ankle, it is also essential to be aware of specific entrapment neuropathies in this area that can cause symptoms such as numbness, tingling, and muscle wasting. Tarsal tunnel syndrome, resulting from posterior tibial nerve compression, causes plantar pain, paresthesia, and numbness. Deep peroneal nerve entrapment, often beneath the inferior extensor retinaculum, is linked to trauma and causes dorsum foot pain extending to the first web space. Superficial peroneal nerve entrapment occurs when the nerve exits the lateral compartment, leading to calf pain and numbness on the foot's dorsum. Palpating the common peroneal nerve around the fibula's neck is essential when deep or superficial peroneal nerve entrapment is suspected. Other entrapment neuropathies, such as sural nerve entrapment and lateral plantar nerve branch entrapment, should be considered if common causes are ruled out.

1.7.6 Examination of Adjacent Joints

It is imperative to examine joints neighboring the affected joint as part of a comprehensive evaluation. The primary pathology can either impact multiple joints or stem from an adjacent joint. For instance, hip pain can result in flexion deformities in the hip and knee, while pain in the knee may actually be referred from the hip and pain in the leg may originate from a prolapsed intervertebral disc. Rheumatoid arthritis may result in flexion deformities of the hip and knee, which can alter the foot and ankle biomechanics. In essence, examining adjacent joints is not merely a suggestion but a crucial component in ensuring precise diagnoses and formulating effective treatment strategies in orthopedic practice.

1.8 Conclusion

In conclusion, when faced with foot and ankle pathology, a thorough examination is imperative for accurate orthopedic diagnosis. While diagnostic tests and imaging techniques have their place, they should complement, not replace, clinical assessments. It is vital to engage patients in the examination process, considering their overall health and not just focusing on the affected area. The proposed scheme of examination, covering gait analysis, footwear assessment, general physical examination, systemic examination, and musculoskeletal examination, provides a comprehensive approach. Orthopedic surgeons must develop their own personalized examination methods while ensuring that they do not overlook essential examination points. Overall, a meticulous examination is essential to unravel the intricate complexities of foot and ankle pathologies and guide effective treatment strategies.

References

1. Manganaro D, Dollinger B, Nezwek TA, Sadiq NM. Anatomy, bony pelvis and lower limb, foot joints. Treasure Island, FL: StatPearls Publishing; 2022.
2. Dy CJ, Mackinnon SE. Ulnar neuropathy: evaluation and management. Curr Rev Musculoskelet Med. 2016;9(2):178–84.
3. Chan KL, Mok CC. Glucocorticoid-induced avascular bone necrosis: diagnosis and management. Open Orthop J. 2012;6:449–57.
4. Rivera-Chavarría IJ, Brenes-Gutiérrez JD. Thromboangiitis obliterans (Buerger's disease). Ann Med Surg (Lond). 2016;7:79–82.
5. Jindal A, Mahapatra AK, Kamal R. Spinal dysraphism. Indian J Pediatr. 1999;66(5):697–705.
6. Malek S, Reinhold EJ, Pearce GS. The Beighton score as a measure of generalised joint hypermobility. Rheumatol Int. 2021;41(10):1707–16.
7. Bubra PS, Keighley G, Rateesh S, Carmody D. Posterior tibial tendon dysfunction: an overlooked cause of foot deformity. J Family Med Prim Care. 2015;4(1):26–9.
8. Brockett CL, Chapman GJ. Biomechanics of the ankle. Orthop Trauma. 2016;30(3):232–8.
9. Singh D. Nils Silfverskiöld (1888-1957) and gastrocnemius contracture. Foot Ankle Surg. 2013;19(2):135–8.

Biomechanics of the Foot and Ankle

2

Bugra Gullu, Omer Subasi, and Lercan Aslan

2.1 Introduction

The foot and ankle play a crucial role in providing stability, adaptability, and efficient locomotion for the human body. This complex system is responsible for absorbing and translating forces while maintaining the whole-body stability. The complexity of these joints arises because their axes cannot be simply explained by coronal, sagittal, and axial movements. Nearly all joints in the foot and ankle have their own axis, thus resulting in the movements becoming combined. Understanding the biomechanics of the foot and ankle is essential for diagnosing and treating common pathologies and improving overall foot health. In this article, we will explore the structural organization and kinematics of the foot and ankle, as well as the passive stability and muscle control that contribute to its function.

B. Gullu
Department of Orthopaedics and Traumatology, Koc University School of Medicine, Istanbul, Turkey

O. Subasi
Foot & Ankle Research and Innovation Laboratory, Department of Orthopaedic Surgery, Massachusetts General Hospital, Harvard Medical School, Boston, MA, USA

L. Aslan (✉)
Department of Orthopaedics and Traumatology, Koc University School of Medicine, Istanbul, Turkey

Foot & Ankle Research and Innovation Laboratory, Department of Orthopaedic Surgery, Massachusetts General Hospital, Harvard Medical School, Boston, MA, USA

S. Sharma et al. (eds.), *Clinical and Radiological Examination of the Foot and Ankle*, https://doi.org/10.1007/978-981-97-4202-8_2

2.2 Functional Anatomy

2.2.1 Talocrural (Ankle) Joint

The talocrural (ankle) joint is a hinge-type joint between the dome of the talus, medial and lateral malleoli, and tibial plafond. The talus is shaped like a truncated cone, or frustum, with the apex directed medially [1]. The talus is 4.2 mm wider anteriorly than posteriorly (Fig. 2.1) [2].

Due to the greater width of the talus in its lateral portion, the diameters of the circles passing through the medial and lateral joint surfaces vary. Consequently, despite ankle joint movement primarily occurring in the sagittal plane, the axis of rotation is not fixed. There are variable rotation axes during plantar flexion and dorsiflexion of the ankle. During ankle dorsiflexion, the axis of rotation is inferior and lateral, whereas as the ankle moves into plantar flexion, it shifts medially [3].

The normal range of dorsiflexion (DF) at the ankle joint is approximately 10°–20°. The normal range of plantar flexion (PF) is around 45°–50°. Although clinically insignificant, there are ankle movements in the transverse and coronal

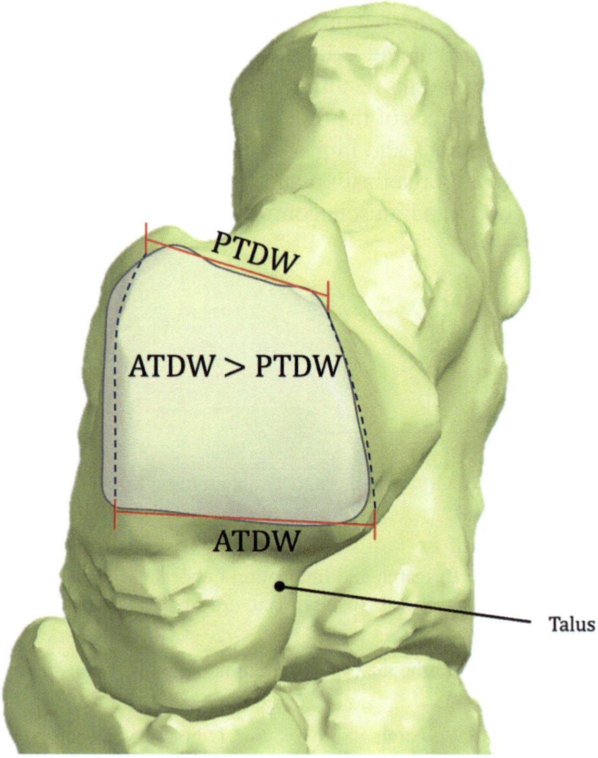

Fig. 2.1 The view of the talar dome that shows the anterior portion is wider than the posterior portion

planes as well. During ankle dorsiflexion, a 5°–6° external rotation is observed in the talus [4]. This rotation is reversed during plantar flexion. In the coronal plane, minimal inversion has been reported during both dorsiflexion and plantar flexion. In addition, the movement of the lateral malleolus also occurs, which is a result of the anterior portion of the talus dome being wider. With ankle dorsiflexion, the fibula translates laterally moving away from the tibia and maintaining the continuity of the mortise. Conversely, during plantar flexion, it moves medially toward the tibia.

2.2.2 Subtalar Joint

The subtalar joint is situated between the talus and the calcaneus: the anterior, middle, and posterior facets on the plantar surface of the talus and corresponding facets of the dorsal surface of the calcaneus make up the joint contact areas. Talus' convex head articulates with the concave surface of the calcaneus, allowing for complex multiplanar motion (Fig. 2.2).

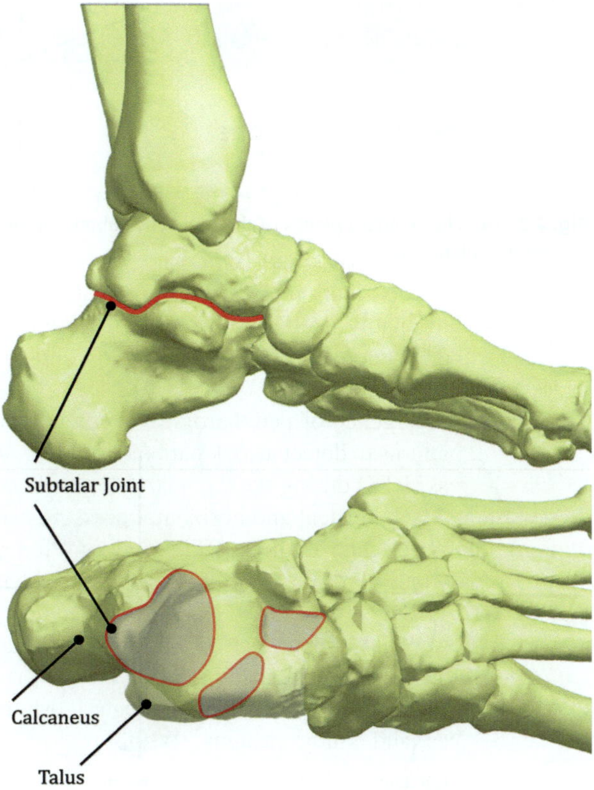

Fig. 2.2 Articulations between talus and calcaneus, consisting of anterior, middle, and posterior facets

The subtalar joint axis is often described as an "oblique axis" because it does not align with any one cardinal plane. Instead, it is an inclined axis that runs from posterior–superior to anterior–inferior through the talus. This axis is roughly oriented at an angle of 42° to the horizontal and about 16° to the sagittal plane [1, 5] (Fig. 2.3). This unique orientation facilitates motion in all three planes, with pronation, abduction, and extension occurring concurrently, just as supination, adduction, and flexion occur simultaneously. The range of motion of the subtalar joint is, approximately, 25°–30° for inversion and 5°–10° for eversion [5] (Fig. 2.4).

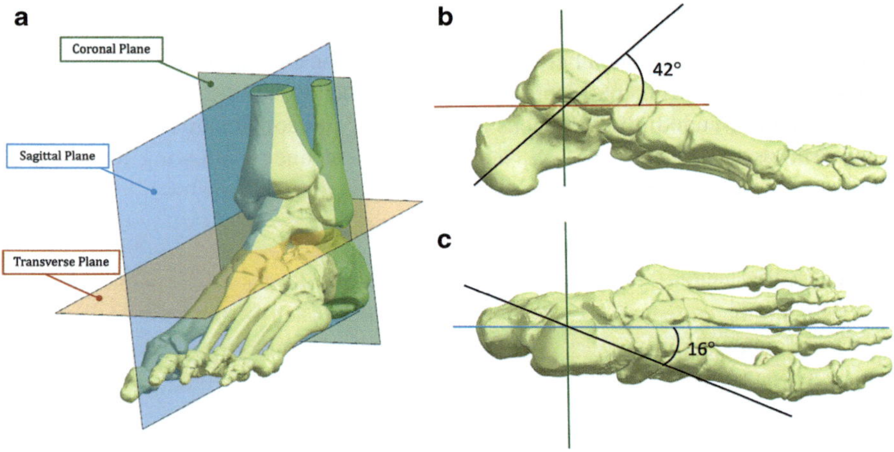

Fig. 2.3 (**a**) The standard planes of the foot. The subtalar joint's axis according to horizontal (**b**) and sagittal planes (**c**)

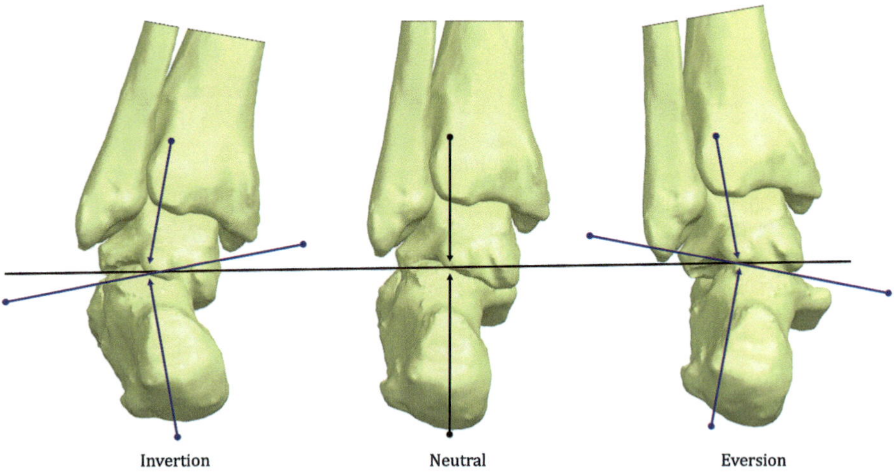

Fig. 2.4 Movement of the subtalar joint

2.2.3 Midtarsal Joint

The foot can be divided into three sections: hindfoot, midfoot, and forefoot (Fig. 2.5). The midtarsal joint, also known as Chopart's joint or transverse tarsal joint, is a complex articulation between hindfoot and midfoot. It consists of talonavicular and calcaneocuboid joints. The calcaneocuboid joint lies on the lateral side of the foot facilitates side-to-side movement and contributes to the transverse arch of the foot. The talonavicular joint is positioned on the medial side and assists in creating the medial longitudinal arch.

The midtarsal joint's movement is characterized by a combination of rotations around two primary axes: the longitudinal axis and the oblique axis. The longitudinal axis of the midtarsal joint runs from posterolateral to anteromedial. This axis slopes upward anteriorly 15° and is medially deviated by 9° [6]. The movement of the longitudinal axis is that of pronation–supination. Some abduction occurs in combination with pronation and, likewise, some adduction with supination. The oblique axis runs from posterolateral to anteromedial as well, more vertically with respect to the longitudinal axis. The axis is oriented at 52° in the sagittal plane and 57° in the coronal plane (Fig. 2.6) [6]. The movement created by the oblique axis is that of plantar flexion–adduction and dorsiflexion–abduction combinations. The range of motion of the midtarsal joint is 8° of rotation around the longitudinal axis and 22° around the oblique axis. The gross motion is 5°–10° for inversion to 10°–15° for eversion.

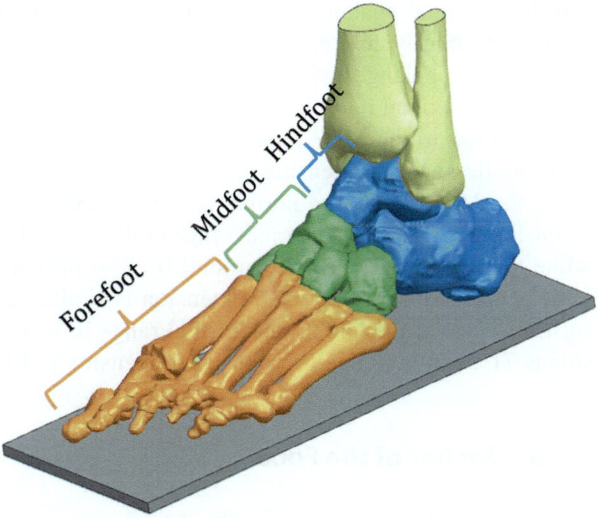

Fig. 2.5 The three sections of foot

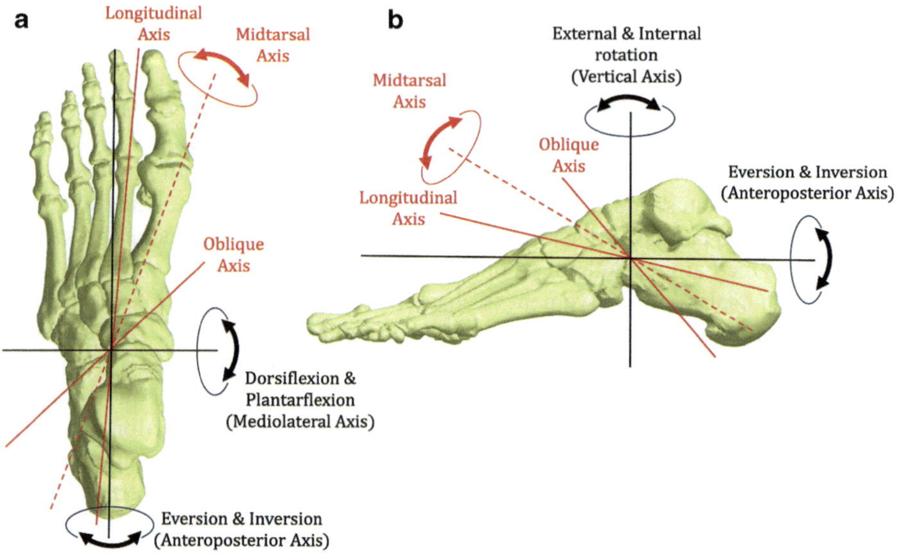

Fig. 2.6 Axes of the midtarsal joint on AP view (**a**) and lateral view (**b**)

2.2.4 Metatarsophalangeal and Interphalangeal Joints

The metatarsophalangeal (MTP) joints are located between the metatarsal bones and the proximal phalanges of the toes. The MTP joints are synovial hinge joints, allowing primarily flexion and extension. The interphalangeal (IP) joints are located between the phalanges of the toes. Each toe, except the hallux, possesses two IP joints: the proximal interphalangeal (PIP) joint and the distal interphalangeal (DIP) joint. Similar to the MTP joints, the IP joints are primarily synovial hinge joints, allowing flexion and extension movements.

At the MTP joints, the normal range of flexion is approximately 40°–60° and extension is 20°–30°. The normal range of flexion at the PIP joint is around 20°–70°, while the extension is usually around 0°. The normal range of flexion at the DIP joint is approximately 25° and extension is limited due to the shape of the joint surfaces [7]. For the hallux, the normal range of active MTP flexion is 23°, extension is 71°; active IP flexion is 46° and extension is 12° [8].

2.2.5 Arches of the Foot

The arches of the foot are formed by the arrangement of bones, ligaments, tendons, and muscles, collectively creating a dynamic and adaptable foundation. The human foot contains three main arches: the medial longitudinal arch, the lateral longitudinal arch, and the transverse arch (Fig. 2.7).

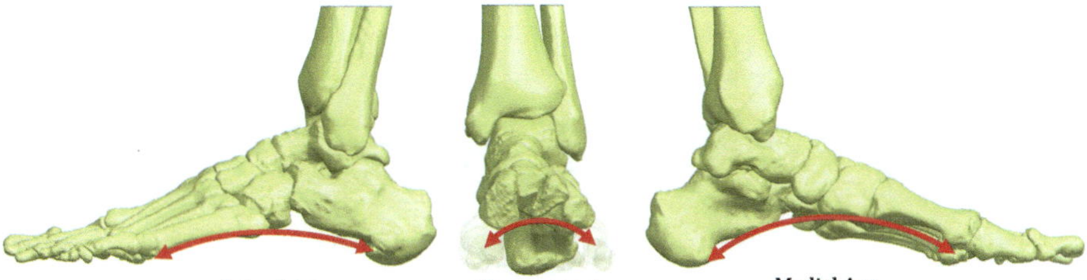

Fig. 2.7 Arches of the foot

The medial longitudinal arch is the most prominent and recognizable arch of the foot. It spans from the heel to the forefoot and consists of several key components. The calcaneus forms the posterior base of the arch, while the talus rests on the calcaneus and acts as the keystone; navicular, cuneiforms, and first, second, and third metatarsals form the rest of the arch. It is supported by the soft tissues of the spring ligament (plantar calcanea navicular ligament), deltoid ligament, posterior tibial tendon, plantar aponeurosis, and flexor hallucis longus and brevis muscles.

Two models have been proposed to explain the medial longitudinal arch's load-bearing mechanism: the Truss and Beam [9]. The Truss model suggests that the arch is a triangular structure, consisting of two columns and a cord at the base. The structure connecting the bottom two corners of the triangle is the plantar fascia, and the point at the top where the force originates from is the talus. When a force is applied, the load is distributed to the columns, causing them to separate. The tension in the plantar fascia prevents this separation and ensures an even distribution of the load. The Beam model, on the other hand, works like a classical Roman arch. With a vertical load, the convex side of the beam creates compressive forces, while on the concave side, tensile forces emerge, and the load is distributed across the beam.

The lateral longitudinal arch is located on the lateral side of the foot and is lower and flatter than the medial arch. It consists of the calcaneus, cuboid, and the fourth and fifth metatarsals. It is described as the only true Roman arch of the foot, the cuboid being the keystone [10].

The existence and nature of the transverse arch of the human foot are subjects of controversy. Some sources suggest that this arch runs horizontally across the midfoot, perpendicular to the longitudinal arch, and is formed by the tarsal bones, while others propose it is formed by the metatarsals [11–13]. Under non-weight-bearing conditions, the midfoot takes on an arcuate structure. However, in weight-bearing situations, this structure is not consistently observed [14]. As a result, the transverse arch of the foot and its functionality remain a subject of debate.

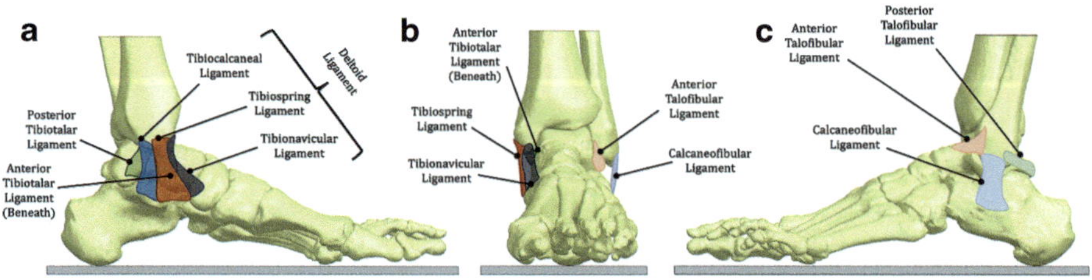

Fig. 2.8 Deltoid ligament components and lateral ligaments of the ankle. (**a**) Medial view. (**b**) Anterior view. (**c**) Lateral view

2.2.6 Ligaments

2.2.6.1 Medial Ligaments

The deltoid ligament complex is a group of ligaments located on the medial side of the ankle. It consists of four main components: tibionavicular ligament, tibiocalcaneal ligament, posterior tibiotalar ligament, and anterior tibiotalar ligament (Fig. 2.8a, b).

The deltoid ligament complex forms a triangular arrangement that supports the medial aspect of the ankle. The ligaments interact and overlap, contributing to a robust and intricate structure; the tibionavicular and tibiocalcaneal ligaments form the base of the triangle, originating from the medial malleolus and extending to the navicular and calcaneus, respectively. The anterior tibiotalar and posterior tibiotalar ligaments extend from the medial malleolus to the talus, with the anterior ligament positioned anteriorly and the posterior ligament situated posteriorly.

2.2.6.2 Lateral Ligaments

These include the anterior talofibular ligament (ATFL), calcaneofibular ligament (CFL), and posterior talofibular ligament (PTFL). ATFL connects the talus to the fibula. It is commonly injured in ankle sprains, where excessive inversion of the foot causes strain or tearing of the ligament. CFL connects the calcaneus to the fibula. This ligament provides stability against inversion and is often involved in lateral ankle sprains. PTFL links the talus to the fibula behind the ankle joint, contributing to overall lateral stability (Fig. 2.8b, c).

The configuration of the lateral ankle ligaments is characterized by their interplay and arrangement: the ATFL, positioned anteriorly, is the weakest of the three ligaments and is often the first to be injured during inversion injuries. The CFL, located between the ATFL and PTFL, provides additional lateral support to the ankle joint. The PTFL, posteriorly positioned, offers stability during extreme plantar flexion and inversion movements.

2.2.6.3 Syndesmotic Ligaments

These include the anterior tibiofibular ligament (ATiFL), interosseous ligament, posterior tibiofibular ligament (PTiFL), and transverse ligament. The syndesmotic ligaments work together to maintain stability and the proper alignment of the tibia

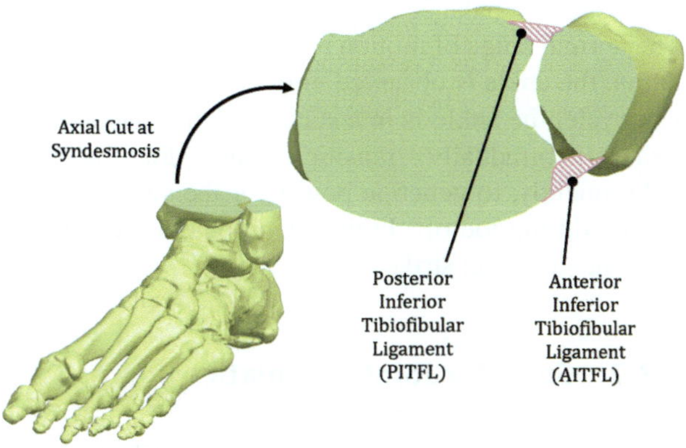

Fig. 2.9 Syndesmotic ligaments

and fibula; The ATiFL and PTiFL, along with the interosseous membrane, stabilize the syndesmosis by holding the tibia and fibula together. The transverse ligament wraps around the distal tibia–fibula joint, providing additional stability and helping to prevent separation (Fig. 2.9).

2.2.7 Plantar Aponeurosis

The plantar aponeurosis is a dense band of connective tissue located on the foot sole, spanning from the calcaneus to the base of the phalanx. It forms a robust, fibrous structure that provides support and contributes to various biomechanical functions. It has central, medial, and lateral components.

The central component starts from the calcaneus and divides into five branches distally, attaching to the metatarsophalangeal complexes. The lateral (peroneal) component starts from the medial calcaneal tubercle and divides into medial and lateral branches beneath the cuboid. The lateral band attaches to the base of the fifth metatarsal, forming the calcaneometatarsal ligament. The medial (tibial) component forms the fascia of the abductor hallucis muscle.

2.3 Biomechanics of the Foot and Ankle

2.3.1 Ankle Kinematics

The ankle joint is a complex component of the lower limb, enabling movement, stability, and weight-bearing. The ankle joint bears the majority of the body's weight during activities such as standing, walking, and running. Its structure and ligaments are designed to withstand substantial forces.

The ankle joint's dorsiflexion at heel strike and plantar flexion during toe-off are essential for normal gait patterns. These movements aid in shock absorption, stability, and propulsion. The joint's ability to control dorsiflexion and plantar flexion contributes to balance and stability during activities that involve uneven surfaces. The normal range of motion is 10°–20° DF and 40°–50° PF. However, during ambulation, the ankle is observed only in 10°–15° of PF and 10° of DF [15]. During the gait cycle, the ankle is in a state of plantar flexion at the heel strike. After the heel strike, it gradually transitions into dorsiflexion through the stance phase. Subsequently, to generate power for the swing phase, the ankle undergoes plantar flexion during toe-off. In the swing phase, the ankle is mostly seen in dorsiflexion until the next heel strike.

2.3.2 Subtalar Joint Kinematics

The subtalar joint actively absorbs and disperses forces during walking and running, reducing the impact on weight-bearing bones and joints. Its range of motion enables the foot to adapt to uneven surfaces, enhancing stability and balance during locomotion. The subtalar joint's motion is integrated into gait mechanics. It pronates during initial contact and midstance for shock absorption and supinates during late stance and toe-off for efficient propulsion. The proper subtalar joint function helps maintain foot and ankle stability, preventing excessive pronation or supination that can lead to overuse injuries.

The motions of the subtalar joint exhibit similarities in both weight-bearing and non-weight-bearing conditions. However, in non-weight-bearing situations, the talus tends to remain static, while the calcaneus moves relative to the talus. Conversely, in weight-bearing scenarios, the calcaneus remains stable, and it is the talus that moves relative to the calcaneus. It is important to note that the fundamental relative motion between the talus and calcaneus remains consistent, but the roles of each bone—active versus stationary—differ between the two conditions.

As previously discussed, the subtalar joint's rotation axis is oblique, allowing motion to occur in all three principal planes. Additionally, the subtalar joint possesses a distinctive characteristic attributed to the articulations between the talus and calcaneus. These facets engage in a screw-like mechanism, causing the talus to translate anteriorly on the calcaneus during supination [5].

2.3.3 Combined Motion of the Ankle, Subtalar, and Midtarsal Joint

The function of the subtalar joint cannot be fully realized without an appreciation for its effect on the ankle, transverse tarsal joints, and subsequent contact forces between the foot and the ground. The axis of rotation in the ankle enables the talus to move freely in the sagittal plane. Nevertheless, in the coronal and transverse planes, the ankle joint has severely limited motion. During ambulation, rotational movements in the pelvis, hip, and knee are transmitted to the foot and ankle through the rotation of the tibia. Given that the primary motions of the subtalar joint are inversion/eversion and abduction/adduction, the combination of the talocrural and subtalar joints affords freedom of motion in all three planes. The talocrural joint provides primarily for forward progression during locomotion, while the subtalar joint provides freedom for the lower leg to rotate in the transverse plane or rock side to side in the coronal plane without requiring the foot to move on the ground. This stable platform allows for essential functions during activities such as forward progression, maintaining balance, changing direction, and navigating uneven surfaces, all while the talus maneuvers around the fixed calcaneus.

The external rotation of the tibia results in the following movements in the subtalar joint: supination of 5.2°, internal rotation of 7.4°, and plantar flexion of 6.4°. Similarly, in the talonavicular joint, the external rotation of the tibia leads to the following movements: supination of 11.5°, internal rotation of 12°, and plantar flexion of 7.8° [16].

The external rotation of the tibia causes the calcaneus to assume a valgus position, aligning the axes of the subtalar and midtarsal joints in parallel, creating an open-packed position. This open-packed position allows for proper supination of the foot by facilitating midtarsal joint movements. On the other hand, internal rotation of the leg induces pronation in the subtalar joint, resulting in a varus position of the calcaneus. When the calcaneus is in varus, the axes of the subtalar and midtarsal joints are not parallel, and they are in a closed-packed position. This restricts midtarsal joint movements.

2.3.4 Tarsal and Tarsometatarsal Mechanisms

This open-packed position allows for proper supination of the foot by facilitating the tarsal mechanism described by Huson [17]. On the other hand, internal rotation of the leg induces inversion in the subtalar joint, resulting in an open-packed position of the calcaneocuboid joint. Considering the severely constrained mobility of the cubonavicular joint, the movements at the calcaneocuboid joint also affect the motion of the navicular. Inversion–abduction–plantar flexion of the calcaneocuboid joint, which occurs with the inversion of the calcaneus, leads to inversion–abduction–plantar flexion of the navicular bone as well. As a result of subtalar and midtarsal inversion, the medial longitudinal arch elevates and the lateral arch lowers, enabling the extremity to move upward.

The tarsometatarsal joint formed as the second metatarsal enters into the medial and intermediate cuneiforms like a socket [18]. Due to the bony congruity and strong ligamentous support, the mobility of the joint is highly constrained. Therefore, the movements of the tarsal bones occur around the second metatarsal. Inversion of the midfoot causes inversion of the second metatarsal as well. The inversion of the second metatarsal results in dorsiflexion at the third, fourth, and fifth tarsometatarsal joints on the lateral side. On the medial side, the first metatarsal comes into plantar flexion to maintain the continuity of the rising medial longitudinal arch with the inversion of the midfoot. Supination movement of the forefoot is achieved through this mechanism. Following the eversion of the midfoot, there is an eversion of the second metatarsal with plantar flexion in the lateral metatarsals and dorsiflexion in the first metatarsal. The tarsometatarsal mechanism is responsible for the pronation and supination movements of the forefoot.

2.3.5 Kinetics and Weight-Bearing

The forces on the ankle during walking are generated by a combination of body weight and the contraction of the muscles responsible for ankle movement. According to Stauffer's study, it has been proposed that the primary muscle group responsible for generating compressive forces in the ankle is the gastrocnemius–soleus complex [19]. It has been reported that the load on the talar dome of the foot is equal to the vector sum of the compressive forces generated by the gastrocnemius–soleus and the ground reaction force (Fig. 2.10).

The ankle joint is subjected to approximately five times the body weight during walking, and this load can increase to up to 13 times the body weight during running [20]. In the ankle joint, the tibia and talus are in constant contact, but the surface area in contact can vary depending on the ankle's position. The total weight-bearing surface of the ankle joint is reported to be 11–13 cm^2 [21]. The broader conical structure of the talus, particularly in the anterior part, may contribute to this variability. The load-bearing tibiotalar joint surface is reported to be approximately 4–5 cm^2 [22, 23].

Some studies reported that dorsiflexion of the ankle increases the contact surface area, while others suggest that it increases with plantar flexion [6, 23]. There are opposing views regarding the relationship between ankle position and the contact surface area of the tibiotalar joint. The changes in ankle position and the variations in pressure within the four compartments of the ankle joint have been studied by Michelson et al. [24]. An increase in pressure proportional to ankle dorsiflexion was observed between the medial malleolus and talus and lateral malleolus. In the medial tibiotalar joint, there was a relatively constant pressure observed from 20° plantar flexion to 10° dorsiflexion, with the pressure decreasing as dorsiflexion increased. In the lateral tibiotalar joint, on the other hand, there was increasing pressure observed as the ankle moved from plantar flexion to dorsiflexion.

It has been demonstrated by Lambert that one-sixth of the ankle load is shared by the fibula and the tibiofibular joint [21]. According to Wang et al., this share decreases with ankle varus and plantar flexion and increases with valgus and dorsiflexion [25].

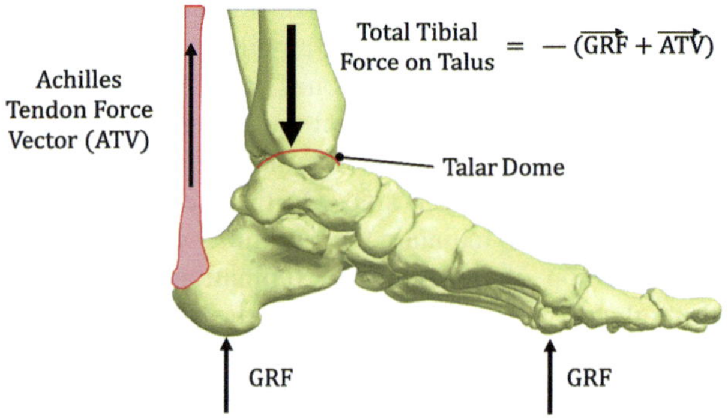

Fig. 2.10 Force vector sum on the talar dome

The tibia's rotation and the position of the ankle joint lead to different conformational changes in the foot under vertical loading. Loading during internal rotation of the leg results in valgus positioning of the calcaneus. This also results in the closed-pack position of the calcaneocuboid joint along with the eversion of the midfoot. The eversion of the midfoot leads to a lowering and elongation of the medial longitudinal arch. The tensioned plantar aponeurosis turns the foot into a rigid lever arm.

Under the load caused by standing, the distribution of pressure on the foot sole has been reported in Cavanaugh's study as 60% of the load on the heel, 28% on the forefoot, and 8% on the midfoot [26]. Historically, it has been proposed that the foot supports weight through a tripod consisting of the heel and the first and fifth metatarsal heads, with a transverse arch at the level of the metatarsal heads contributing to this weight distribution [13]. As previously mentioned, the existence of a functional transverse arch is a subject of debate. All metatarsal heads play a role in bearing weight. In the measurements by Jones, the second, third, fourth, and fifth metatarsal heads shared equal weight, while the first metatarsal head was reported to bear twice as much load [27]. Indeed, the positions of the leg and ankle can alter the distribution of load on the metatarsal heads. Ankle dorsiflexion increases the load on the first metatarsal head, while plantar flexion increases the load on the second through fifth metatarsal heads. Inversion resulting from hip adduction increases the load on the first metatarsal head, whereas eversion due to hip abduction increases the load on the second through fifth metatarsal heads.

The medial longitudinal arch, in both beam and truss models, effectively functions in weight-bearing. Under load, there is a tendency for the medial arch to flatten and both models' components resist this flattening. The effects of these models have been demonstrated by Hicks [28]. The distribution of a 100 lb. load was measured under three conditions: normal (without transections), with plantar aponeurosis transection, and with both plantar aponeurosis and abductor hallucis transection. Without transections, the load was distributed 60 lb. in the truss and 24 lb. in the beam, while 16 lb. were unaccounted for. With the plantar aponeurosis transected, 13 lb. were in the truss and 55 lb. were in the beam, while 32 lb. were unaccounted for. Lastly, when both the plantar aponeurosis and the abductor hallucis muscle were transected, 6 lb. were in the truss and 81 lb. were in the beam, while 13 lb. were unaccounted for. Plantar aponeurosis' significance in bearing weight is evident. Furthermore, in situations where the center of mass, such as during forward bending, moves to the anterior part of the ankle, the load-bearing mechanisms of the arches differ. The flattening effect on the arches and the strain on the plantar aponeurosis increase.

2.3.6 Stability of Ankle and the Foot

The ankle joint's stability relies on dynamic and static factors. Passive stabilizers of the ankle joint are the bony congruity of the tibiotalar joint, ligamentous support, and other soft tissues. Dynamic stabilizers are the musculature, ground reactions, and gravity. The position of maximum stability for the ankle is dorsiflexion. In

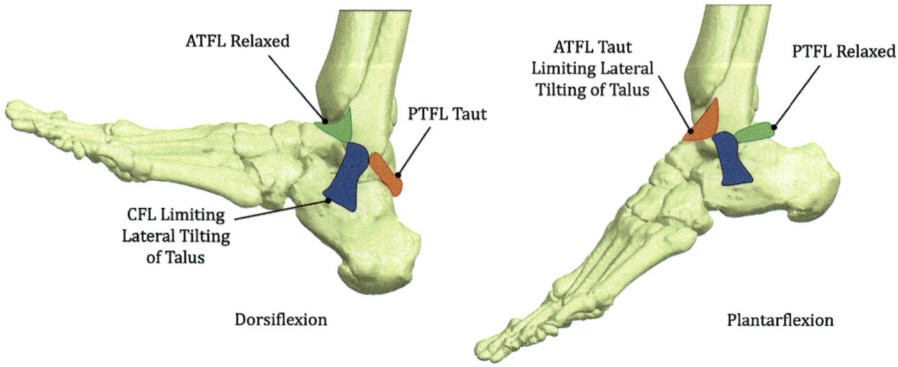

Fig. 2.11 Lateral ligaments of the ankle in plantar flexion and dorsiflexion

non-weight-bearing situations, coronal plane stability is provided by the malleoli and the lateral and medial ligaments. In the sagittal plane, balance depends solely on the ligaments. The fibrous tendon sheaths of the crural muscles also contribute to stability.

Dorsiflexion of the ankle causes the broader anterior joint surface of the talus to occupy the tibiotalar joint. This tends to separate the tibia and fibula. Syndesmotic ligaments limit this separation to 1 mm [25]. The contribution of syndesmotic components to stability is as follows: ATiFL 35%, deep PTFL 32%, interosseous ligament 21%, and superficial PTiFL 8% [29].

The lateral ligaments of the ankle collectively restrict ankle inversion and limit the lateral movement of the talus (Fig. 2.11). The ATFL is taut in plantar flexion and relaxed in dorsiflexion, serving as the primary ligament preventing anterior translation of the talus. The PTFL is taut in dorsiflexion and relaxed in plantar flexion, primarily restricting dorsiflexion, and preventing external rotation of the talus.

The calcaneofibular ligament (CFL) stabilizes both the ankle and subtalar joints. It is taut in dorsiflexion and relaxed in plantar flexion. The tension in the CFL varies based on its structural variations, as studied by Ruth, with different shapes such as oblique, horizontal, vertical, or fan-shaped [30]. When the CFL is horizontally oriented, it limits the valgus position of the calcaneus, while vertically oriented CFL limits the varus. When the ligament is obliquely oriented, its tension remains consistent throughout motion. CFL and ATFL are the primary restrictors against lateral tilt. CFL prevents lateral tilting of the talus in dorsiflexion, and ATFL prevents the lateral tilt of the talus in plantar flexion.

The deltoid ligament complex is the primary restrictor of ankle eversion. The deep deltoid fibers are the strongest ligaments and prevent lateral translation of the talus. They also restrict ankle dorsiflexion. The talocalcaneal component of the deltoid ligament limits eversion at the subtalar joint. The tibionavicular ligament and the anterior tibiotalar ligament restrict the plantar flexion of the foot and ankle. The anterior tibiotalar ligament also plays a role in transmitting the internal rotation of the tibia to the talus and resists external rotation. The posterior tibiotalar ligament is the major component of the deltoid ligament and is the strongest ligamentous structure among them. It serves as the primary restrictor of ankle dorsiflexion and posterior translation of the talus.

The load distribution of ankle stabilizers has been studied by Stormont and his colleagues [31]. The major contributions of ligaments and joint surfaces to stability vary based on ankle rotations, sagittal plane position, and weight-bearing conditions. Resistance to internal rotation is provided by ATFL (56%) and deltoid (30%). Resistance to external rotation is provided by CFL (65%) and PTFL (>20%). Resistance to inversion is provided by CFL (50%), ATFL (30%), and PTFL (%15). Resistance to eversion is provided mostly by the deltoid (%83). When under load, inversion and eversion resistance only come from the articular contact. Transection of the tibiocalcaneal fascicle of the superficial deltoid ligament results in a 43% decrease in the talar contact area, a 30% increase in peak pressures, and a 4-mm lateral shift according to Earll et al. [32]. This can increase stress forces in the talar joint surface and cause arthritis in a clinical setting.

The subtalar joint transitions to a closed-pack position with the valgus position of the calcaneus. In the closed-pack position, the posterior facet joint surface is in maximum contact, providing stability through bone congruence. The cervical ligament restricts the internal rotation of the calcaneus. The major ligament contributing to the stability of the subtalar joint is the talocalcaneal ligament of the canalis tarsi. This ligament primarily restricts the external rotation of the calcaneus. In a study where it was transected, the increase in subtalar joint movements was as follows: 21% in the horizontal plane (internal–external rotation), 16% in the frontal plane (pronation–supination), 57% in the sagittal plane (dorsiflexion–plantar flexion), where 43% of the increase was in dorsiflexion and 14% in plantar flexion. The talocalcaneal ligament of the canalis tarsi is considered to have a more significant stabilizing role than simply limiting external rotation [33].

The stability of the tarsometatarsal joints is provided by a network of ligaments and the joint surfaces. Dorsal, plantar, and interosseous ligaments connect the metatarsal bones to each other and to the tarsal bones. Among the interosseous ligaments, the Lisfranc ligament is the strongest. The Lisfranc ligament is the ligament between the medial cuneiform and the base of the second metatarsal and serves as the major stabilizer of the tarsometatarsal joint. The bony congruence of the second cuneometatarsal joint plays an important role both in transmitting movements from the midfoot to the forefoot and in maintaining stability. In a study conducted by Lakin et al., it was shown that the second and third tarsometatarsal joints bear the majority of the load [34].

2.3.7 Muscle Control

The tendons situated behind the axis of motion at the talocrural joint are responsible for ankle plantar flexion. Whereas the tendons positioned in front of the axis of the talocrural joint serve as the dorsiflexors of the ankle. During walking, the plantar

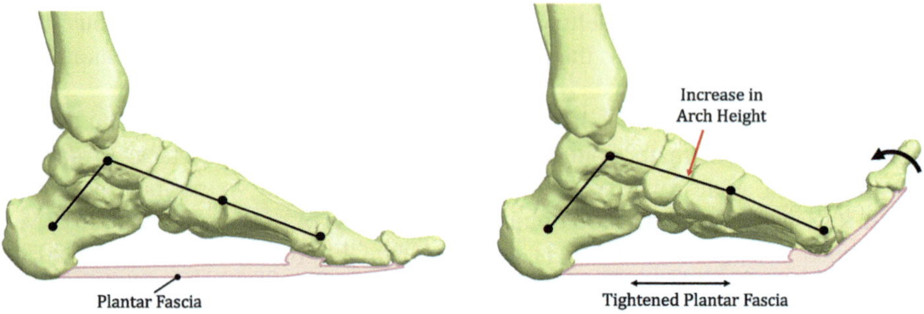

Fig. 2.12 Windlass mechanism

flexor muscles generate four times more energy compared to the dorsiflexor muscles. The muscles responsible for plantar flexion movement of the ankle are the gastrocnemius–soleus (triceps surae), flexor hallucis longus, flexor digitorum longus, tibialis posterior, peroneus longus, and brevis muscles. Approximately 90% of the energy for plantar flexion is provided by the triceps surae. The dorsiflexors of the ankle are the tibialis anterior, extensor digitorum longus, extensor hallucis longus, and peroneus tertius muscles. According to Lang and Wachsmuth, triceps surae have a 16.4-m kg work capacity, whereas the stronger dorsiflexor tibialis anterior has a 2.5-m kg [33].

The toes play a role in tensing the plantar aponeurosis and the skin on the sole during the push-off phase of walking. The motor movements of the toes play a role in the windlass mechanism, as described by Hicks [9]. In this mechanism, the metatarsal head acts as the drum, the proximal phalanx serves as the handle, and the cable wound around the drum represents the plantar aponeurosis. The stretching force generated on the plantar aponeurosis by the hyperextension of the hallux leads to the elevation of the medial arch, resulting in the flexion of the forefoot. This mechanism also operates in the opposite direction. When the arch lowers, it releases the windlass, and increased tension on the plantar fascia causes flexion at the metatarsophalangeal joints (Fig. 2.12).

2.3.8 Gait Cycle

The gait cycle begins with the heel contact and ends with the subsequent heel contact of the same foot (Fig. 2.13). Within this cycle, there are two phases: the stance phase occurs while the foot is in contact with the ground and accounts for about 60% of the gait cycle and the swing phase occurs while the foot is not in contact with the ground and approximately accounts for 40% of the cycle. In the swing phase, the ankle is in 20° of dorsiflexion with slight supination, and the ankle goes into a neutral position until the heel contact.

The stance phase can be further examined by dividing it into intervals. It begins with heel contact initially and continues with the loading response as the other

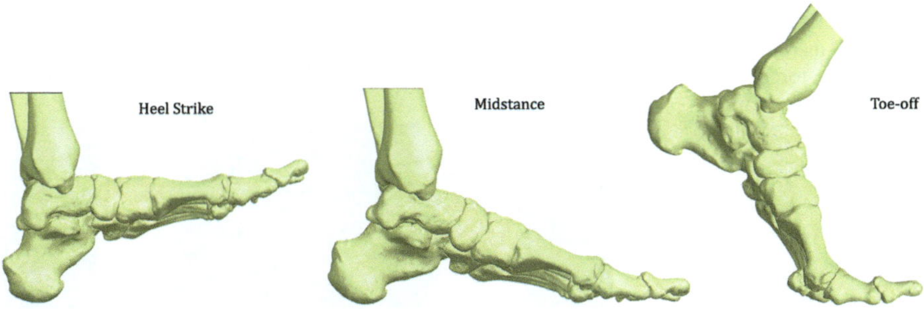

Fig. 2.13 Stance phase progression

extremity swings. As the center of mass aligns with the foot, midstance is reached, which is followed by the terminal stance with the lift-off of the heel and pre-swing with the lift-off of the toes.

During walking, after heel contact, the lateral aspect of the foot initially contacts the ground. Following loading, compressive forces on the cuboid stabilize the lateral longitudinal arch. Progressively, the entire sole is in contact with the ground, and with the internal rotation of the leg, the pronation of the heel places the calcaneocuboid joint in a closed-pack position. Simultaneously, dorsiflexion of the ankle stretches the plantar ligaments and plantar aponeurosis, which stabilize the tarsal joints. In this way, the foot creates a rigid lever arm just before heel-off.

The rise of the heel from the ground marks the emergence of static and dynamic stabilizing mechanisms. The windlass mechanism, facilitated by MTP hyperextension, stabilizes the medial arch. Additionally, the contraction of the peroneus longus and tibialis posterior muscles stabilizes the intertarsal and tarsometatarsal joints.

As the leg prepares for push-off, it undergoes external rotation, leading to subtalar joint inversion and the open-packed position of the calcaneocuboid joint. The foot becomes capable of supination for a powerful push-off. The ankle moves from a dorsiflexion of 15° to a plantar flexion of 20°, leading to toe-off, and the stance phase is completed.

2.4 Biomechanics in a Clinical Setting

2.4.1 Physical Examination

Assessing biomechanics is of great importance in the physical examination of the foot and ankle. Information related to biomechanics can be gathered at all stages of the examination. Footwear patterns should be inquired about, and whether they provide a proper fit should be evaluated. Skin inspection can reveal the presence of corns, as well as issues related to pressure distribution. The presence of structural abnormalities is assessed. The alignment of the entire lower extremity should be evaluated. The presence of deformities is assessed. Arch heights and dynamics are

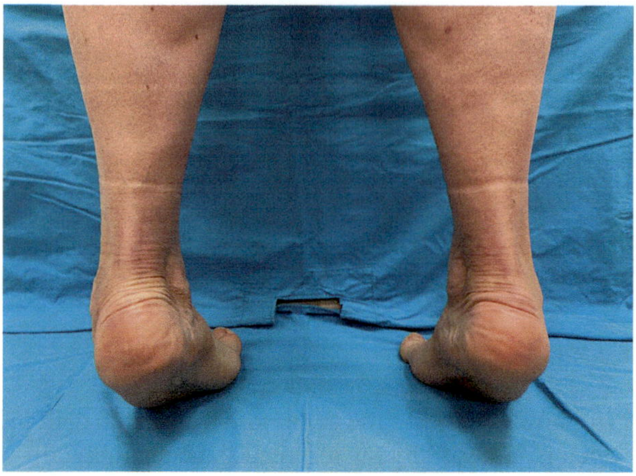

Fig. 2.14 When standing on tiptoe, medial arch height is restored, and the heel comes into varus

tested. The joint range of motion and the presence of instability are evaluated. Muscle strength is tested in the neurological examination.

The examination of midfoot supination, as described by Huson, provides insights into foot biomechanics [17]. Two physical examination findings in this triad include the observation/palpation of the "step" between the calcaneus and cuboid and the palpation of the talus head on the lateral aspect of the foot dorsum. The third finding is the widening of the tarsal sinus, which can be assessed through radiographic examination. The presence of the inversion triad is a normal finding in the subtalar joint and midfoot supination.

Arch abnormalities, such as pes planus, are frequently encountered pathologies in clinical practice. When evaluating a patient presenting with pes planus, one important assessment involves observing whether the heel everts when the patient rises onto their toes, indicating whether the condition is flexible or rigid. The single-leg heel raise test provides information about arch dynamics and can be indicative of tibialis posterior insufficiency. Another sign of pes planus is the "too many toes sign," which is observed from the rear and refers to the visibility of three or more toes beyond the lateral border of the foot [35]. Additionally, the "Jack's toe raise test" is an examination that assesses the windlass mechanism by hyperextending the toes (Fig. 2.14) [36]. Normally, the medial arch should rise during the test.

2.4.2 Biomechanical Disorders and Injuries

Pes planus occurs when the arches of the foot collapse, causing the entire sole of the foot to contact the ground. The majority of these deformities are seen in children. Most of the cases are flexible flatfeet. In some children, the development of the medial arch is observed until the age of 10. This condition can result from various factors, including genetic predisposition, weak arch-supporting muscles, or ligament laxity. Dysfunction of the joints or soft tissues of the medial longitudinal arch

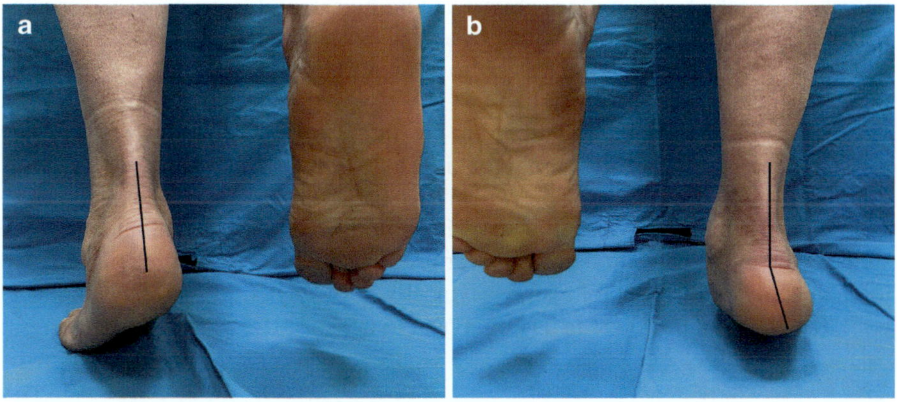

Fig. 2.15 In the case of tibialis posterior tendon deficiency, the heel cannot come into varus and stays in the valgus due to the excessive tension in the triceps surae. (**a**) Normal heel varus. (**b**) Tibialis posterior tendon deficiency

may result in acquired pes planus. Causes of the acquired flat foot deformity include excessive tension in the triceps surae, obesity, posterior tibial tendon dysfunction, or ligamentous laxity in the spring ligament, plantar fascia, or other supporting plantar ligaments. It may also result from a tight Achilles tendon or calf muscle contracture (Fig. 2.15).

In contrast to pes planus, pes cavus is an excessive curvature of the arches, leading to reduced surface area in contact with the ground. This biomechanical variation can lead to poor shock absorption and increased pressure on specific areas of the foot, potentially resulting in issues such as metatarsalgia and stress fractures. The exact pathophysiology is not known, but it is believed to be caused by untreated contractions or contractures of the peroneal muscles. It can also occur due to varus deformities of the calcaneus. In terms of etiology, neurological problems (such as cerebral palsy and hereditary neuropathies such as Charcot-Marie-Tooth disease), untreated pes equinovarus, or traumatic injuries can contribute to its development.

Certain biomechanical abnormalities have been associated with various overuse injuries. External tibial torsion is linked to Achilles peritendinitis, foot pronation abnormalities are associated with tibialis posterior tendinitis, pes planus/cavus is linked to flexor hallucis longus tendinitis, and forefoot valgus/varus is often associated with plantar fasciitis.

2.4.3 Clinical Implications

Understanding the biomechanics of the foot and ankle can help optimize treatments in clinical practice. For example, in ankle arthrodesis, excessive internal rotation of the ankle can lead to stress on the subtalar and midtarsal joints, causing pain. On the other hand, excessive external rotation can result in the foot turning on its medial edge. Similarly, in cases of excessive varus of the ankle, the patient may walk on the lateral edge of the foot, limiting subtalar and midtarsal joint movements and causing stress. The position of the ankle in sagittal is also important. Sufficient dorsiflexion should be achieved during arthrodesis to allow the foot to clear the ground during the swing phase. Excessive dorsiflexion, however, can reduce the plantigrade contact area and lead to pain. The optimum position for ankle arthrodesis is typically considered to be 5°–10° of external rotation, 5° of valgus, and neutral dorsiflexion. Similar to ankle arthrodesis, achieving proper fusion with correct alignment is crucial in the isolated diseases of the subtalar and midtarsal joints. Malalignment in the subtalar or midtarsal joint can lead to arthritis in the other joint. For the subtalar joint, the optimal position for fusion is typically 5° of valgus, while for the midtarsal joint, fusion should be done in a neutral position.

2.4.4 Future of Biomechanics

Foot and ankle biomechanics remain a subject with unresolved issues and controversial concepts. The challenges of assessing biomechanics under load in research studies, as well as the presence of anatomical variations, can lead to opposing views. Advances in technology have provided new tools that offer opportunities to better understand foot and ankle biomechanics. Finite element analyses are increasingly improving models of foot and ankle pathologies and providing data that can be useful in clinical practice.

Weight-bearing computed tomography (WBCT) is another example. WBCT offers the ability to obtain high-resolution images with the patient bearing weight, the possibility of three-dimensional reconstruction, and rapid image acquisition. With the advancement of technology, it also provides a low rate of radiation exposure, compact device size, and cost-effectiveness compared to conventional CT scans. Most importantly, WBCT allows for the three-dimensional examination of the foot and ankle under load.

Key Points

- The normal range of dorsiflexion at the ankle joint is approximately 10°–20°. The normal range of plantar flexion (PF) is around 45°–50°. During ankle dorsiflexion, a 5°–6° external rotation is observed in the talus.
- The subtalar joint axis is often described as an "oblique axis" because it does not align with any one cardinal plane. Instead, it is an inclined axis that runs from posterior–superior to anterior–inferior through the talus.
- The medial longitudinal arch is the most prominent and recognizable arch of the foot. It spans from the heel to the forefoot and consists of several key components. The calcaneus forms the posterior base of the arch while the talus rests on the calcaneus and acts as the keystone; navicular, cuneiforms, and first, second, and third metatarsals form the rest of the arch.
- The gait cycle begins with the heel contact and ends with the subsequent heel contact of the same foot. Within this cycle, there are two phases: the stance phase occurs while the foot is in contact with the ground and accounts for about 60%

of the gait cycle and the swing phase occurs while the foot is not in contact with the ground and approximately accounts for 40% of the cycle.
- The rise of the heel from the ground marks the emergence of static and dynamic stabilizing mechanisms. The windlass mechanism, facilitated by MTP hyperextension, stabilizes the medial arch. Additionally, the contraction of the peroneus longus and tibialis posterior muscles stabilizes the intertarsal and tarsometatarsal joints.

References

1. Inman VT. The joints of the ankle. Philadelphia: Williams & Wilkins; 1976.
2. Sarrafian SK. Anatomy of the foot and ankle: descriptive, topographic, functional. Philadelphia: Lippincott; 1993.
3. Barnett CH, Napier JR. The axis of rotation at the ankle joint in man; its influence upon the form of the talus and the mobility of the fibula. J Anat. 1952;86(1):1–9.
4. Close JR. Some applications of the functional anatomy of the ankle joint. J Bone Joint Surg Am. 1956;38-a(4):761–81.
5. Manter JT. Movements of the subtalar and transverse tarsal joints. Anat Rec. 1941;80(4):397–410.
6. Kelikian AS, Sarrafian SK. Sarrafian's anatomy of the foot and ankle: descriptive, topographic, functional. Philadelphia: Wolters Kluwer Health/Lippincott Williams & Wilkins; 2011. p. 507–643.
7. Myerson MS, Shereff MJ. The pathological anatomy of claw and hammer toes. J Bone Joint Surg Am. 1989;71(1):45–9.
8. Joseph J. Range of movement of the great toe in men. J Bone Joint Surg Br. 1954;36-B(3):450–7.
9. Hicks JH. The mechanics of the foot. II. The plantar aponeurosis and the arch. J Anat. 1954;88(1):25–30.
10. MacConaill MA, Basmajian JV. Muscles and movements: a basis for human kinesiology. Philadelphia: Williams & Wilkins Company; 1969.
11. Anthony D'Antoni GDM. Ankle and Foot. In: Standring S, editor. Gray's anatomy. Amsterdam: Elsevier; 2021. p. 1430–66.
12. Asghar A, Naaz S. The transverse arch in the human feet: a narrative review of its evolution, anatomy, biomechanics and clinical implications. Morphologie. 2022;106(355):225–34.
13. Kapandji IA. The physiology of the joints, vol. 2. Lower Limb: Churchill Livingstone; 1974.
14. Daentzer D, Wülker N, Zimmermann U. Observations concerning the transverse metatarsal arch. Foot Ankle Surg. 1997;3(1):15–20.
15. Stauffer RN, EYS C, Brewster RC. Force and motion analysis of the Normal, diseased, and prosthetic ankle joint. Clin Orthop Relat Res. 1977;127:189–96.
16. Lundberg A, Svensson OK, Bylund C, Selvik G. Kinematics of the ankle/foot complex—Part 3: influence of leg rotation. Foot Ankle. 1989;9(6):304–9.
17. Huson A. Functional anatomy of the foot. In: Janss MH, editor. Disorders of the foot and ankle: medical and surgical treatment. 1. Philadelphia: Saunders; 1991. p. 408–31.
18. Kelikian H, Valgus H. Allied deformities of the forefoot and Metatarsalgia. Philadelphia: Saunders; 1965.
19. Stauffer RN, Chao EY, Brewster RC. Force and motion analysis of the normal, diseased, and prosthetic ankle joint. Clin Orthop Relat Res. 1977;127:189–96.
20. Burdett RG. Forces predicted at the ankle during running. Med Sci Sports Exerc. 1982;14(4):308–16.
21. Lambert KL. The weight-bearing function of the fibula. A strain gauge study. J Bone Joint Surg Am. 1971;53(3):507–13.
22. Kimizuka M, Kurosawa H, Fukubayashi T. Load-bearing pattern of the ankle joint. Contact area and pressure distribution. Arch Orthop Trauma Surg. 1980;96(1):45–9.

23. Macko VW, Matthews LS, Zwirkoski P, Goldstein SA. The joint-contact area of the ankle. The contribution of the posterior malleolus. J Bone Joint Surg Am. 1991;73(3):347–51.
24. Michelson JD, Checcone M, Kuhn T, Varner K. Intra-articular load distribution in the human ankle joint during motion. Foot Ankle Int. 2001;22(3):226–33.
25. Wang Q, Whittle M, Cunningham J, Kenwright J. Fibula and its ligaments in load transmission and ankle joint stability. Clin Orthop Relat Res. 1996;330:261–70.
26. Cavanagh PR, Rodgers MM, Iiboshi A. Pressure distribution under symptom-free feet during barefoot standing. Foot Ankle. 1987;7(5):262–76.
27. Jones RL. The human foot. An experimental study of its mechanics, and the role of its muscles and ligaments in the support of the arch. American. J Anat. 1941;68(1):1–39.
28. Hicks JH. The foot as a support. Acta Anat (Basel). 1955;25(1):34–45.
29. Ogilvie-Harris DJ, Reed SC, Hedman TP. Disruption of the ankle syndesmosis: biomechanical study of the ligamentous restraints. Arthroscopy. 1994;10(5):558–60.
30. Ruth CJ. The surgical treatment of injuries of the fibular collateral ligaments of the ankle. JBJS. 1961;43(2):229–39.
31. Stormont DM, Morrey BF, An KN, Cass JR. Stability of the loaded ankle. Relation between articular restraint and primary and secondary static restraints. Am J Sports Med. 1985;13(5):295–300.
32. Earll M, Wayne J, Brodrick C, Vokshoor A, Adelaar R. Contribution of the deltoid ligament to ankle joint contact characteristics: a cadaver study. Foot Ankle Int. 1996;17(6):317–24.
33. Lang J, von Lanz T, Wachsmuth W. Bein und Statik. Berlin Heidelberg: Springer; 1972.
34. Lakin RC, DeGnore LT, Pienkowski D. Contact mechanics of normal tarsometatarsal joints. J Bone Joint Surg Am. 2001;83(4):520–8.
35. Young CC, Niedfeldt MW, Morris GA, Eerkes KJ. Clinical examination of the foot and ankle. Prim Care. 2005;32(1):105–32.
36. Harris EJ, Vanore JV, Thomas JL, Kravitz SR, Mendelson SA, Mendicino RW, et al. Diagnosis and treatment of pediatric flatfoot. J Foot Ankle Surg. 2004;43(6):341–73.

Clinical Anatomy of the Foot and Ankle

3

Ulaş Yavuz, Arın Celayir, and Bedri Karaismailoglu

3.1 Tibiotalar (Talocrural) Joint

Also known as the ankle joint, the talotibial joint is a hinge-type joint that is formed by talus and the distal end of the tibia and fibula. The normal range of dorsiflexion is approximately 10° when the knee is extended and 30° when the knee is flexed. Plantar flexion has a normal range up to 30°. The blood supply to the ankle joint is provided by the posterior and anterior tibial arteries, as well as the fibular arteries. Innervation of the joint is carried out by the deep fibular nerve, saphenous nerve, sural nerve, tibial nerve, and plantar nerves (Fig. 3.1) [1].

The medial and lateral ligament complexes play a crucial role in maintaining the stability of the ankle joint. Weakness or dysfunction of the fibular muscles can lead to varus instability of the joint, while posterior tibial tendon dysfunction may result in valgus instability.

3.1.1 Medial Collateral Ligament Complex of the Ankle Joint

The medial collateral ligament, also known as the deltoid ligament, is a robust triangular ligament that connects to the anterior and posterior borders of the medial

U. Yavuz · A. Celayir
Department of Orthopaedics and Traumatology, Istanbul University-Cerrahpasa, Istanbul, Turkey

B. Karaismailoglu (✉)
Department of Orthopaedics and Traumatology, Istanbul University-Cerrahpasa, Istanbul, Istanbul, Turkey

FARIL - Foot and Ankle Research and Innovation Laboratory, Mass General Hospital - Harvard Medical School, Boston, USA

CAST - Cerrahpasa Research Simulation and Design Laboratory, Istanbul University-Cerrahpasa, Istanbul, Turkey
e-mail: bedrikio@iuc.edu.tr

S. Sharma et al. (eds.), *Clinical and Radiological Examination of the Foot and Ankle*, https://doi.org/10.1007/978-981-97-4202-8_3

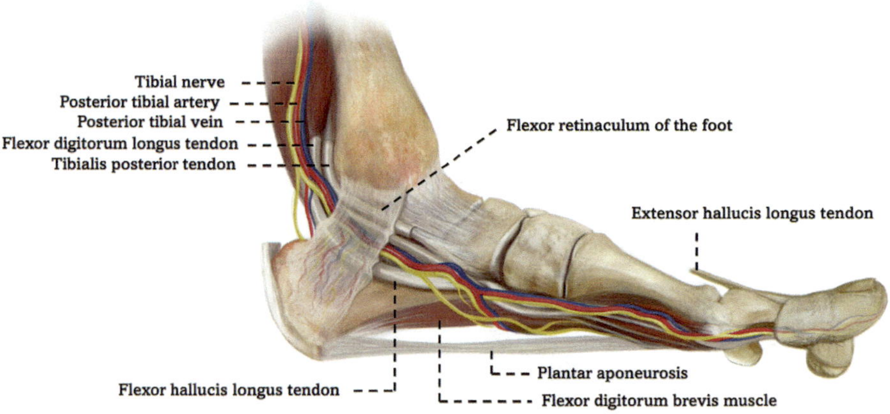

Fig. 3.1 Medial view of the ankle and foot showing the posteromedial neurovascular and tendon structures

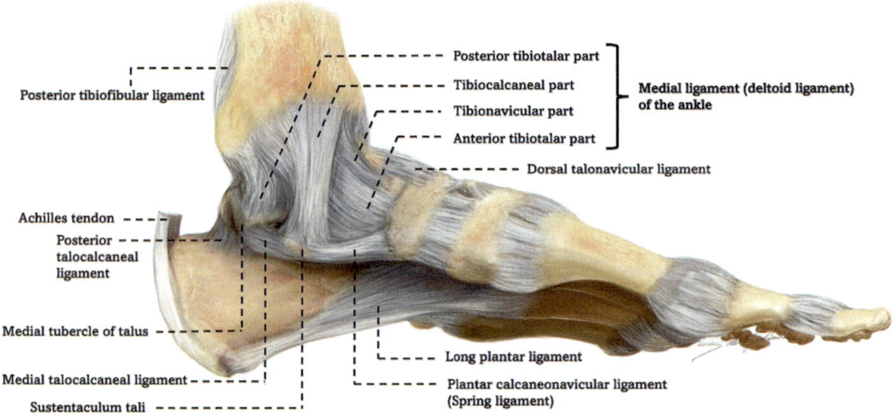

Fig. 3.2 Medial view of ankle showing ligamentous structures

malleolus. Its primary branches include the tibiocalcaneal, tibionavicular, anterior tibiotalar, and posterior tibiotalar ligaments (Fig. 3.2). It is composed of two sets of fibers: superficial and deep. The superficial layer consists of the tibiocalcaneal, tibionavicular, and a part of the posterior tibiotalar components, while the deep layer includes the anterior tibiotalar and a portion of the posterior tibiotalar ligaments. Functionally, the superficial layer prevents the talus from shifting into a valgus position, thus inhibiting the outward rotation of the hindfoot. On the other hand, the deep layer primarily acts as a barrier, preventing the external rotation of the talus [2]. The tibialis posterior and flexor digitorum longus tendons pass over this ligament. Tears of this ligament are commonly associated with fractures of the lateral malleolus and distal fibula [3–5].

A Pott fracture-dislocation of the ankle happens when there is a forceful eversion of the foot. This exerts tension on the remarkably strong medial ligament, often leading to the fracture of the medial malleolus or deltoid ligament rupture. Consequently, the talus bone moves laterally, resulting in fibular fracture. In some cases, if the tibia is pushed forward, the posterior edge of the distal end of the tibia is also sheared off by the talus, resulting in a trimalleolar fracture.

3.1.2 Lateral Collateral Ligament Complex of the Ankle Joint

The lateral collateral ligament of the ankle joint consists of three parts: the anterior talofibular ligament, posterior talofibular ligament, and calcaneofibular ligament (Fig. 3.3). Among these, the anterior talofibular ligament is the most commonly injured during sprains, while injuries to the posterior talofibular ligament are rare. Inversion sprains are the most common mechanism of injury for the ankle joint [6–8]. The anterior talofibular ligament has two fascicles: superior and inferior. Injury to the superior fascicle is associated with chronic symptoms due to limited healing capacity, given its intra-articular location [9].

Lateral ligament sprains are frequently observed in sports that involve running and jumping, with basketball being a notable example. The lateral ligament is more susceptible to injury compared to the medial ligament and plays a crucial role in resisting inward twisting at the talocrural joint. Avulsion fractures can occur below the ankle joint, where a fragment of bone is pulled off due to the force exerted by an attached ligament.

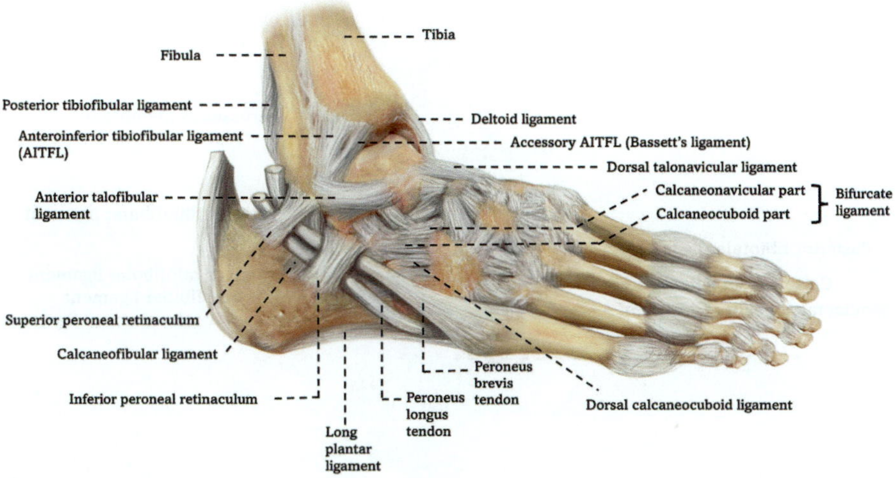

Fig. 3.3 Lateral view of the ankle and foot showing the ligamentous structures

3.1.3 Osteochondral Lesions of the Talus

Osteochondral lesions of the talus (OLT) frequently develop following ankle injuries or repetitive minor injuries, although the precise underlying causes are not fully understood. The primary symptoms often involve localized pain in the inner or outer ankle areas, which intensify during activities that involve bearing weight. This pain is often accompanied by tenderness and swelling.

Several studies have indicated that the most common location of the osteochondral lesions is situated at the medial rim of the talus, primarily within the middle third or posterior third of the anterior–posterior axis. A smaller portion of the lesions are observed at the lateral talar rim, predominantly within the middle third of the anterior-posterior direction [10].

3.2 The Inferior Tibiofibular Joint

The inferior tibiofibular joint is classified as a syndesmosis. It consists of anterior and posterior tibiofibular ligaments, transverse, and intraosseous ligaments (Figs. 3.3 and 3.4). The articulation occurs between the distal end of the tibia and fibula. Unlike other joints, the inferior tibiofibular joint lacks a fibrous capsule. In some individuals, an accessory anteroinferior tibiofibular ligament can exist, which might sometimes be symptomatic, causing anterolateral impingement of the ankle due to its pressure on the anterolateral talus (Fig. 3.3).

The anterior tibiofibular ligament descends laterally between the edges of the tibia and fibula [11]. The interosseous ligament is a continuation of the interosseous

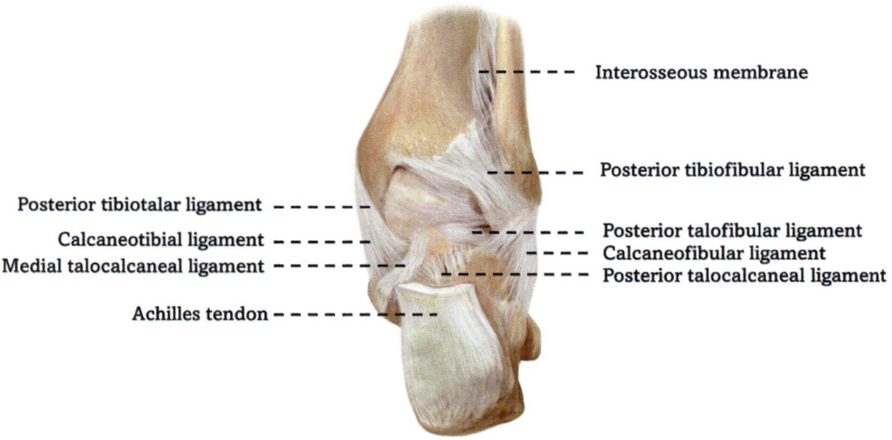

Fig. 3.4 Posterior view of the ankle and foot showing ligamentous structures

membrane. The posterior tibiofibular ligament is the strongest ligament of the joint. The transverse ligament is a round ligament that follows a horizontal path between the posterior tibia and the outer border of the distal lateral malleolar fossa, situated beneath the posterior tibiofibular ligament [2].

Arterial supply to the joint comes from the peroneal artery and the lateral malleolar branches of the posterior tibial artery. Innervation of the joint is provided by the deep peroneal and sural nerves. There are no significant muscles directly associated with this joint.

Anterior tibiofibular ligament avulsion fractures can occur at its attachment site on the tibia, which is also known as Chaput fractures. This ligament is important for the rotational stability of the ankle [2, 7, 12].

Posterior tibiofibular ligament is important for stabilizing against posterior translation. Interosseous ligament allows slight separation of the tibia and fibula during ankle dorsiflexion and limits excessive separation [7].

3.3 The Talocalcaneal Joint

It is formed by the articulation between the calcaneus and talus bones. It is a multiaxial joint, and there is a fibrous capsule between the bones. The main soft tissue support for this joint includes the lateral, medial, and interosseous talocalcaneal ligaments. The medial talocalcaneal ligament is connected to the deltoid ligament [13]. Innervation of the joint is provided by the tibial, medial plantar, and sural nerves. The posterior tibial artery, along with the tibial nerve and associated veins, runs posteriorly to the joint (Fig. 3.1) [14]. The fibularis longus and fibularis brevis muscles pass along the lateral border of the joint, with both tendons passing behind the lateral malleolus. The sural nerve is behind these two tendons (Fig. 3.5).

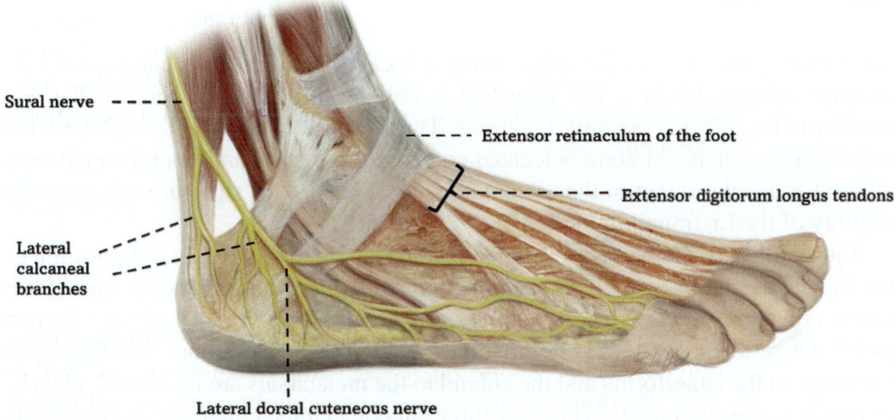

Fig. 3.5 Lateral view of the ankle and foot showing the sural nerve and its branches

3.4 Hindfoot

The calcaneus is the largest bone in the foot. Its long axis is directed anteriorly, superiorly, and laterally. The calcaneal pitch, lateral and anteroposterior talocalcaneal angles can be abnormal in deformities such as pes planus, pes cavus, and congenital clubfoot.

The Achilles tendon is the conjoint tendon of the gastrocnemius and soleus muscles. It attaches to the posterior surface of the calcaneus (Fig. 3.4). The Achilles tendon is responsible for plantar flexion of the ankle. The plantaris tendon originates in close relation to the lateral head of the gastrocnemius. It runs obliquely distally and medially and merges with the medial the Achilles tendon. Its insertion can vary, as it may blend with the Achilles tendon or attach to adjacent areas of the calcaneus. The retrocalcaneal bursa is located between the Achilles tendon and the posterior surface of the calcaneus. Inflammation of this bursa can cause symptoms such as pain and swelling, which is known as retrocalcaneal bursitis. A bony prominence can exist at the back and upper part of the heel referred as Haglund's deformity and is often associated with retrocalcaneal bursitis, a significant contributor to pain in the posterior heel region.

While the Achilles tendon holds the distinction of being both the strongest and largest tendon in the body, it also stands as the tendon most frequently susceptible to rupture. The Achilles tendon lacks a conventional tendon sheath; instead, it possesses a well-vascularized paratenon that serves as a conduit for the tendon's blood supply and enables smooth movement between the subcutaneous tissue and the posterior fascia. Blood supply to the tendon's proximal and distal regions is provided by the posterior tibial artery, while the midsection is nourished by the peroneal artery. Given its comparably limited blood supply, the midsection is particularly susceptible to degeneration and rupture [15].

3.5 Midfoot

The tarsometatarsal joint (Lisfranc joint) is located between the cuneiform bones and the metatarsal bases. The joint line is oblique, progressing from medial and anterior to lateral and posterior directions. The presence of an indentation where the intermediate cuneiform bone is located compared to medial and lateral cuneiforms allows for the placement of the second metatarsal head. This structure enhances the stability of the Lisfranc joint [16, 17].

The stability of the bones is established through the trapezoidal configuration of the metatarsal bases, which, in conjunction with their corresponding cuneiform bones, creates a sturdy arch commonly referred to as a "transverse arch" or "Roman arch," where the second TMT joint serves as the central keystone. The ligaments that connect the cuneiforms and the cuboid to the metatarsals are the dorsal, plantar, and interosseous ligaments. There are three cuneometatarsal interosseous ligaments located in the cuneometatarsal spaces. The strongest one starts from the lateral surface of the medial cuneiform and extends to the second metatarsal. This ligament is also known as the Lisfranc ligament and is crucial for tarsometatarsal joint stability [2].

3.6 Ligaments of the Foot

3.6.1 Calcaneonavicular Joint Ligaments and Acetabulum Pedis

The talar head articulates with a socket-like structure called the acetabulum pedis [18]. The components that form this structure are the posterior articular surface of the navicular, the anterior and middle articular surfaces of the calcaneus, the calcaneonavicular component of the bifurcate ligament, the superomedial calcaneonavicular ligament, and the plantar calcaneonavicular ligament. The acetabulum pedis provides flexibility during the relative movements of the talus, calcaneus, and navicular.

3.6.2 Superomedial Calcaneonavicular Ligament

It originates from the medial aspect of the sustentaculum tali and attaches to the superomedial part of the navicular bone [19]. It supports the head of the talus and prevents its inferomedial translation [20].

3.6.3 Inferior Calcaneonavicular Ligament (Spring Ligament)

It is aligned with the nonarticular inferior segment of the talus head. It originates from the inferior surface of the calcaneus (between the sustentaculum tali and the cuboid articulation surface) and attaches to the plantar surface of the navicular bone and helps to maintain stability [21].

3.6.4 Bifurcate Ligament (Chopart Ligament)

The V-shaped bifurcate ligament is formed by the calcaneonavicular and calcaneocuboid ligaments. It starts from the superior surface of the calcaneus, originating from the dorsolateral aspect (Fig. 3.3). The lateral calcaneocuboid ligament attaches to the dorsolateral surface of the navicular bone, while the medial calcaneocuboid ligament attaches to the dorsomedial surface of the cuboid bone [22]. The bifurcate ligament is a strong ligament and plays an important role in the stabilization of the midtarsal joints [23].

3.6.5 Talocalcaneonavicular Joint Ligaments

This structure comprises several ligaments, including the cervical ligament, tarsal canal ligament, lateral talocalcaneal ligament, posterior talocalcaneal ligament, medial talocalcaneal ligament, talonavicular ligament, and dorsolateral calcaneocuboid ligament (Fig. 3.4).

3.7 Tibialis Anterior and Posterior Tendons

3.7.1 Tibialis Anterior Tendon

The muscle originates from the lateral tibial condyle and the proximal one-third of the lateral tibial shaft [24]. The tendon begins at the level of the lower to middle one-third of the tibia [25]. It passes beneath the inferior extensor retinaculum [26]. It travels from the anteromedial part of the ankle to the medial border of the foot, where it makes a turn. It attaches to the inferomedial part of the first metatarsal base and the medial aspect of the first cuneiform [24]. It is encased within its own tunnel formed by the inferior extensor retinaculum. The tibialis anterior tendon is innervated by the deep fibular nerve. It functions to perform dorsiflexion and inversion movements [24].

3.7.2 Tibialis Posterior Tendon

It originates from the proximal to middle one-third of the tibia and the posterior surface of the interosseous membrane at this level [24]. The tendon divides into anterior, middle, and posterior components just anterior to the tubercle of the navicular bone. The major component is the anterior one, which attaches to the tubercle of the navicular bone, the inferior aspect of the medial cuneonavicular joint capsule, and the undersurface of the medial cuneiform bone (Fig. 3.1). The middle component continues distally on the plantar aspect of the foot and attaches to the middle and lateral cuneiforms, the cuboid bone, and over the peroneal canal. From there, the metatarsal extension of the tendon continues, passing dorsally to the peroneus longus tendon and attaching to the bases of the second, third, fourth, and sometimes fifth metatarsals. The posterior branch separates from the tendon before attaching to the tubercle of the navicular bone and attaches to the anterior aspect of the sustentaculum tali. The complex attachment pattern of the posterior tibial tendon allows it to firmly anchor to the plantar aspect of the foot [27]. It is innervated by the tibial nerve. Tibialis posterior is the primary muscle responsible for inversion of the foot. Tibialis posterior is the primary stabilizer of the medial longitudinal arch and in cases of its insufficiency, acquired flatfoot deformity may occur [24].

The flexor digitorum longus tendon courses from the posteromedial side of the talus and along the medial edge of the sustentaculum tali. The flexor hallucis longus tendon traverses the posterior talar groove and passes beneath the inferior aspect of the sustentaculum tali. The flexor digitorum longus tendon is adjacent to the tibialis posterior tendon. All three tendons then angle anteriorly, pass through the tarsal tunnel, and reach the plantar aspect of the foot (Fig. 3.1) [28].

Posteromedial tibial neurovascular bundle (tibial nerve and posterior tibial vessels) is located in a compartment adjacent to these tendons, in the intertendinous interval between the tibialis posterior, flexor digitorum longus tendons, and flexor hallucis longus tendon (Fig. 3.1) [29].

3.8 Peroneal Tendons

3.8.1 Peroneus Longus

Peroneus longus muscle originates from the proximal half of the fibula, including the fibular head and lateral aspect, as well as the deep fascia and anterior and posterior intermuscular septa. It courses distally along the lateral aspect of the fibula, posterior to the lateral malleolus. It passes through a tunnel formed by the superior peroneal retinaculum, located behind the malleolus (retromalleolar groove), alongside the peroneus brevis tendon (Fig. 3.3). It then turns anteroinferiorly in the lateral malleolar region. It passes below the trochlear process of calcaneus, and at the level of this process, it passes beneath the inferior peroneal retinaculum. It changes direction again, inferomedially, and passes along the lateral border of the foot, between the cuboid and the base of the fifth metatarsal. It then makes its third turn and enters the plantar aspect of the foot, attaching to the first metatarsal base and the first cuneiform. It is innervated by the superficial peroneal nerve and functions in plantar flexion of the ankle and eversion of the foot. It also supports the transverse and longitudinal arches of the foot [30]. In addition, peroneus longus tendon assists in stabilization of the medial column during the stance [31].

3.8.2 Peroneus Brevis

Peroneus brevis muscle originates from the lateral surface of the distal half of the fibula, as well as the anterior and posterior intermuscular septa. Its tendon travels along the posterior surface of the lateral malleolus and glides through the retromalleolar groove. It passes beneath the lateral malleolus and angles anteriorly. It then proceeds inferiorly, anteriorly, and laterally, passing over the trochlear process of the calcaneus [32]. It traverses the tunnel formed by the inferior peroneal retinaculum and attaches to the styloid process of the fifth metatarsal (Fig. 3.3). It is innervated by the superficial fibular nerve and functions in eversion of the foot. Together with the peroneus longus muscle, by contracting against a sudden inversion stress, it acts as a stabilizer of the ankle joint following a lateral sprain [31].

3.8.3 Peroneus Tertius

Peroneus tertius muscle originates from the distal one-third of the fibula. Its tendon lies lateral to the extensor digitorum longus. It passes beneath the superior and inferior extensor retinaculum. It courses anteriorly and laterally and attaches to the dorsal aspect of the fifth metatarsal base. It may be present in 88% of individuals. It is innervated by the deep fibular nerve, and it functions in dorsiflexion and eversion of the ankle [31].

3.8.4 Peroneal Tendon Pathology

Pathology of peroneal tendons may be associated with lateral hindfoot pain and functional aberrations. The pathologic conditions of these tendons may be divided into two groups: tendinopathy (including tendinosis, splits, and tears) and instability (including subluxation and dislocation). Tendinopathy may develop due to impingement by nearby bony structures such as the malleolar tip or lateral calcaneal wall. Tendinopathy may progress to degeneration of the tendons, resulting in splits or tears. Peroneal tendon instability usually develops due disruption of the superior peroneal retinaculum following a forceful ankle dorsiflexion and hindfoot inversion. These conditions are more commonly encountered in patients with coexisting ankle and foot pathology such as malalignment or patients with frequent athletic activities. In addition, pathology of the accessory bone os peroneum, a low-lying peroneus brevis muscle and the accessory peroneus quartus muscle, may contribute to peroneal tendon pathology [31].

3.9 Important Neurovascular Structures of the Foot Ankle

3.9.1 Superficial Peroneal Nerve

Superficial peroneal nerve courses through the anterolateral compartment of the leg, and in the lower third of the leg, it divides into the medial and intermediate dorsal cutaneous nerves of the foot dorsum. The cutaneous branches may anastomose with the lateral sural nerve. Its largest branch, the medial dorsal cutaneous nerve, travels medially toward the inner side of the foot. It further divides into lateral, middle, and medial branches, which provide innervation to the dorsum regions of the foot. Superficial peroneal nerve innervates the peroneus longus and brevis muscles and also provides the cutaneous sensation of the lateral aspect of lower leg. A superficial peroneal nerve lesion may result in weakness in foot eversion and sensory loss on the lateral aspect of the lower leg and dorsum of the foot. The nerve can be affected by compartment syndrome of the lower leg [33].

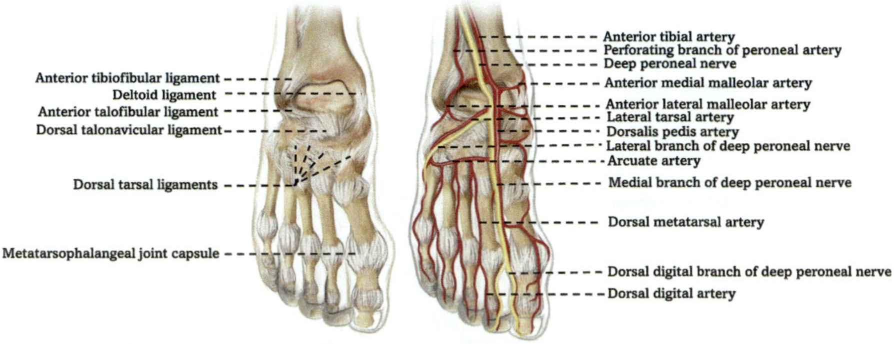

Fig. 3.6 Dorsal view of the ankle and foot showing the ligamentous structures on the left and neurovascular structures on the right

3.9.2 Deep Peroneal Nerve

In the leg, deep peroneal nerve pierces through the extensor digitorum longus muscle and joins the anterior tibial artery along its course. It runs between the extensor digitorum longus and tibialis anterior muscles. It remains lateral to the anterior tibial artery both proximally and distally. Above the ankle joint, it divides into medial and lateral terminal branches (Fig. 3.6). The medial branch is located medially to the dorsalis pedis artery and runs between the extensor hallucis brevis muscle and the hallucis longus tendon, reaching the first intermetatarsal space. Some of the superficial peroneal branches join the deep peroneal nerve and course toward the first web space. The lateral branch of the deep fibular nerve runs anterolaterally and innervates the extensor digitorum brevis muscle. The deep peroneal nerve innervates tibialis anterior, extensor hallucis longus, and digitorum longus muscles. This nerve may also be affected by the compartment syndrome of the lower leg. It may also be insulted by a local mass occupying lesion (e.g., intraneural ganglion). Deep peroneal nerve lesions cause weakness of ankle dorsiflexion and extension of all toes [33].

3.9.3 Tibial Nerve

In the crural region the tibial nerve runs medially to the Achilles tendon and remains posterior to the posterior tibial artery. It then extends into the tarsal tunnel (Fig. 3.1). Inside the tunnel, proximal to the medial malleolus, it divides into the medial plantar nerve and the lateral plantar nerves (Fig. 3.7). Its cutaneous branches are distributed to the medial malleolar area and the inner part of the heel [34]. Its terminal calcaneal branches connect with the sural nerve laterally. The medial plantar nerve descends at the intersection of the flexor hallucis longus and flexor digitorum longus

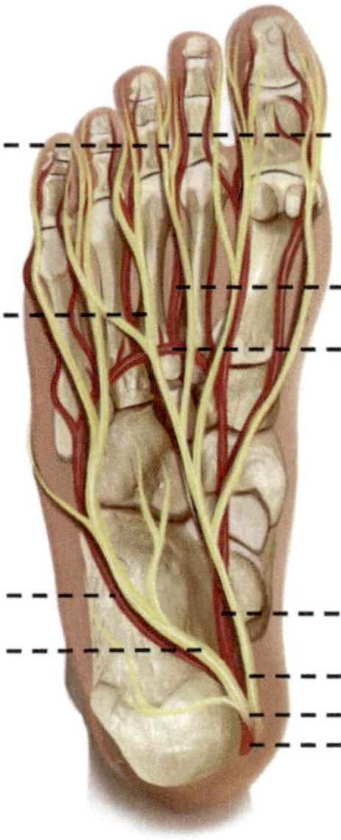

Fig. 3.7 Plantar view of the foot showing neurovascular structures

tendons. It gives off cutaneous, muscular, and articular branches. It provides motor branches to both heads of the flexor hallucis brevis muscle. The lateral plantar nerve enters the medial plantar space, which is situated between the quadratus plantae and flexor digitorum brevis muscles. It approaches the base of the fifth metatarsal and divides into its terminal branches. The lateral plantar nerve and its branches gives motor branches to the quadratus plantae muscle, opponens of the fifth toe, some of the lumbrical and interossei muscles, and innervates tarsal and metatarsal joints. Tibial nerve may be affected by the compartment syndrome of the lower leg or direct trauma. The nerve or its branches may be affected by entrapment in tarsal tunnel, the condition termed as tarsal tunnel syndrome [33].

3.9.4 Dorsalis Pedis Artery

The dorsalis pedis artery is the continuation of the anterior tibial artery at the level of the ankle joint. It follows a course along a line between the first intermetatarsal space and the middle of the transverse malleolar line (Fig. 3.6). Along its pathway, the artery passes by and provides blood supply to the talus, navicular, second cuneiform, and the second metatarsal. It then penetrates the first intermetatarsal region (Armen, [35]). The dorsalis pedis artery eventually joins with the lateral plantar artery to complete the deep plantar arch. Important relations of dorsalis pedis artery include talus, navicular bone, intermediate cuneiform, inferior extensor retinaculum, and extensor hallucis brevis. During its course in the dorsum of the foot, it is lateral to extensor hallucis longus tendon and medial to extensor digitorum longus tendon and terminal branch of deep peroneal nerve [36].

3.9.5 Posterior Tibial Artery

Posterior tibial artery is one of the two branches of the tibioperoneal trunk. It descends between tibialis posterior muscle and triceps surae. Along its course it is accompanied by tibial nerve. The artery then becomes superficial and proceeds into the tarsal canal. It gives its terminal branches, the medial and lateral plantar arteries (Fig. 3.7) [33, 36].

3.10 Plantar Fascia

The plantar fascia is a strong, thick, and longitudinally oriented ligament (Fig. 3.8). Repetitive tensile microtrauma may induce acute or chronic changes in plantar fascia. Plantar fasciitis includes both degenerative changes such as disorganized fibrous tissue (fasciosis) and inflammation of the plantar fascia and/or surrounding tissue. The repeated trauma is most notable at the calcaneal origin of the fascia [37, 38].

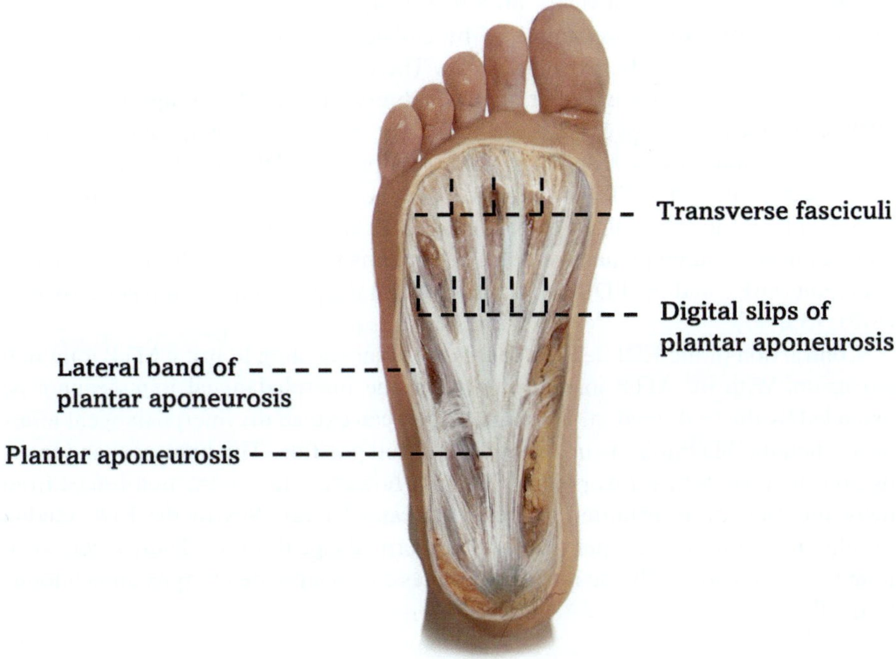

Fig. 3.8 Plantar view of the foot showing plantar fascia and its orientation

Calcaneus, midtarsal joint and the metatarsal bones form a triangular arch, and the plantar fascia acts as a tie-rod connecting the calcaneus to the metatarsals. The plantar fascia resists the force generated by body weight, which tends to flatten the arch during weight bearing. Such flattening effect is also produced by ground reaction forces as these forces are applied to the calcaneus and metatarsal heads. The tension of the plantar fascia is altered during different phases of gait. During the propulsion phase, dorsiflexion causes plantar fascia to be tensioned, thus decreasing the distance between the calcaneus and metatarsal and increasing the height of the arch. This is called the "windlass mechanism," the biomechanical model used to describe how plantar fascia supports the arch of the foot during weight-bearing and how its tension changes during the gait by dorsiflexion of the great toe. Coexisting abnormalities of the foot may lead to the development and progression of plantar fasciitis by causing changes in this mechanism. Structural deformities such as forefoot malalignment and increased or decreased arch can be contributing factors due to altered foot biomechanics [37, 38].

3.11 Balance of Intrinsic and Extrinsic Foot Muscles and Related Deformities

3.11.1 Intrinsic and Extrinsic Muscles of Lesser Toes

The metatarsophalangeal (MTP), proximal interphalangeal (PIP), and distal interphalangeal (DIP) joints are stabilized by collateral ligaments, plantar plate, joint capsule, and dynamically, by the tendons. The extrinsic muscles include extensor digitorum longus (EDL), extensor digitorum brevis (EDB), flexor digitorum longus (FDL), and these muscles extends the MTP joint and flex the interphalangeal joints. The intrinsic muscles include flexor digitorum brevis (FDB), lumbrical, interosseous, and they flex the MTP joint and extend the interphalangeal joints. The FDB courses plantar to the FDL, it divides into two slips on the proximal phalanx and attaches to the middle phalanx. The FDL performs flexion at the distal interphalangeal joint (DIPJ) and the FDB performs flexion at the proximal interphalangeal joint (PIPJ) [39, 40].

Contraction of the EDL tendon causes the extensor sling to bring MTP joint into extension. With the MTP joint in extension, the interphalangeal joints cannot be extended by the EDL tendon. The EDL tendon can extend the interphalangeal joints only when the MTP joint is in a neutral or flexed position. The lumbrical and interosseous tendons act as flexors on MTP joint; however, the contraction forces from these muscles are transmitted to the medial and lateral slips of the EDL tendon distally, resulting in the extension of the interphalangeal joints. Thus, when MTP joint is in extension, only the intrinsic muscles can extend the interphalangeal joints [38, 39].

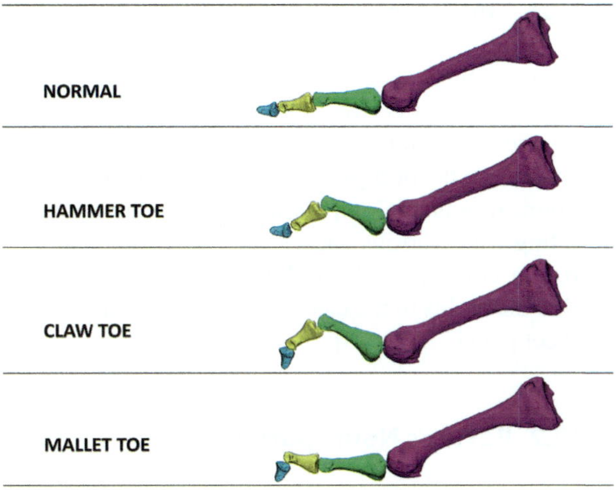

Fig. 3.9 Sagittal plane deformities of lesser toe

3.11.2 Deformities of the Lesser Toes

Deformities of the lesser toes may be caused by an imbalance between the extension and flexion forces acting on metatarsophalangeal and interphalangeal joints (Fig. 3.9) [39]. Mallet toe is characterized by flexion deformity at the DIP joint. The PIP and MTP joints remain in neutral position. It may be caused by external pressure applied by a hard shoe, which may result in DIP flexion contracture and a tight FDL. It can also be caused by injury to the EDL at the level of DIP joint. Hammer toe is the flexion deformity of the PIP joint with the extension of the DIP joint extended or neutral MTP joint [41]. In this deformity, extensors are not able to correct the flexion deformity of the PIP joint and cause extension at the MTP joint [39]. Claw toe is the concurrent flexion deformities of the PIP and DIP joints with hyperextension at the MTP joint. In claw toe, stronger extrinsic forces overcome the intrinsic forces. Similar to claw toe deformity, curly toe is the flexion deformity of both the DIP and PIP joints but in contrast to claw toe deformity the MTP is either neutral or in flexion [41].

3.12 Nerve Entrapment Syndromes of Ankle and Foot

Entrapment of any nerve may be a result of compression by surrounding structures (e.g., vessels, muscles) or space-occupying lesions. The manifestation is usually a burning or tingling sensation in innervation area of the affected nerve [42].

3.12.1 The Tarsal Tunnel and Tarsal Tunnel Syndrome

The tarsal tunnel is an osseo-fibrous structure located at the medial aspect of the foot. It is situated between the flexor retinaculum and the medial borders of distal tibia, calcaneus, and talus. The tibial nerve, posterior tibial vein and artery, posterior tibial tendon, flexor digitorum, and flexor hallucis longus tendons pass through this anatomical tunnel. Compression of the tibial nerve or its branches in this region, resulting from various causes such as anatomical anomalies or tenosynovitis, can lead to a clinical condition known as tarsal tunnel syndrome. The area from the medial malleolus to the calcaneus is affected and the most common manifestation is the heel pain [27, 42, 43].

3.12.2 Baxter's Neuropathy

Baxter's neuropathy is caused by compression of the inferior calcaneal nerve, a branch of the lateral plantar nerve arising within the tarsal tunnel. The compression of the nerve may be due to various conditions including entrapment by a hypertrophied abductor hallucis muscle or compression by surrounding soft. The lateral plantar nerve is a terminal branch of the posterior tibial nerve that supplies intrinsic muscles of the foot, including the abductor digiti minimi, quadratus plantae, flexor digiti minimi brevis, adductor hallucis, the interossei muscles, and the lumbricals. It also provides cutaneous innervation of the area sensation from the lateral sole and from the fifth toe and a part of the fourth toe. Baxter's neuropathy may manifest with heel pain, numbness along the lateral sole of the foot, and weakness of the abductor digiti minimi [42].

A less common cause of pain in the anteroinferior area of the heel is the entrapment of the medial calcaneal nerve (MCN). MCN is one of the nerves arising from the tibial nerve. The MCN provides cutaneous sensation of the anteromedial plantar surface of the calcaneus. The most common origin of the MCN is where tibial nerve exits the tarsal tunnel. However, its origin can vary. This nerve travels superficially along the anteromedial edge of the calcaneus and to provides sensation to the anteromedial plantar surface of the heel. Consequently, along with tibial and plantar nerves, it should always be included in the differential diagnosis of the heel pain [44].

Key Points

- The tibial muscles are found on the medial (inner) aspect of the lower leg and are primarily responsible for ankle inversion and plantar flexion. The primary peroneal muscles include the peroneus longus and peroneus brevis, which run along the outside of the calf and help with ankle eversion and stabilizing the ankle.
- Ligaments such as the anterior talofibular ligament (ATFL) and calcaneofibular ligament (CFL) are commonly injured in ankle sprains, requiring careful examination and rehabilitation.
- Knowledge of the arches of the foot is essential for diagnosing conditions such as flat feet (pes planus) or high arches (pes cavus).

- The plantar fascia, a thick band of connective tissue on the sole of the foot, is frequently involved in conditions such as plantar fasciitis, necessitating focused evaluation.
- The posterior tibial artery travels behind the medial malleolus and gives off branches that supply blood to the plantar surface of the foot. The anterior tibial artery passes anteriorly between the tibia and fibula, supplying the dorsum of the foot.
- Innervation of the foot and ankle involves several nerves, including the tibial nerve and peroneal nerves. Entrapment or injury of these nerves may cause sensory of motor deficits.

Acknowledgments We thank *Buket Serdar Yörükçüler* (Medical Illustrator, Istanbul University-Cerrahpasa) for drawing the illustrations for this chapter.

References

1. Lynn L. Subtalar joint. In: Clinical kinesiology and anatomy. Philadelphia: F. A. Davis; 2011.
2. Stranding S, Anand N, Birch R, Collins P. Ligaments of the foot and ankle. In: Gray's anatomy. London: Elsevier; 2016. p. 1418–21.
3. Guerra-Pinto F, Fabian A, Mota T, Díaz T, Monzo M, Martin Oliva X. The tibiocalcaneal bundle of the deltoid ligament—prevalence and variations. Foot Ankle Surg. 2021;27:138–42. https://doi.org/10.1016/j.fas.2020.03.014.
4. Parson SH. Clinically oriented anatomy. J Anat. 2009;215:474. https://doi.org/10.1111/j.1469-7580.2009.01136.x.
5. Savage-Elliott I, Murawski CD, Smyth NA, Golanó P, Kennedy JG. The deltoid ligament: an in-depth review of anatomy, function, and treatment strategies. Knee Surg Sports Traumatol Arthrosc. 2013;21:1316–27. https://doi.org/10.1007/s00167-012-2159-3.
6. Golanó P, Vega J, de Leeuw PAJ, Malagelada F, Manzanares MC, Götzens V, van Dijk CN. Anatomy of the ankle ligaments: a pictorial essay. Knee Surg Sports Traumatol Arthrosc. 2010a;18:557–69. https://doi.org/10.1007/s00167-010-1100-x.
7. Kelikian A. Ligaments of the foot and ankle. In: Sarrafian's anatomy of the foot and ankle. Chicago, IL: Lippincott Williams & Wilkins; 2011a. p. 163–223.
8. Standring S. Ankle joint. In: Grays anatomy. London: Elsevier; 2016a. p. 1430–1.
9. Vega J, Malagelada F, Dalmau-Pastor M. Ankle microinstability: arthroscopic findings reveal four types of lesions to the anterior talofibular ligament's superior fascicle. Knee Surg Sports Traumatol Arthrosc. 2021;29:1294–303. https://doi.org/10.1007/s00167-020-06089-z.
10. Bruns J, Habermann C, Werner M. Osteochondral lesions of the talus: a review on talus osteochondral injuries, including osteochondritis Dissecans. Cartilage. 2021;13:1380S–401S. https://doi.org/10.1177/1947603520985182.
11. Mróz I, Kurzydło W, Bachul P, Jaworek J, Konarska M, Bereza T, Walocha K, Mazur M, Kuniewicz M, Depukat P, Mizia E, Chmielewski P, Warchoł Ł. Inferior tibiofibular joint (tibiofibular syndesmosis)—own studies and review of the literature. Folia Med Cracov. 2015;55:71–9.
12. Golanó P, Vega J, de Leeuw PAJ, Malagelada F, Manzanares MC, Götzens V, van Dijk CN. Anatomy of the ankle ligaments: a pictorial essay. Knee Surg Sports Traumatol Arthrosc. 2010b;18:557–69. https://doi.org/10.1007/s00167-010-1100-x.
13. Espinosa N. The subtalar joint. Foot Ankle Clin. 2018;23:11–2. https://doi.org/10.1016/j.fcl.2018.05.001.

14. Bartoníček J, Rammelt S, Naňka O. Anatomy of the subtalar joint. Foot Ankle Clin. 2018;23:315–40. https://doi.org/10.1016/j.fcl.2018.04.001.
15. Szaro P, Nilsson-Helander K, Carmont M. MRI of the Achilles tendon—a comprehensive pictorial review. Part two. Eur J Radiol Open. 2021;8:100343. https://doi.org/10.1016/j.ejro.2021.100343.
16. Chen J, Sagoo N, Panchbhavi VK. The Lisfranc injury: a literature review of anatomy, etiology, evaluation, and management. Foot Ankle Spec. 2021;14:458–67. https://doi.org/10.1177/1938640020950133.
17. Siddiqui NA, Galizia MS, Almusa E, Omar IM. Evaluation of the Tarsometatarsal joint using conventional radiography, CT, and MR imaging. Radio Graphics. 2014;34:514–31. https://doi.org/10.1148/rg.342125215.
18. Epeldegui T, Delgado E. Acetabulum pedis. Part I: Talocalcaneonavicular joint socket in normal foot. J Pediatr Orthop B. 1995;4:1–10.
19. Tryfonidis M, Jackson W, Mansour R, Cooke PH, Teh J, Ostlere S, Sharp RJ. Acquired adult flat foot due to isolated plantar calcaneonavicular (spring) ligament insufficiency with a normal tibialis posterior tendon. Foot Ankle Surg. 2008;14:89–95. https://doi.org/10.1016/j.fas.2007.11.005.
20. Richie DH. Biomechanics and clinical analysis of the adult acquired flatfoot. Clin Podiatr Med Surg. 2007;24:617–44. https://doi.org/10.1016/j.cpm.2007.07.003.
21. Steffensmeier AM, Matar R, Chung D, Dixon T, Laughlin R. Fracture of the posteromedial tubercle of the talus: surgical approach with case examples. Tech Foot Ankle Surg. 2021;20:175–81. https://doi.org/10.1097/BTF.0000000000000328.
22. Melão L, Canella C, Weber M, Negrão P, Trudell D, Resnick D. Ligaments of the transverse tarsal joint complex: MRI–anatomic correlation in cadavers. Am J Roentgenol. 2009;193:662–71. https://doi.org/10.2214/AJR.08.2084.
23. Kafka RM, Aveytua IL, Choi PJ, DiLandro AC, Tubbs RS, Loukas M, Mintz DN, Baidya R, Kumar S, Sangari SK, Mtui EP, D'Antoni AV. Anatomico-radiological study of the bifurcate ligament of the foot with clinical significance. Cureus. 2019;11:e3847. https://doi.org/10.7759/cureus.3847.
24. Standing S, Anand N, Birch R, Collins P. Muscles of the foot and ankle. In: Gray's anatomy. London: Elsevier; 2016. p. 1439–44.
25. Lee MH, Chung CB, Cho JH, Mohana-Borges AV, Pretterklieber ML, Trudell DJ, Resnick D. Tibialis anterior tendon and extensor retinaculum: imaging in cadavers and patients with tendon tear. Am J Roentgenol. 2006;187:W161–8. https://doi.org/10.2214/AJR.05.0073.
26. Maquirriain J, Sammartino M, Ghisi JP, Mazzuco J. Tibialis anterior tenosynovitis: avoiding extensor retinaculum damage during endoscopic debridement. Arthroscopy. 2003;19:1–3. https://doi.org/10.1053/jars.2003.50059.
27. Kohls-Gatzoulis J, Angel JC, Singh D, Haddad F, Livingstone J, Berry G. Tibialis posterior dysfunction: a common and treatable cause of adult acquired flatfoot. BMJ. 2004;329:1328–33. https://doi.org/10.1136/bmj.329.7478.1328.
28. Kelikian A. Muscles of ankle and foot. Chicago, IL: Lippincott Williams & Wilkins; 2011b. p. 223–92.
29. Sora M-C, Jilavu R, Grübl A, Genser-Strobl B, Staykov D, Seicean A. The posteromedial neurovascular bundle of the ankle: an anatomic study using plastinated cross sections. Arthroscopy. 2008;24:258.e1–7. https://doi.org/10.1016/j.arthro.2007.08.030.
30. Standring S. Arches of the foot. In: Gray's anatomy. London: Elsevier; 2016b. p. 1438–40.
31. Davda K, Malhotra K, O'Donnell P, Singh D, Cullen N. Peroneal tendon disorders. EFORT Open Rev. 2017;2:281–92. https://doi.org/10.1302/2058-5241.2.160047.
32. Walt J, Massey P. Peroneal tendon syndromes. Treasure Island, FL: StatPearls Publishing; 2023.
33. Stranding S. Muscles of foot. In: Grays anatomy. London: Elsevier; 2016. p. 1440–5.
34. Tonogai I, Tsuruo Y, Sairyo K. Implications of dorsalis Pedis artery anatomical variants for dorsal midfoot surgery. Foot Ankle Int. 2022;43:942–7. https://doi.org/10.1177/10711007221081527.
35. Kelikian A. Arteries of the foot and ankle. In: Sarrafian's anatomy of the foot and ankle. Chicago, IL: Lippincott Williams & Wilkins; 2011. p. 302–70.

36. Standing S. Innervation of the foot. In: Gray's anatomy. London: Elsevier; 2016. p. 1444–7.
37. Bolgla LA, Malone TR. Plantar fasciitis and the windlass mechanism: a biomechanical link to clinical practice. J Athl Train. 2004;39:77–82.
38. Schwartz EN, Su J. Plantar fasciitis: a concise review. Perm J. 2014;18:e105. https://doi.org/10.7812/TPP/13-113.
39. Malhotra K, Davda K, Singh D. The pathology and management of lesser toe deformities. EFORT Open Rev. 2016;1:409–19. https://doi.org/10.1302/2058-5241.1.160017.
40. Soysa A, Hiller C, Refshauge K, Burns J. Importance and challenges of measuring intrinsic foot muscle strength. J Foot Ankle Res. 2012;5:29. https://doi.org/10.1186/1757-1146-5-29.
41. Shirzad K, Kiesau CD, DeOrio JK, Parekh SG. Lesser toe deformities. Am Acad Orthopae Surg. 2011;19:505–14. https://doi.org/10.5435/00124635-201108000-00006.
42. Beltran L, Bencardino J, Ghazikhanian V, Beltran J. Entrapment neuropathies III: lower limb. Semin Musculoskelet Radiol. 2010;14:501–11. https://doi.org/10.1055/s-0030-1268070.
43. Nelson SC. Tarsal tunnel syndrome. Clin Podiatr Med Surg. 2021;38:131–41. https://doi.org/10.1016/j.cpm.2020.12.001.
44. Diers DJ. Medial calcaneal nerve entrapment as a cause for chronic heel pain. Physiother Theory Pract. 2008;24:291–8. https://doi.org/10.1080/09593980701738392.

Radiological Assessment of Foot and Ankle

4

Sarthak Sharma, Ankit Dadra, Mandeep S. Dhillon, and Siddhartha Sharma

4.1 Introduction

The foot and ankle are anatomically and functionally complex structures. The pathological processes involving this region are diverse. Knowledge of imaging modalities as well as radiographic projections is essential to obtain optimal diagnostic information in patients undergoing evaluation for foot and ankle disorders. The various diagnostic imaging modalities serve to complement each other in many clinical scenarios. Radiographs continue to be the most commonly used initial imaging modality [1]. Computed tomography (CT) offers additional diagnostic value in anatomically complex areas and traumatic cases. Magnetic resonance imaging (MRI) enables diagnosis of a wide variety of ligamentous, tendinous, cartilaginous, and other soft tissue pathology [2]. Ultrasonography is useful in the evaluation of superficial soft tissue structures, especially tendons and ligaments.

S. Sharma
Jammu Healthcare & Diagnostics Pvt. Ltd., Jammu, Jammu and Kashmir, India

A. Dadra (✉)
Foot and Ankle Biomechanics, Experimentation & Research (FABER) Laboratory, Department of Orthopedics, Postgraduate Institute of Medical Education and Research, Chandigarh, India

M. S. Dhillon
Department of Orthopedics, Fortis Hospital, Mohali, Punjab, India

S. Sharma
Foot and Ankle, Biomechanics, Experimentation & Research (FABER) Laboratory, Postgraduate Institute of Medical Education and Research, Chandigarh, Chandigarh, India

Foot and Ankle Research and Innovation Laboratory, Massachusetts General Hospital, Harvard Medical School, Boston, USA

S. Sharma et al. (eds.), *Clinical and Radiological Examination of the Foot and Ankle*, https://doi.org/10.1007/978-981-97-4202-8_4

4.2 Radiographs

Radiographs are the most commonly used initial imaging modality for a variety of suspected disease processes, including (but not limited to) trauma, degeneration, infection, deformities, or tumors [1]. They are widely available, inexpensive, and can be performed rapidly.The primary advantage of radiographs is assessment of bony detail, including cortex and medullary cavity, joint space, and joint alignment. The disadvantages include exposure to ionizing radiation as well as limited ability to identify and characterize soft tissue pathology.

4.2.1 Radiographic Projections

There are a number of variably oriented bones and articulations in the foot and ankle, making radiographic evaluation of this region more complex as compared to the proximal extremities. As the ankle and foot are oriented in different anatomical planes, their radiographic projections also differ [3]. Hence, it is impossible to perform a "combined" foot and ankle anteroposterior projection.

The choice of projection depends on the clinical scenario. Both non-weight-bearing and weight-bearing radiographs can be obtained. However, weight-bearing radiographs provide the added benefit of assessing osseous as well as articular alignment in the functional state, which is especially useful in functional assessment of deformities.

Due to the multiplanar nature of articular alignments in the foot and ankle, oblique views are needed in a significant number of cases to depict articular congruency as well as to remove overlap of osseous structures, which can potentially obscure significant pathology.

A knowledge of radiographic projections used in foot and ankle imaging is essential in clinical practice. These are enumerated in the following tables. A brief, clinically relevant overview of the most common projections is also provided.

4.2.2 Radiographic Projections: Ankle

Important radiographic projections for evaluation of the ankle, along with their indications, are enumerated in Table 4.1. A brief account of the most commonly used projections is also provided.

The tibiotalar joint and medial mortise are well delineated on AP view (Fig. 4.1), whereas the lateral mortise is obscured on AP view and visualized on mortise view (Fig. 4.2). The normal tibiotalar joint space measures 3–4 mm.

On the AP view (Fig. 4.1), the following parameters are used to evaluate the integrity of mortise and inferior tibiofibular syndesmosis:

Medial clear space (MCS): indicative of medial mortise, interval between medial malleolus and medial border of talus, measured 5 mm inferior to tibial plafond, normally measures <4 mm [4].

Table 4.1 Radiographic projections—ankle

Projection	Anatomy/indication
Anteroposterior	Tibiotalar joint, medial mortise, distal tibiofibular syndesmosis, medial/lateral malleoli
Anteroposterior with 15–20° internal rotation (mortise view)	Mortise, distal tibiofibular syndesmosis, talar dome, talofibular joint, lateral malleolus
Lateral	Tibiotalar joint, talar dome, posterior malleolus, distal tibial and fibula, posterior subtalar joint, Kager's fat pad
Lateral oblique, medial oblique	Medial and lateral malleoli, subtalar joint
Varus stress view Valgus stress view	Suspected ligamentous injury or insufficiency

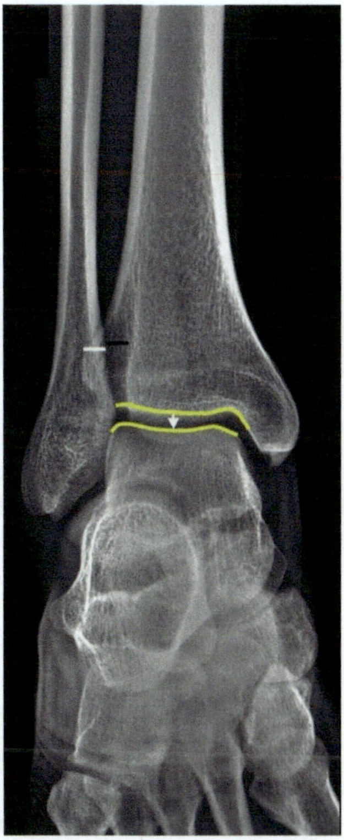

Fig. 4.1 AP view ankle. Curvilinear yellow lines depict tibiotalar articular congruency. The medial clear space is well visualized. Short white line depicts tibiofibular overlap, short black line depicts total clear space, indicative of syndesmotic integrity

Tibiofibular clear space (TCS) and tibiofibular overlap (TFO): reflect integrity of the tibiofibular syndesmosis.

TCS: distance between the posterior border of tibia and the medial border of fibula, measured 1 cm above the tibial plafond, normally measures <6 mm [5].

TFO: distance between lateral border of tibia and medial border of the fibula, measured 1 cm above the tibial plafond, normally measures >6 mm on AP view and > 1 mm on mortise view [5].

On the lateral view (Fig. 4.3), the anteroposterior extents of the tibiotalar joint and talar dome as well as the posterior malleolus are well delineated. The posterior

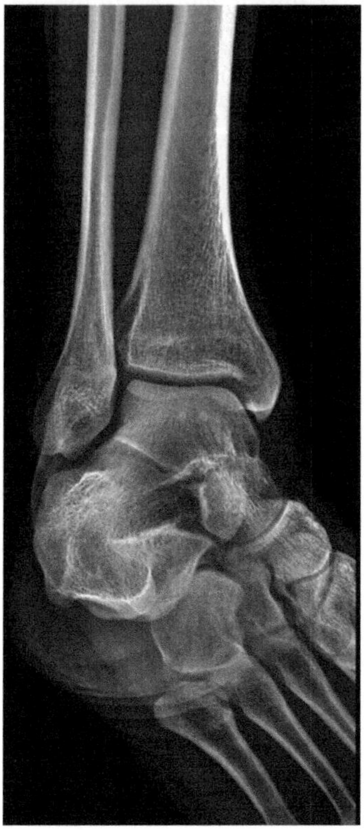

Fig. 4.2 Mortise view ankle. The ankle mortise, especially the lateral aspect, is well depicted

aspect of the subtalar joint can also be visualized. The lateral view can also demonstrate significant tibiotalar joint effusions as well as provide indirect evidence of soft tissue abnormalities related to Kager's fat pad (such as Achilles peritendinitis).

Although stress views of the ankle are infrequently performed in clinical practice, due to availability of MRI for assessment of ligamentous insufficiency, they do allow assessment of medial and lateral ankle stability. Valgus and varus stress views are performed in plantar flexion and neutral position, preferably under fluoroscopic guidance. Evaluation of bilateral ankles allows comparison with the normal side. Despite variations in criteria for instability reported in literature, an interval difference of >3 mm and a talar tilt angle difference of >5° between the symptomatic and normal sides is considered suggestive of ligamentous injury/insufficiency [6].

4.2.3 Radiographic Projections: Foot

Important radiographic projections for evaluation of the foot, along with their indications, are enumerated in Table 4.2. A brief account of the most commonly used projections is also provided.

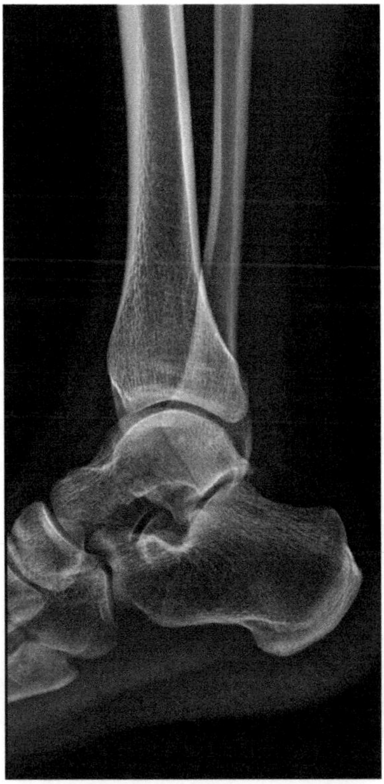

Fig. 4.3 Lateral view ankle

Table 4.2 Radiographic projections—foot

Projection	Anatomy/indication
Anteroposterior (dorsoplantar), anteroposterior medial oblique	Metatarsals, tarsometatarsal joints, talonavicular joint, calcaneocuboid joint, cyma line cuneiforms, sinus tarsi
Anteroposterior (dorsoplantar) lateral (reverse) oblique	First–second cuneiform interval, first–second metatarsal interval
Lateral	Tibiotalar joint, cyma line, subtalar joint, navicular, metatarsals, medial cuneiform
Axial view calcaneum	Transverse view of calcaneus
Lateral view calcaneus	Anteroposterior view of calcaneus
Subtalar oblique	Subtalar joint
Canale view	Talar neck
Sesamoid view	Tangential view of sesamoid bones of first metatarsal
Anteroposterior toes	Phalangeal tufts, interphalangeal space, subungual region
Oblique toes	Phalanges, dislocation
Weight-bearing anteroposterior/ lateral	Deformities, tarsal coalition, functional measurements of foot
Hindfoot alignment view (weight-bearing)	Suspected hindfoot malalignment

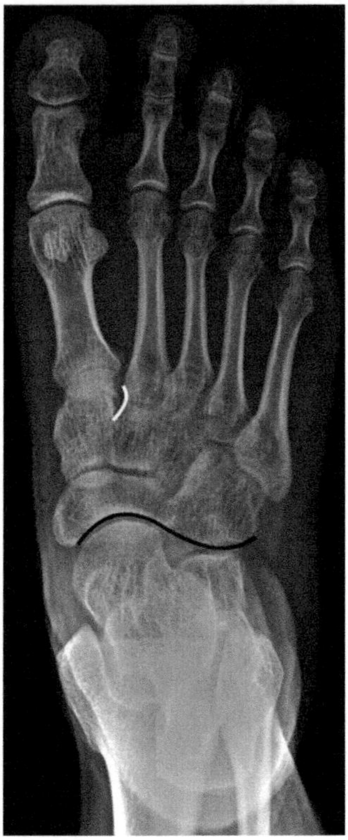

Fig. 4.4 AP view foot. Short white lines depict normal second tarsometatarsal joint alignment. Curved black line depicts the cyma line. The third, fourth, and fifth tarsometatarsal relationships are obscured due to overlap

On AP view (Fig. 4.4), the midfoot and forefoot bones as well as tarsometatarsal, metatarsophalangeal, and interphalangeal joints can be demonstrated. The third, fourth, and fifth metatarsophalangeal joints as well as lateral tarsals show overlap on AP view, necessitating the use of an AP medial oblique view (Fig. 4.5), which demonstrates these structures free of overlap. It is important to assess alignment of tarsometatarsal joints and widening of the interval between proximal ends of first and second metatarsals, the latter may be the only initial evidence of Lisfranc injury [7].

On AP and lateral views, the "lazy S"-shaped cyma line formed by smooth union of talonavicular and calcaneocuboid (mid-tarsal) joint lines can be demonstrated, reflecting congruency of the Chopart joint. Disruption of this line can occur in Chopart injury and pes planus [8].

A lateral view (Fig. 4.6) of the foot supplements the AP view. The subtalar joint, hindfoot, Chopart joint (cyma line), navicular, medial cuneiform, and metatarsal shaft can be assessed. Dorsal subluxation of the base of the second metatarsal on lateral view may be a clue to Lisfranc injury. Evaluation of Bohler and Gissane angles is also performed on lateral view (Fig. 4.7).

On a lateral oblique view of the foot, evaluation of intervals between medial and intermediate cuneiforms as well as proximal ends of first and second metatarsals can be performed.

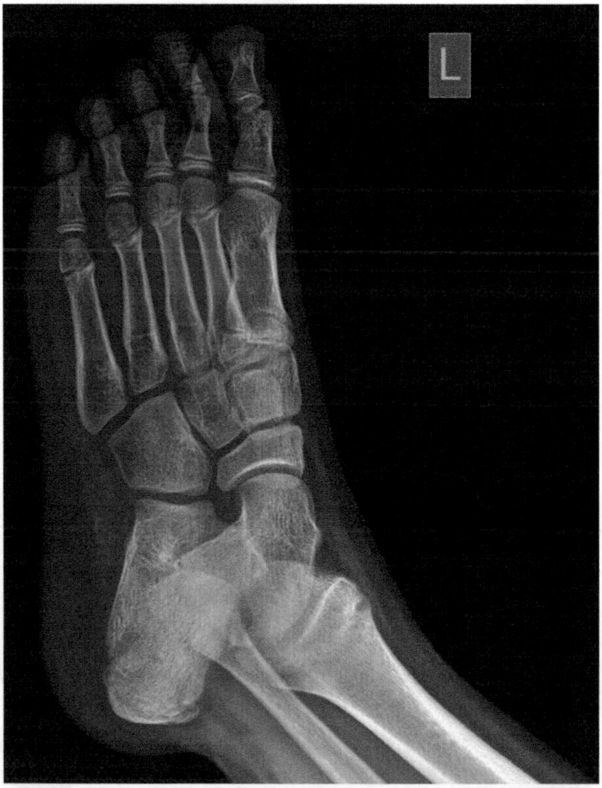

Fig. 4.5 AP medial oblique view foot. The third, fourth, and fifth tarsometatarsal relationships are well defined, free of overlap

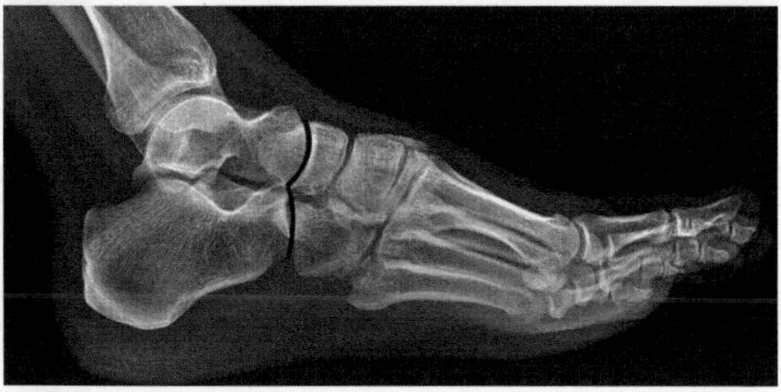

Fig. 4.6 Lateral view foot. Curved black line depicts the cyma line. The subtalar joint, calcaneum, navicular, medial cuneiform, and metatarsals are depicted

Weight-bearing views (Figs. 4.8, 4.9, 4.10 and 4.11) are essential in assessing articular alignment and performing functional measurements, since subtle structural changes may not be evident on non-weight-bearing views.

An overview of the most commonly assessed measures of alignment on weight-bearing views of the foot is provided in Table 4.3 and Figs. 4.8, 4.9, 4.10 and 4.11. However, this is not an exhaustive list. A thorough description of these measurements can be found in some excellent existing publications, the details of which are provided in the references section of this chapter.

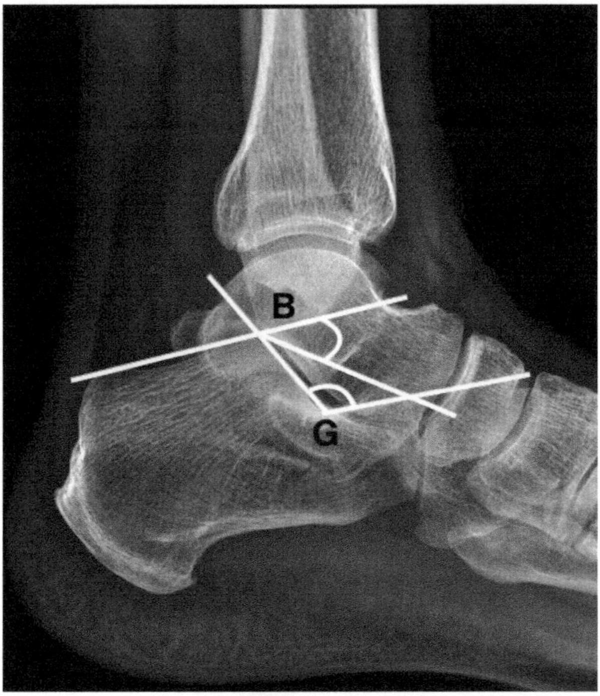

Fig. 4.7 Lateral view hindfoot. *B* Bohler's angle, *G* Gissane's angle

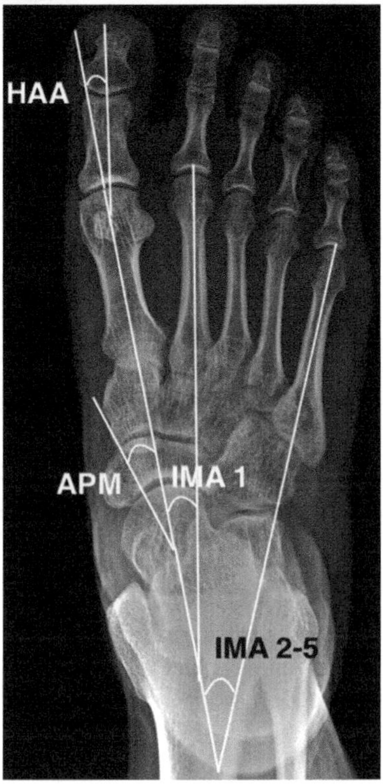

Fig. 4.8 AP weight-bearing view of the foot. *APM* anteroposterior Meary's angle, *HAA* hallux valgus angle, *IMA 1* first intermetatarsal angle, *IMA 2–5* second–fifth intermetatarsal angle

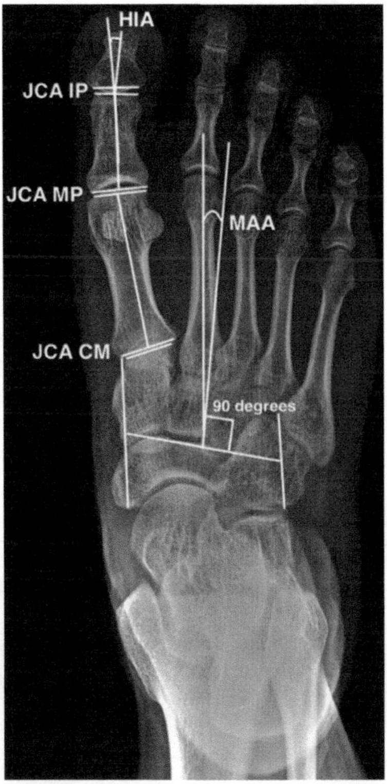

Fig. 4.9 AP weight-bearing view of the foot. *MAA* metatarsus adductus angle, *HIA* hallux interphalangeal angle, *JCA IP* joint line convergence angle of interphalangeal joint of hallux, *JCA MP* joint line convergence angle of first metatarsophalangeal joint, *JCA CM* joint line convergence angle of medial cuneiform-first metatarsal joint

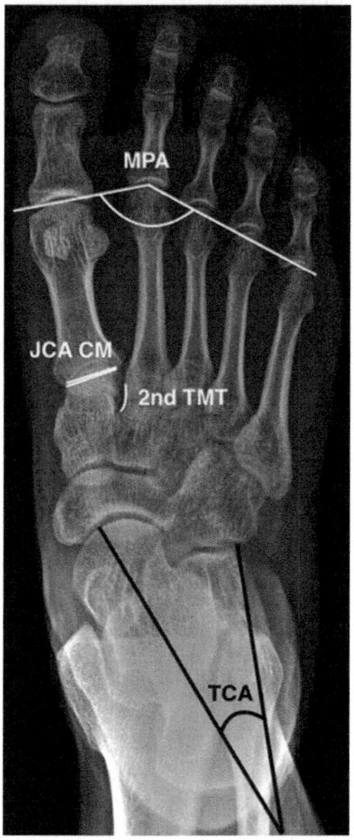

Fig. 4.10 AP weight-bearing view of the foot. *MPA* metatarsal parabola angle, *second TMT* articular alignment of second tarsometatarsal joint, *TCA* talocalcaneal angle (Kite's angle)

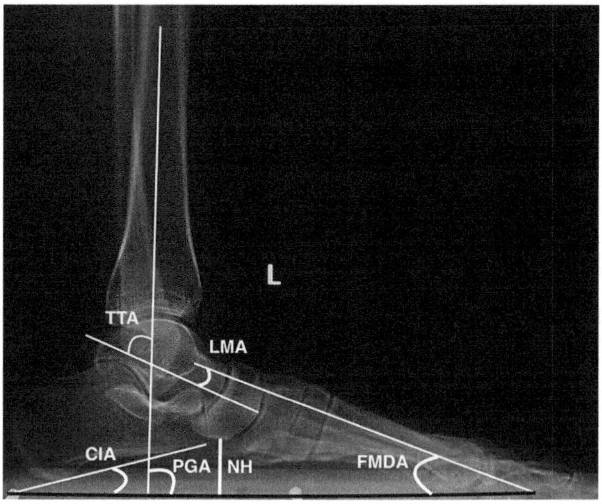

Fig. 4.11 Lateral weight-bearing view of the foot. Black line is parallel to the weight-bearing surface. *TTA* tibiotalar angle, *LMA* lateral Meary's angle, *CIA* calcaneal inclination angle, *FMDA* first metatarsal declination angle, *PGA* plantigrade angle, *NH* navicular height

Table 4.3 Radiographic parameters assessed on weight bearing anteroposterior (AP) and lateral views of the foot

Parameter	Indication
Weight-bearing view foot—Anteroposterior (dorsoplantar)	
Anteroposterior Meary's angle	Midfoot abduction/ adduction
Hallux valgus angle	Hallux valgus
Hallux interphalangeal angle	Hallux valgus
Joint line convergence angle first tarsometatarsal joint	Hallux valgus
Joint line convergence angle first metatarsophalangeal joint	Hallux valgus
Joint line convergence angle first interphalangeal joint	Hallux valgus
First intermetatarsal angle	Metatarsus primus varus, hallux abductus
Metatarsus adductus angle	Metatarsus adductus
Metatarsal parabola angle	Relative shortening of first metatarsal
Talocalcaneal (Kite's) angle	Hindfoot varus/ valgus
Weight-bearing view foot—Lateral	
Lateral Meary's angle	Pes planus, pes cavus
Tibiotalar angle	Ankle–hindfoot malalignment
First metatarsal declination angle	Metatarsus primus elevatus
Calcaneal inclination angle	Pes planus, pes cavus

4.3 Computed Tomography (CT)

Present-day multidetector CT scanners enable fast scanning with high spatial resolution. Excellent bony detail as well as the ability to acquire isotropic volumetric datasets and perform multiplanar reconstructions makes CT invaluable in the anatomically complex region of foot and ankle. The most common indication is evaluation of traumatic injury. CT provides detailed information with regard to fracture geometry, fracture fragment alignment, as well as healing response. CT is also invaluable in the evaluation of arthritis and tarsal coalitions.

Three-dimensional volumetric images as well as surface reconstructions enable depiction of structures with different densities, such as bone and soft tissues to be displayed using different color maps, increasing the utility of CT in surgical planning. The use of maximum intensity projection (MIP) reduces beam hardening artifacts from orthopedic fixation devices. Additionally, CT has a role in evaluating bone tumors or tumor-like conditions in patients with contraindications for MRI.

For evaluation of the ankle and hindfoot, the patient's feet are positioned perpendicular to the CT table with the knees extended; the use of a foot board may be beneficial in this situation. For evaluation of the midfoot and forefoot, the patient's feet are positioned flat on the CT table with the knees flexed. In post-traumatic cases, the ability to manipulate acquired volumetric image data into any desired plane may compensate for suboptimal patient positioning.

The disadvantages include exposure to ionizing radiation and relative insensitivity in assessment of soft tissue pathology. Image degradation resulting from beam hardening artifacts due to orthopedic implants can be offset to some extent by adjusting position of the scanned part so that the implant presents the least cross-sectional area to the X-ray beam, adjusting scan parameters such as peak voltage and tube charge, using thinner sections for scanning, modifying reconstruction kernel, and using maximum intensity projection images.

4.4 Ultrasonography

Modern-day ultrasound transducers use high frequencies, permitting a high level of soft tissue resolution. Ultrasound has a wide variety of applications in the foot and ankle, including evaluation of tendons, ligaments, muscles, as well as joint effusions, synovitis, suspected foreign bodies, and initial evaluation of superficial soft tissue lesions. Ultrasound also provides real-time guidance for a variety of diagnostic and therapeutic procedures such as aspirations, biopsies, and therapeutic injections.

Addition of color Doppler and power Doppler imaging provides additional information regarding soft tissue perfusion and locoregional vascular flow.

The advantages include wide availability, reduced cost burden, fast image acquisition, lack of ionizing radiation, and the ability to perform real-time, dynamic assessment of structures, especially tendons. The latter is especially useful in assessment of suspected tendon subluxation and evaluation of widening of tendinous gap in tears of the Achilles tendon [9].

The disadvantages include a high degree of dependence on operator skill as well as relative difficulty in assessing deep structures.

4.5 Magnetic Resonance Imaging (MRI)

Advances in coil technology and design as well as improvements in pulse sequence design over the past decade have dramatically improved MR image quality. Superior contrast resolution and multiplanar imaging capability make MRI an indispensable tool in the evaluation of osseous as well as soft tissue structures in the foot and ankle. Although MRI imaging at field strength of 1.5 T provides adequate diagnostic information in most scenarios, the use of higher field strength (3 T) has the advantage of increasing signal-to-noise ratio as well as better depiction of soft tissue abnormalities, especially those related to cartilage [10].

T1-weighted sequences provide an assessment of normal anatomy, bone marrow signal, and tendons. T2-weighted sequences or proton density (PD)-weighted sequences acquired with or without fat suppression techniques and short tau inversion recovery (STIR) sequence provide the bulk of diagnostic information with regard to soft tissues, including ligaments, tendons, joint capsule, synovium, and periosseous soft tissues [11].

The use of fat suppression techniques combined with T2-or PD-weighted sequences enables detection of bone marrow oedema as well as soft tissue injury, infection, and inflammation [12].

Addition of gadolinium-based contrast sequences may be necessary in cases of suspected inflammation, infection, and neoplasm.

Commonly used cartilage-specific sequences include dual echo steady state (DESS) and 3D-gradient echo [13]. The use of T2 relaxometry and delayed Gadolinium enhanced MRI of cartilage (dGEMRIC) techniques enables articular cartilage mapping (cartography) with color-coded maps, providing differentiation between normal and diseased cartilage [14].

The selection of the MRI imaging sequence and imaging plane depends heavily on the suspected area of abnormality. The use of alternative imaging sequences as well as modifying the scan plane may be necessary to improve visualization of pathology and eliminate imaging artifacts. A common example of such an artifact is the magic angle artifact pertaining to the evaluation of flexor and peroneal tendons adjacent to the ankle.

Due to the obliquity of most of the ligamentous structures with respect to standard anatomical planes, the use of oblique planes parallel and/or orthogonal to the course of the ligament may be needed to adequately visualize the ligament [15].

Common indications for MRI include osteomyelitis, soft tissue infections, osteochondral injury, ligamentous injury, tendinopathy, arthropathy, synovitis, neuropathy, diabetic foot assessment, impingement syndromes, and tumors.

The most significant disadvantages of MRI include contraindications precluding its use (claustrophobia, permanent pacemaker, aneurysm coils, and other contraindications of MRI in general) and a higher cost burden. Image degradation may result from preexisting orthopedic implants, although metallic artifact reduction sequences may mitigate image degradation to some extent. The inability to perform a real-time dynamic assessment of tendons with MRI is also a disadvantage when compared with ultrasonography.

4.6 Summary

Knowledge of imaging modalities and various radiographic projections is essential in diagnosing and managing the wide range of foot and ankle disorders. The use of special radiographic projections may be warranted in certain clinical scenarios to detect pathology that is obscured on standard radiographic projections. Weight-bearing radiographic views are essential in diagnosing and quantifying the extent of deformity and malalignment, if present. The use of CT allows multiplanar assessment of fractures and other osseous abnormalities, especially in areas that are obscured on radiography. MRI provides detailed assessment of soft tissue pathology. Ultrasonography is useful in assessment of superficial soft tissue pathology and dynamic assessment of tendons.

References

1. Hughes P, Miranda R, Doyle AJ. MRI imaging of soft tissue tumours of the foot and ankle. Insights Imaging. 2019;10(1):60.
2. Gorbachova T. Magnetic resonance imaging of the ankle and foot. Pol J Radiol. 2020;85:e532–49.
3. Lamm BM, Stasko PA, Gesheff MG, Bhave A. Normal foot and ankle radiographic angles, measurements, and reference points. J Foot Ankle Surg. 2016;55(5):991–8.
4. Murphy JM, Kadakia AR, Schilling PL, Irwin TA. Relationship among radiographic ankle medial clear space, sex, and height. Orthopedics. 2014;37(5):e449–54.
5. Hermans JJ, Wentink N, Beumer A, Hop WC, Heijboer MP, Moonen AF, Ginai AZ. Correlation between radiological assessment of acute ankle fractures and syndesmotic injury on MRI. Skeletal Radiol. 20121;41(7):787–801.
6. Gribble PA. Evaluating and differentiating ankle instability. J Athl Train. 2019;54(6):617–27.
7. Moracia-Ochagavía I, Rodríguez-Merchán EC. Lisfranc fracture-dislocations: current management. EFORT Open Rev. 2019;4(7):430–44.
8. Meyr AJ, Sansosti LE, Ali S. A pictorial review of reconstructive foot and ankle surgery: evaluation and intervention of the flatfoot deformity. J Radiol Case Rep. 2017;11(6):26–36.
9. Hodgson RJ, O'Connor PJ, Grainger AJ. Tendon and ligament imaging. Br J Radiol. 2012;85(1016):1157–72.
10. Novakofski KD, Pownder SL, Koff MF, Williams RM, Potter HG, Fortier LA. High-resolution methods for diagnosing cartilage damage in vivo. Cartilage. 2016;7(1):39–51.
11. McMahon KL, Cowin G, Galloway G. Magnetic resonance imaging: the underlying principles. J Orthop Sports Phys Ther. 2011;41(11):806–19.
12. Martín Noguerol T, Luna A, Gómez Cabrera M, Riofrio AD. Clinical applications of advanced magnetic resonance imaging techniques for arthritis evaluation. World J Orthop. 2017;8(9):660–73.
13. Gold GE, Chen CA, Koo S, Hargreaves BA, Bangerter NK. Recent advances in MRI of articular cartilage. AJR Am J Roentgenol. 2009;193(3):628–38.
14. Potter HG, Black BR, Chong le R. New techniques in articular cartilage imaging. Clin Sports Med. 2009;28(1):77–94.
15. Hermans JJ, Ginai AZ, Wentink N, Hop WC, Beumer A. The additional value of an oblique image plane for MRI of the anterior and posterior distal tibiofibular syndesmosis. Skeletal Radiol. 2011;40(1):75–83.

Basics of Pedobarography

5

Atta Taseh, Nour Nassour, and Jiyong Ahn

5.1 Introduction

Pedobarography is the study of foot pressure distribution when it is in contact with a supporting surface [1]. Foot pressure distribution has been studied for a long time, evolving from using simple ink to using modern complex platforms with pressure sensors [2, 3]. It is now increasingly used in the evaluation of several foot and ankle pathologies [4]. Using this method, we can gain invaluable knowledge regarding the condition of the foot and ankle during different loading phases. Medical conditions in the lower extremity change the contact pressure distribution of the foot, rendering it a key aspect of the overall evaluation of such conditions.

5.2 The Fundamentals

5.2.1 Historical Background and Evolution of Pedobarographic Techniques

The concept of studying the foot pressure has been around since the 1930s, since then, several devices have come into play in how to measure the different aspects of

A. Taseh (✉) · N. Nassour
Foot and Ankle Research and Innovation Lab, Harvard Medical School, Massachusetts General Hospital, Boston, MA, USA
e-mail: ataseh@mgh.harvard.edu

J. Ahn
Foot and Ankle Research and Innovation Lab, Harvard Medical School, Massachusetts General Hospital, Boston, MA, USA

Department of Orthopaedic Surgery, College of Medicine, The Catholic University of Korea, Seoul, Republic of Korea

S. Sharma et al. (eds.), *Clinical and Radiological Examination of the Foot and Ankle*, https://doi.org/10.1007/978-981-97-4202-8_5

the pressure distribution of the foot [3]. There are two main approaches to studying foot pressure distribution patterns: static and dynamic. One of the earliest devices to study this pressure distribution was the kinetograph, a rubber pressure mat with a reflecting fluid on a glass plate, to measure pressure change during walking motion [5, 6]. Early studies primarily focused on clinical applications, such as understanding foot deformities and pathological gait patterns. As technology progressed, the emergence of electronic sensors and computing systems in the late twentieth century transformed pedobarography. These technologies allowed for capturing and analyzing more comprehensive and accurate data, leading to a better understanding of foot biomechanics [7]. With the advent of new pedobarography devices, the scope of the applications moved beyond clinical settings to sports performance, and footwear and orthotic design boosting the overall efficiency of individuals with foot deformities.

Various technologies are available for collecting plantar pressure data, ranging from floor-mounted plantar pressure devices to in-shoe pressure sensor systems. These technologies enable the measurement of plantar pressure variations while individuals engage in various activities such as walking, running, and cycling. Various pedobarography devices exist, each offering data collection based on distinct sensors and technologies. Piezoelectric, capacitive, resistive, piezoresistive, and optical sensors are the frequently used ones [8, 9]. Each sensor type has distinct characteristics that make it suitable for specific applications such as sports activities that require a larger pressure range and capturing frequency [8].

5.2.2 Types of Pedobarographic Systems

Plantar pressure measurement can be conducted through two primary approaches: platform-based and insole-based systems. While neither method has been definitively established as superior, each possesses distinct applications and serves specific purposes.

Platform systems are made from horizontally placed pressure platforms embedded with pressure sensors. Such platforms allow for static or dynamic pressure analysis. The concentration of sensors, operating frequency, and sensing area of the sensors determine the accuracy of the data received. Such platforms should be attached to a computer system that will receive and analyze the data. Smaller platform systems are used for static gait analysis (Fig. 5.1), while larger ones are used for both static and dynamic analyses (Fig. 5.2), be it a standing, walking, or running one. One limitation of such systems is that the foot has to land on predetermined areas, which could affect the natural gait pattern in a patient undergoing motion analysis [7].

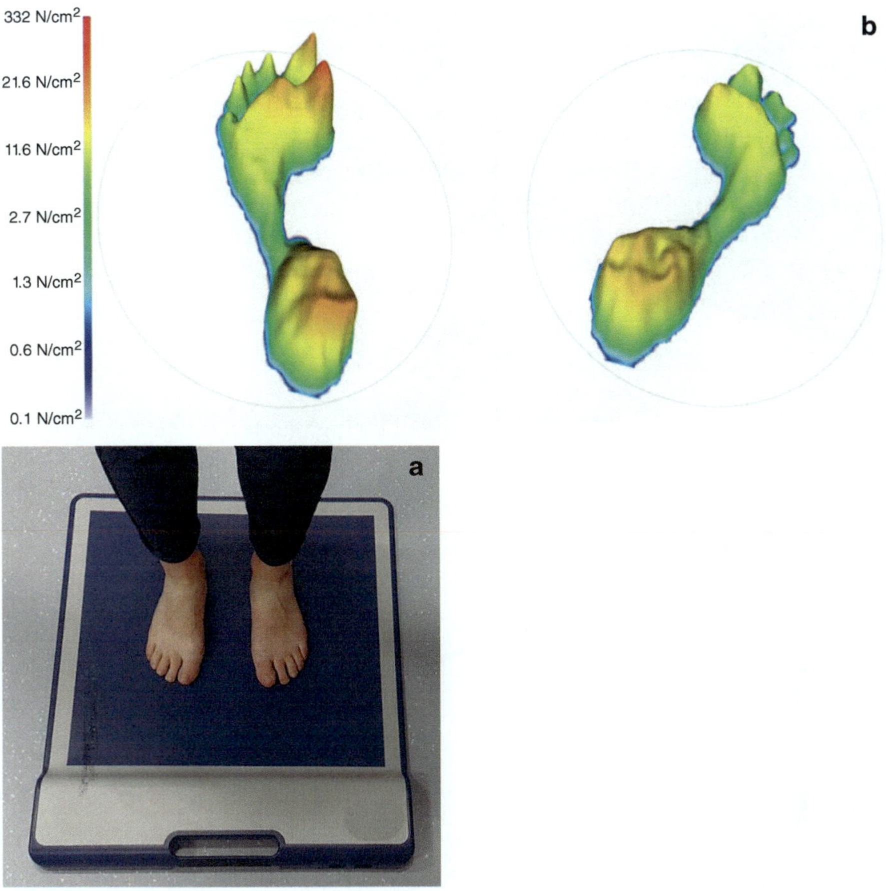

Fig. 5.1 Platform-based static pedobarography (**a**). Three-dimensional presentation of the pressure distribution (**b**)

Pressure-sensitive insoles are thin shoe-lining inserts that incorporate pressure sensors capable of capturing pressure variations between the sole of the foot and the shoe. This method is particularly favored for assessing footwear ergonomics and facilitates testing in a more natural environment for patients. As a portable solution, it is suitable for both indoor and outdoor activities, allowing patients to walk or run with a more natural gait pattern. However, a major limitation of this method is the changing sensitivity of the sensors as it depends on how they are integrated within either the shoe or socks [7, 8, 10, 11]. Other limitations include sensor slippage during movement, the limited number of sensors, and the trapped heat of this closed system, which could alter the validity of the results [8, 12, 13].

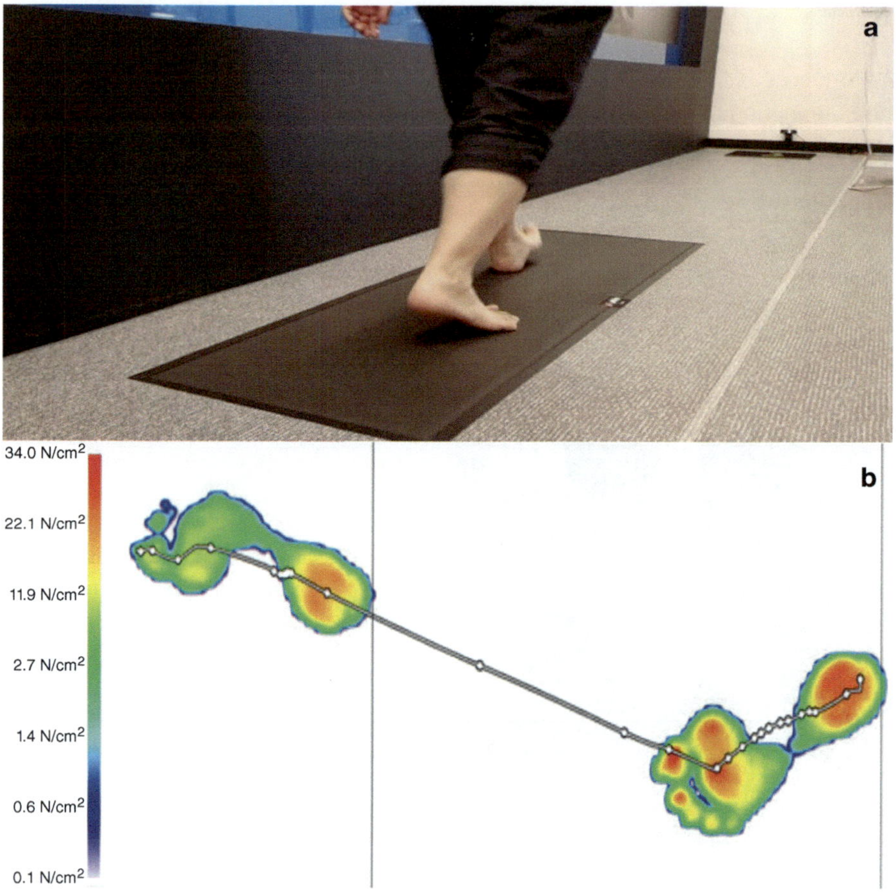

Fig. 5.2 Platform-based dynamic pedobarography (**a**). The pressure data Is presented as a heatmap with the white line denoting the trajectory of center of pressure (**b**)

5.3 Conducting a Pedobarographic Examination

5.3.1 Examination Protocol

With either static or dynamic measurements, it is crucial to calibrate the device before starting the examination. Calibration is important in establishing the validity and reliability of the results obtained [14]. Static weight calibration is one of the most frequently used methods, which includes placing a known weight and associating the weight with the measured pressure values. Often devices come with a protocol and software calibration capability, which allow adjustments through the interface. Regardless of the calibration method, it is very important to place the platform on a level surface.

Depending on the purpose of the examinations, various approaches exist regarding how to conduct the procedure. When a static examination is required, first, the pressure-sensing platform such as the pressure plate is placed on a level surface of the examination room. After connecting the platform to its corresponding hardware/software acquisition system, the examiner will have to position the patient upright at the center of the platform, with feet shoulder width and aligned parallel to one another [7]. When performing the measurements in challenging poses such as the Romberg's test, or single heel raise test, it is important to take safety precautions to prevent potential fall incidents.

When conducting a dynamic examination such as walking, patients are advised to maintain their natural walking pattern and pace. It is necessary to avoid restrictive clothing, as this can affect the range of motion of the joints and subsequently the walking pattern. They should be instructed to walk along the designated path and follow the technician's cues on starting and stopping. It is necessary that the patient communicates any pain or discomfort after each trial, which could help with the interpretation of the data.

When conducting an examination on a platform, various protocols involve, asking the patient to either walk a certain number of steps or a certain distance. The latter is called the "mid-gait" protocol, during which the measurements are done excluding the first and last steps [15]. It is best practice to conceal the platform in the middle of the walkway to avoid any potential gait alterations due to the patients targeting the platform [16]. The mid-gait protocol is less favorable in special patient populations such as stroke or diabetic neuropathy because the fatigue could cause alterations in the gait pattern [15, 17]. On the contrary, the two-step protocol entails starting the measurements on the second step. This approach has been shown to generate comparable results as the mid-gait one and to be more suitable for the aforementioned patient populations [18–20].

For both protocols, the examination needs to be repeated multiple times to ensure consistency and to catch probable variations in gait patterns [7, 19]. It is a common approach to do two to six walking trials and then average the values, to increase the reliability of measurements. On the contrary, the in-shoe systems allow for more steps to be taken and included in the measurements as they are not limited by the size of the platform. In such settings, an average of 12 steps is considered sufficient to produce valid data. However, measuring some parameters such as the force-time integral may require a larger number of steps [21].

5.3.2 Parameters and Data Analysis

The primary data that is delivered by a pedobarographic device is the absolute pressure/force values. All other parameters are inferred from this raw data including metrics related to contact area, foot pressure distribution, center of pressure (COP), and gait. Each one of these categories can then be further analyzed for a more detailed description of the condition depending on the purpose of the examination. If it is intended to find an overloaded region of interest (ROI), as it is in diabetic

Table 5.1 Practical pedobarographic parameters in foot and ankle assessments

Parameter (unit of measurement)	Definition
Peak plantar pressure (kPa or N/cm^2)	The highest-pressure value measured within a specific area of the foot during the process of walking or running [22]
Maximum force (N or % of body weight)	The greatest amount of force recorded in a specific area of the foot; distinct from pressure as it is not influenced by the size of the area [22]
Force time integral (N.s)	Serves as a metric for quantifying the accumulated exposure to force over a period of time within a specific region of the foot [22]
Pressure time integral (N.s/cm^2)	Quantifies the total pressure experienced over a period of time in a specific area of the foot. It is determined by calculating the area beneath the pressure–time curve [22]
Contact area (cm^2)	The contact area of the foot with the floor refers to the portion of the foot that is in direct contact with the surface or ground during weight-bearing activities such as walking, running, or standing [22, 23]
Contact time (s)	Contact time refers to the duration for which the foot remains in contact with the ground during each step or stride
Arch index (%)	Midfoot area divided by the total foot area excluding the toes [24]
Relative midfoot index	It represents the depth of the midfoot valley of the force–time curve in relation to the amount of maximal force (MF) in the hindfoot and forefoot [25]
Center of pressure excursion index (%)	The assessment of how much the center of pressure deviates from a reference line during the stance phase of gait [26]

neuropathy patients, various pressure parameters including peak pressure, mean pressure, pressure gradient, and pressure–time integral are essential. If the purpose is to evaluate a patient's balance, as in a multiple sclerosis patient, then different COP metrics will be generated, including COP sway velocity, trajectory length, ellipse area, and excursion index. The gait parameters include, but are not limited to, walking speed, stride length/time, step width/length, and cadence. They come into play especially when assessing sports performance or footwear functionality. Table 5.1 provides a set of parameters that are useful in the realm of foot and ankle surgery.

There are no universally established reference range values for the above-mentioned parameters. The reason partly being that the values generated by different devices are not comparable and partly that multiple patient-related factors such as age, body mass index, sex, and disease status can affect the final results [27]. This makes pedobarography a challenging option when it comes to diagnosis. Clinicians and researchers often rely on either comparing the pressure differences between the injured leg and the normal contralateral leg or assessing the trend in pressure changes before and after the initiation of a treatment or an intervention. The first approach is not frequently used due to concerns about the healthy side being also affected as a result of the changes in weight distribution. The second one is favorable in treatment selection, monitoring, rehabilitation, and footwear design. However, across the literature, some authors have found different cut-offs that can point to some sort of foot abnormality. For instance, a relative midfoot index of <0.8 showed a positive predictive value of 80% and sensitivity of 78% to detect an unhealthy foot condition such as end-stage ankle osteoarthritis [25]. To design insoles for offloading ROIs, it is often common to use a peak plantar pressure (PPP) of 200 kPa as a threshold for detecting these regions [28].

While the entire foot can be included in the pedobarographic analysis, often examiners are interested in a specific location of the foot and choose to do a location-wise analysis, leaving out other parts. Plantar pressure measurements involve dividing the foot into predefined regions to analyze pressure distribution referred to as"masks" [22]. Generally, the foot is divided into three regions including hindfoot, midfoot, and forefoot based on a percentage of the length of the foot, then each area can be further subdivided to provide more detailed information and fulfill the aims of the examination [29]. Masking is usually automated and the partitioning is done automatically through specific software during the examination [16]. Examiners can also create personalized masks that are focused on specific ROIs, including particular joints, bony prominences, or other specific regions. However, a more detailed mask requires more advanced devices that offer a higher resolution of sensors (Fig. 5.3). Based on the mask selected, the computer will generate a

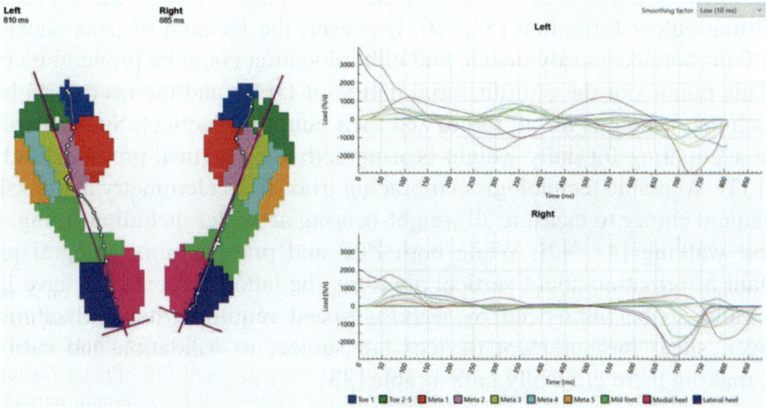

Fig. 5.3 Location-wise analysis of the plantar pressure. On the left, a mask with ten regions is shown. The graphs show the changes in the plantar pressure over time, corresponding to each specific region

three-dimensional topographic pressure map of the foot that reveals regions of both heightened and reduced pressures [30, 31] (Fig. 5.1).

5.4 Clinical Applications

5.4.1 Another Aspect of Foot and Ankle Ailments

5.4.1.1 Diabetic Foot Ulcer (DFU)

Plantar pressure distribution is altered in patients with diabetic peripheral neuropathy. While neuropathy itself is considered the main cause of the alteration, other factors such as foot deformities, increased BMI, and soft tissue changes may further aggravate the condition [32].

Since the emergence of pedobarography researchers have been striving to find meaningful patterns to detect at-risk patients for DFU. Theoretically speaking, plantar tissue stress (PTS) during the day is the factor that can best help with screening these patients [33]. Vertical and horizontal pressures applied to the feet, along with their cumulative effects, all affect PTS, rendering it a rather difficult measurement. Barefoot walking PPP of the entire foot, as one of the most studied vertical loading parameters, is frequently reported as being elevated in these patients. It is also correlated to the severity of neuropathy [34, 35]. Values ranging from 6 to 12.3 kg/cm^2 with varying sensitivity and specificity have been proposed as a cut-off point for PPP [36]. Yet, no consensus exists in this regard, primarily due to the heterogeneity in the devices and study designs. While region-specific measurements show increased pressure in the midfoot and forefoot in these patients, a more meticulous assessment reveals an association between the history of DFU and higher midfoot pressures [35, 37].

Peak shear stress as a horizontal pressure also plays a role in tissue breakdown and ultimate ulcer formation [38, 39]. However, the location of peak shear stress and PPP do not necessarily match, and either location could be prone to ulceration [40]. This points out the multifactorial nature of DFUs and the need for a holistic approach that accounts for all forces and their cumulative effect. Such calculations require accounting for daily weight-bearing activities, vertical pressure, and shear stress [33]. Wearable technologies embracing triaxial accelerometry are considered a convenient choice to measure all weight-bearing activities including sitting, standing, and walking [41, 42]. While both PPP and pressure–time integral provide important information about vertical pressure, the latter is deemed to have limited added value to be influenced by walking speed requiring standardization [43]. Moreover, shear measurement devices are subject to validation and calibration issues, making them clinically unfavorable [33].

5.4.1.2 Foot Posture

Changes in the foot posture, especially the medial longitudinal arch, change the contact area and as a result the pressure distribution. As faulty foot posture is a risk factor for some other lower extremity injuries, early detection and correction of these deformities is of clinical value [44]. Arch index is an important metric based on the foot contact area that is used for foot posture classification. Values between 0.21 and 0.26 denote neutral posture, while lesser values (<0.21) and greater ones (>0.26) classify the foot as planus and cavus, respectively [24]. The center of pressure excursion index (CPEI) is another metric for detecting planus deformity. While no established cutoff exists, values below 18.57–19.4%, depending on the device being used, are considered to be useful for detecting planus deformity [29, 45].

Pedobarographic studies of the planus foot deformity reveal an increased medial midfoot contact area along with elevated PPP in the hallux, second toe, and metatarsophalangeal (MTP) joint area [29]. Pressure–time integral values are higher in the medial midfoot and lower in the fifth MTP joint area [29]. The COP sway is also altered with the COP being deviated medially [29, 46]. On the contrary, higher PPP is observed in the heel, and lateral side of the forefoot in the cavus foot and higher contact area in the fourth and fifth MTP joints [23, 29]. Higher pressure–time integral is detected in the whole-foot measurements, along with the rearfoot and forefoot ones [23]. Moreover, the COP laterally deviates in these patients [29, 46].

As foot posture is related to several foot and ankle ailments, assessing the relationship between static foot posture and foot function using dynamic pedobarography is also another interesting research avenue. Attempts to assess the predictability of the dynamic values from the static ones have yielded some preliminary results [47, 48]. Variables such as foot area, arch height, forefoot to rearfoot angle, individual's height and weight, calcaneal inclination, and rear foot eversion have been shown to significantly account for some level of variability in the whole-foot maximum force and midfoot PPP [22].

5.4.1.3 Chronic Ankle Instability

The condition is characterized by the recurrent giving way of the lateral side of the foot, mainly because of a ligamentous laxity. Dynamic pedobarographic studies demonstrate an increased PPP in the lateral side of the foot in these patients [49, 50]. Pressure–time integral values are also elevated. There is some evidence on this increased PPP in the forefoot area is correlated to medial talar cartilage deformation [51]. This altered joint contact stress may explain the higher rate of osteoarthritis in these patients [52]. Measures of balance show a shift of COP to the lateral side of the foot along with an increased COP sway velocity, indicating less stability compared to healthy individuals [17, 53].

5.4.2 Orthotic Design

Orthopedic insoles have been around for years with the primary purpose of correcting anatomical malalignments and thereby relieving symptoms. With the emergence of pedobarography devices, researchers started investigating the effect of these insoles on the distribution of plantar pressures. In light of these assessments, plantar pressure evaluation is now incorporated into the process of insole design and manufacturing.

Different materials and design elements have been tested in terms of their offloading capacity so far. Among the materials, polyurethane and ethylene vinyl acetate (EVA) show promising offloading effects by decreasing the PPP, especially under the lesser metatarsi and medial heel [54]. It looks like these materials conform easily to the plantar surface, increase the contact area, and decrease the PPP. Patients with rheumatoid arthritis, for example, benefit from the softer variations of these materials by offloading the forefoot, which is an ROI in these patients [55]. It is difficult to draw conclusions about specific design elements due to the convention of using more than one element in insoles [56]. The frequently used design elements including medial longitudinal arch support, heel cup, metatarsal pad, and medial/lateral postings/wedges have shown various pedobarographic results. However, using the plantar pressure input to fine-tune the design shows interesting results [57]. For instance, using this data to design custom insoles for flatfoot patients seems to offload the lateral metatarsals when compared to conventional prefabricated insoles primarily by redistributing the pressure [58].

The altered pressure distribution in patients with diabetic neuropathy causes elevated pressure values in specific regions commonly in the forefoot. Offloading this area is the main purpose of prescribing footwear and insoles for the population. Despite the conflicting literature, the cutoff point to identify ROIs in the forefoot has frequently been considered to be 200 kPa [28]. Various insole specifications have been proposed and used for offloading purposes based on pedobarographic assessments. Among those medial, longitudinal arch support and metatarsal pads have been shown to decrease the PPP between 35 and 40 kPa in the ROIs [59]. Creating a cut-out through removable square plugs is another strategy to eliminate the pressure on these ROIs [60].

5.4.3 Treatment Selection and Monitoring

As the interpretation complexity and lack of sufficient evidence on specific pathologies challenge the applicability of the method, it barely is incorporated in treatment guidelines as a standard of care in the field of orthopedics. However, due to the invaluable information, it provides researchers continue to modify the method and make it more clinically relevant. The following section focuses on presenting some of these studies to provide some clues about the potential of the method.

Paralytic cavovarus deformity is a progressive one during which the high arch gradually progresses to hindfoot equinovarus and metatarsalgia with callosities. Treatment aims to increase the range of motion and stability and decrease the symptoms [61–63]. The choice of treatment, whether it be a soft tissue release, tendon transfer, or osteotomy, depends on the hindfoot flexibility and age at the time of surgery [61, 64, 65]. Given our knowledge about the plantar pressure distribution in cavus foot deformity, pedobarography looks like a reasonable choice to objectively evaluate the choice of treatment. A study result demonstrated that the plantar pressure distributions are closer to normal individuals with the combination of soft tissue release and first ray extension osteotomy when compared to soft tissue release alone [66]. These changes entailed a significant increase in the forefoot pressures, especially under the first metatarsal head. Although no definite conclusions could be drawn from these results, dynamic pedobarography may be considered in addition to radiographic parameters when choosing a surgical approach for patients with paralytic cavovarus deformity.

Subtalar extra-articular screw arthroereisis (SESA) is a minimally invasive procedure for the correction of juvenile flexible flatfoot. To track the post-operative changes following a SESA operation, pedobarography has shown promising results. In a study, researchers followed patients up to 1 month after the surgery [67]. Static pedobarography assessments showed an immediate increase in vertical ground force on the lateral side of the foot and a decrease on the medial side. They revealed that the changes on the medial side started to abate, and 1 month after the surgery, 37% of the changes were maintained. A similar trend with CPEI was observed, increasing from 17% preoperatively to 39% 3 days after the surgery, while after 1 month it decreased to 25%. Similarly, the outcome of calcaneo-cuboid-cuneiform osteotomies was evaluated for the correction of planus feet in cerebral palsy patients [68]. While a meaningful shift of the plantar pressure to the lateral side of the foot was observed, the radiological parameters including talonavicular coverage and anteroposterior talo-first metatarsal angles were significantly correlated to the pressure changes in the midfoot. Moreover, correlating subjective outcomes of disability (foot function index) to plantar pressure in patients with idiopathic flatfoot deformity revealed less disability with medially shifted hindfoot, laterally shifted midfoot peak pressures, and larger hindfoot pressure [69].

Intra-articular calcaneal fracture is a common foot and ankle fracture often leading to long-term pain, disability, and reduced quality of life [70]. Studies regarding the plantar pressure alterations in these patients showed elevated midfoot and forefoot pressure, a medially shifted COP, and decreased hindfoot, first and second toe pressure. These findings were meaningfully correlated to functional outcomes when the whole-foot pressure values were included but not the location-wise ones. In a study the velocity of the center of pressure was measured over 2 years after surgery was done [71]. When compared to the healthy side, the increased values of the terminal-stance were normalized after 2 years, and the decreased values of the mid-stance were normalized after a year. These values were associated with the biomechanical evaluations of the range of motion but not with pain scores. Overall, pedobarography could be considered to monitor treatment outcomes in these patients, not to mention that the pressure data could also be utilized for prescribing insoles in this population.

5.5 Limitations and Considerations

5.5.1 Lack of Standardization

There is a lack of universal standardization in examination, sensor calibration, data collection, and analysis methods and protocols. This is a flaw that can lead to variations in results when comparing studies from different research groups.

5.5.2 Inter-Device Discrepancy

The variations of the values between different plantar pressure measurement devices make it difficult to produce universal values that could be incorporated in clinical guidelines.

5.5.3 Limited Data Interpretation

The lack of reference values, as well as the broad range of parameters, renders the data interpretation a challenging one, which requires additional expertise and context.

5.5.4 Calibration

Pedobarography platforms and insoles often require regular calibration to maintain accuracy. Failure to calibrate the system properly can lead to inaccurate or unreliable data.

5.6 Future Directions in Pedobarography

Given the vast amount of data generated by pedobarography devices, trying to find relevant parameters by performing clinical trials and statistical analyses does not seem to be the best approach. As with many other fields, pedobarography could benefit from big data and artificial intelligence (AI) applications. Training AI algorithms on large amount of plantar pressure data could potentially lead to more precise diagnostic capabilities and treatment recommendations.

As telemedicine is evolving rapidly, pedobarography devices could potentially benefit the patients, especially when it comes to treatment monitoring. Patients could perform pressure measurements at home using portable devices, and healthcare professionals could review the data remotely, providing timely feedback.

Key Points

- The two types of pedobarography devices include platform-based and insole-based, which allow for either static or dynamic measurements of the plantar pressure.
- When conducting a dynamic examination, the choice of number of steps and when to do the measurements depends on the parameters being measures and the aim of the examination.
- While various parameters can be inferred from the raw pressure data, choosing the right ones that fulfill the purpose of the examination is a crucial step.
- Plantar tissue stress consists of vertical and horizontal loads along with their cumulative effects during the day. While the vertical load can be measure by all

devices, measuring the horizontal ones (i.e., shear stress) faces validation challenges.
- While several location-wise pressure readings provide invaluable knowledge on foot posture, arch index is a simple and practical one used for foot classification.
- Pedobarography inputs help designing insoles to redistribute the plantar pressure and offload the regions of interest.
- Despite the limited evidence, plantar pressure measurement could provide important information to track the course of healing in a number of orthopedic injuries including foot posture deformities and calcaneal fractures.

References

1. Hughes J. The clinical use of pedobarography. Acta Orthop Belg. 1993;59:10.
2. Klimiec E, Zaraska K, Piekarski J, et al. Durable sensors for measurement of foot plantar pressure with piezoelectric polyvinylidene fluoride foil. Sensors Actuators A Phys. 2016;247:504–13.
3. Soames RW. Foot pressure patterns during gait. J Biomed Eng. 1985;7:120–6.
4. Metaxiotis D, Accles W, Pappas A, Doederlein L. Dynamic Pedobarography (DPB) in operative management of cavovarus foot deformity. Foot Ankle Int. 2000;21:935–47. https://doi.org/10.1177/107110070002101108.
5. Elftman H. A cinematic study of the distribution of pressure in the human foot. Anat Rec. 1934;59:481–91. https://doi.org/10.1002/ar.1090590409.
6. Morton DJ. Structural factors in static disorders of the foot. Am J Surg. 1930;9:315–28. https://doi.org/10.1016/S0002-9610(30)91100-2.
7. Zulkifli SS, Loh WP. A state-of-the-art review of foot pressure. Foot Ankle Surg. 2020;26:25–32. https://doi.org/10.1016/j.fas.2018.12.005.
8. Abdul Razak AH, Zayegh A, Begg RK, Wahab Y. Foot plantar pressure measurement system: a review. Sensors. 2012;12:9884–912. https://doi.org/10.3390/s120709884.
9. Klöpfer-Krämer I, Brand A, Wackerle H, et al. Gait analysis—available platforms for outcome assessment. Injury. 2020;51(Suppl 2):S90–6. https://doi.org/10.1016/j.injury.2019.11.011.
10. Burnfield JM, Few CD, Mohamed OS, Perry J. The influence of walking speed and footwear on plantar pressures in older adults. Clin Biomech. 2004;19:78–84. https://doi.org/10.1016/j.clinbiomech.2003.09.007.
11. Mei Q, Fernandez J, Fu W, et al. A comparative biomechanical analysis of habitually unshod and shod runners based on a foot morphological difference. Hum Mov Sci. 2015;42:38–53. https://doi.org/10.1016/j.humov.2015.04.007.
12. Putti AB, Arnold GP, Cochrane L, Abboud RJ. The Pedar® in-shoe system: repeatability and normal pressure values. Gait Posture. 2007;25:401–5. https://doi.org/10.1016/j.gaitpost.2006.05.010.
13. Woodburn J, Helliwell P. Observations on the F-scan in-shoe pressure measuring system. Clin Biomech. 1996;11:301–4. https://doi.org/10.1016/0268-0033(95)00071-2.
14. Collins SH, Adamczyk PG, Ferris DP, Kuo AD. A simple method for calibrating force plates and force treadmills using an instrumented pole. Gait Posture. 2009;29:59–64. https://doi.org/10.1016/j.gaitpost.2008.06.010.
15. Orlin MN, McPoil TG. Plantar pressure assessment. Phys Ther. 2000;80:399–409.
16. Lorkowski J, Gawronska K, Pokorski M. Pedobarography: a review on methods and practical use in foot disorders. Appl Sci. 2021;11:11020. https://doi.org/10.3390/app112211020.
17. Morrison KE, Hudson DJ, Davis IS, et al. Plantar pressure during running in subjects with chronic ankle instability. Foot Ankle Int. 2010;31:994–1000. https://doi.org/10.3113/FAI.2010.0994.

18. Bryant A, Singer K, Tinley P. Comparison of the reliability of plantar pressure measurements using the two-step and midgait methods of data collection. Foot Ankle Int. 1999;20:646–50.
19. Bus SA, de Lange A. A comparison of the 1-step, 2-step, and 3-step protocols for obtaining barefoot plantar pressure data in the diabetic neuropathic foot. Clin Biomech (Bristol, Avon). 2005;20:892–9. https://doi.org/10.1016/j.clinbiomech.2005.05.004.
20. Cavanagh PR, Ulbrecht JS. Clinical plantar pressure measurement in diabetes: rationale and methodology. Foot. 1994;4:123–35. https://doi.org/10.1016/0958-2592(94)90017-5.
21. Arts MLJ, Bus SA. Twelve steps per foot are recommended for valid and reliable in-shoe plantar pressure data in neuropathic diabetic patients wearing custom made footwear. Clin Biomech. 2011;26:880–4. https://doi.org/10.1016/j.clinbiomech.2011.05.001.
22. Buldt AK, Allan JJ, Landorf KB, Menz HB. The relationship between foot posture and plantar pressure during walking in adults: a systematic review. Gait Posture. 2018;62:56–67. https://doi.org/10.1016/j.gaitpost.2018.02.026.
23. Burns J, Crosbie J, Hunt A, Ouvrier R. The effect of pes cavus on foot pain and plantar pressure. Clin Biomech. 2005;20:877–82. https://doi.org/10.1016/j.clinbiomech.2005.03.006.
24. Cavanagh PR, Rodgers MM. The arch index: a useful measure from footprints. J Biomech. 1987;20:547–51. https://doi.org/10.1016/0021-9290(87)90255-7.
25. Arno F, Roman F, Martin W, et al. Facilitating the interpretation of pedobarography: the relative midfoot index as marker for pathologic gait in ankle osteoarthritic and contralateral feet. J Foot Ankle Res. 2016;9:47. https://doi.org/10.1186/s13047-016-0177-y.
26. Song J, Hillstrom HJ, Secord D, Levitt J. Foot type biomechanics. Comparison of planus and rectus foot types. J Am Podiatr Med Assoc. 1996;86:16–23. https://doi.org/10.7547/87507315-86-1-16.
27. Telfer S, Bigham JJ. The influence of population characteristics and measurement system on barefoot plantar pressures: a systematic review and meta-regression analysis. Gait Posture. 2019;67:269–76. https://doi.org/10.1016/j.gaitpost.2018.10.030.
28. Lin T-L, Sheen H-M, Chung C-T, et al. The effect of removing plugs and adding arch support to foam based insoles on plantar pressures in people with diabetic peripheral neuropathy. J Foot Ankle Res. 2013;6:29. https://doi.org/10.1186/1757-1146-6-29.
29. Hillstrom HJ, Song J, Kraszewski AP, et al. Foot type biomechanics part 1: structure and function of the asymptomatic foot. Gait Posture. 2013;37:445–51. https://doi.org/10.1016/j.gaitpost.2012.09.007.
30. Skopljak A, Muftic M, Sukalo A, et al. Pedobarography in diagnosis and clinical application. Acta Inform Med. 2014;22:374–8. https://doi.org/10.5455/aim.2014.22.374-378.
31. Žvorc M. Dijagnostički postupci kod promjena na stopalu. Acta Med Croatica. 2011;64:15–25.
32. Sutkowska E, Sutkowski K, Sokołowski M, et al. Distribution of the highest plantar pressure regions in patients with diabetes and its association with peripheral neuropathy, gender, age, and BMI: one Centre study. J Diabetes Res. 2019;2019:e7395769. https://doi.org/10.1155/2019/7395769.
33. Lazzarini PA, Crews RT, van Netten JJ, et al. Measuring plantar tissue stress in people with diabetic peripheral neuropathy: a critical concept in diabetic foot management. J Diabetes Sci Technol. 2019;13:869–80. https://doi.org/10.1177/1932296819849092.
34. Abri H, Aalaa M, Sanjari M, et al. Plantar pressure distribution in diverse stages of diabetic neuropathy. J Diabetes Metab Disord. 2019;18:33–9. https://doi.org/10.1007/s40200-019-00387-1.
35. Sacco ICN, Hamamoto AN, Tonicelli LMG, et al. Abnormalities of plantar pressure distribution in early, intermediate, and late stages of diabetic neuropathy. Gait Posture. 2014;40:570–4. https://doi.org/10.1016/j.gaitpost.2014.06.018.
36. de Castro JPW, Ferreira FC, Vargas JGF, et al. Accuracy of foot pressure measurement on predicting the development of foot ulcer in patients with diabetes: a systematic review and meta-analysis. J Diabetes Sci Technol. 2021;17:70–8. https://doi.org/10.1177/19322968211043550.
37. Bacarin TA, Sacco ICN, Hennig EM. Plantar pressure distribution patterns during gait in diabetic neuropathy patients with a history of foot ulcers. Clinics (Sao Paulo). 2009;64:113–20. https://doi.org/10.1590/s1807-59322009000200008.

38. Cavanagh PR, Ulbrecht JS, Caputo GM. New developments in the biomechanics of the diabetic foot. Diabetes Metab Res Rev. 2000;16:S6–S10. https://doi.org/10.1002/1520-7560(200009/10)16:1+<::AID-DMRR130>3.0.CO;2-Z.
39. Yavuz M. Plantar shear stress distributions in diabetic patients with and without neuropathy. Clin Biomech (Bristol, Avon). 2014;29:223–9. https://doi.org/10.1016/j.clinbiomech.2013.11.003.
40. Stucke S, McFarland D, Goss L, et al. Spatial relationships between shearing stresses and pressure on the plantar skin surface during gait. J Biomech. 2012;45:619–22. https://doi.org/10.1016/j.jbiomech.2011.11.004.
41. Crews RT, Yalla SV, Dhatt N, et al. Monitoring location-specific physical activity via integration of Accelerometry and Geotechnology within patients with or at risk of diabetic foot ulcers: a technological report. J Diabetes Sci Technol. 2016;11:899–903. https://doi.org/10.1177/1932296816651631.
42. Sheahan H, Canning K, Refausse N, et al. Differences in the daily activity of patients with diabetic foot ulcers compared to controls in their free-living environments. Int Wound J. 2017;14:1175–82. https://doi.org/10.1111/iwj.12782.
43. de Lange T, Melai T (2009) Foot loading during different walking speeds: is THERE a balance between peak pressures and pressure-time integrals?
44. Tong JWK, Kong PW. Association between foot type and lower extremity injuries: systematic literature review with meta-analysis. J Orthop Sports Phys Ther. 2013;43:700–14. https://doi.org/10.2519/jospt.2013.4225.
45. Cacace LA, Hillstrom HJ, Dufour AB, Hannan MT. The association between pes planus foot type and the prevalence of foot disorders: the Framingham foot study. Osteoarthr Cartil. 2013;21:S166–7. https://doi.org/10.1016/j.joca.2013.02.356.
46. Wong L, Hunt A, Burns J, Crosbie J. Effect of foot morphology on center-of-pressure excursion during barefoot walking. J Am Podiatr Med Assoc. 2008;98:112–7. https://doi.org/10.7547/0980112.
47. Cavanagh PR, Morag E, Boulton AJ, et al. The relationship of static foot structure to dynamic foot function. J Biomech. 1997;30:243–50. https://doi.org/10.1016/s0021-9290(96)00136-4.
48. Mootanah R, Song J, Lenhoff MW, et al. Foot type biomechanics part 2: are structure and anthropometrics related to function? Gait Posture. 2013;37:452–6. https://doi.org/10.1016/j.gaitpost.2012.09.008.
49. Koldenhoven RM, Feger MA, Fraser JJ, et al. Surface electromyography and plantar pressure during walking in young adults with chronic ankle instability. Knee Surg Sports Traumatol Arthrosc. 2016;24:1060–70. https://doi.org/10.1007/s00167-016-4015-3.
50. Schmidt H, Sauer LD, Lee SY, et al. Increased in-shoe lateral plantar pressures with chronic ankle instability. Foot Ankle Int. 2011;32:1075–80. https://doi.org/10.3113/FAI.2011.1075.
51. Torp DM, Thomas AC, Hubbard-Turner T, Donovan L. Plantar pressure profile during walking is associated with talar cartilage characteristics in individuals with chronic ankle instability. Clin Biomech (Bristol, Avon). 2022;95:105656. https://doi.org/10.1016/j.clinbiomech.2022.105656.
52. Valderrabano V, Hintermann B, Horisberger M, Fung TS. Ligamentous posttraumatic ankle osteoarthritis. Am J Sports Med. 2006;34:612–20. https://doi.org/10.1177/0363546505281813.
53. Wan D, Bao H, Wang J, et al. Plantar pressure distribution and posture balance during walking in individuals with unilateral chronic ankle instability: an observational study. Med Sci Monit. 2023;29:e940252. https://doi.org/10.12659/MSM.940252.
54. Gerrard JM, Bonanno DR, Whittaker GA, Landorf KB. Effect of different orthotic materials on plantar pressures: a systematic review. J Foot Ankle Res. 2020;13:35. https://doi.org/10.1186/s13047-020-00401-3.
55. Tenten-Diepenmaat M, Dekker J, Heymans MW, et al. Systematic review on the comparative effectiveness of foot orthoses in patients with rheumatoid arthritis. J Foot Ankle Res. 2019;12:32. https://doi.org/10.1186/s13047-019-0338-x.
56. Herchenröder M, Wilfling D, Steinhäuser J. Evidence for foot orthoses for adults with flatfoot: a systematic review. J Foot Ankle Res. 2021;14:57. https://doi.org/10.1186/s13047-021-00499-z.

57. Hsu C-Y, Wang C-S, Lin K-W, et al. Biomechanical analysis of the FlatFoot with different 3D-printed insoles on the lower extremities. Bioengineering (Basel). 2022;9:563. https://doi.org/10.3390/bioengineering9100563.
58. Jiang Y, Wang D, Ying J, et al. Design and preliminary validation of individual customized insole for adults with flexible flatfeet based on the plantar pressure redistribution. Sensors (Basel). 2021;21:1780. https://doi.org/10.3390/s21051780.
59. Collings R, Freeman J, Latour JM, Paton J. Footwear and insole design features for offloading the diabetic at risk foot—a systematic review and meta-analyses. Endocrinol Diabet Metab. 2021;4:e00132. https://doi.org/10.1002/edm2.132.
60. Arts MLJ, de Haart M, Waaijman R, et al. Data-driven directions for effective footwear provision for the high-risk diabetic foot. Diabet Med. 2015;32:790–7. https://doi.org/10.1111/dme.12741.
61. McCluskey WP, Lovell WW, Cummings RJ. The cavovarus foot deformity. Etiology and management. Clin Orthop Relat Res. 1989;247:27–37.
62. Mubarak SJ, Van Valin SE. Osteotomies of the foot for cavus deformities in children. J Pediatr Orthop. 2009;29:294–9. https://doi.org/10.1097/BPO.0b013e31819aad20.
63. Sherman FC, Westin GW. Plantar release in the correction of deformities of the foot in childhood. J Bone Joint Surg Am. 1981;63:1382–9.
64. Jahss MH. Evaluation of the cavus foot for orthopedic treatment. Clin Orthop Relat Res. 1983;181:52–63.
65. Paulos L, Coleman SS, Samuelson KM. Pes cavovarus. Review of a surgical approach using selective soft-tissue procedures. J Bone Joint Surg Am. 1980;62:942–53.
66. Kwon YU, Kim HW, Hwang JH, et al. Changes in dynamic Pedobarography after extensive Plantarmedial release for paralytic pes Cavovarus. Yonsei Med J. 2014;55:766. https://doi.org/10.3349/ymj.2014.55.3.766.
67. Hagen L, Pape JP, Kostakev M, Peterlein C-D. Pedobarographic changes during first month after subtalar extra-articular screw arthroereisis (SESA) operation of juvenile flexible flatfoot. Arch Orthop Trauma Surg. 2020;140:313–20. https://doi.org/10.1007/s00402-019-03230-7.
68. El-Hilaly R, El-Sherbini MH, Abd-Ella MM, Omran AA. Radiological outcome of calcaneo-cuboid-cuneiform osteotomies for planovalgus feet in cerebral palsy children: relationship with pedobarography. Foot Ankle Surg. 2019;25:462–8. https://doi.org/10.1016/j.fas.2018.02.019.
69. Hösl M, Böhm H, Oestreich C, et al. Self-perceived foot function and pain in children and adolescents with flexible flatfeet—relationship between dynamic pedobarography and the foot function index. Gait Posture. 2020;77:225–30. https://doi.org/10.1016/j.gaitpost.2020.01.014.
70. Hirschmüller A, Konstantinidis L, Baur H, et al. Do changes in dynamic plantar pressure distribution, strength capacity and postural control after intra-articular calcaneal fracture correlate with clinical and radiological outcome? Injury. 2011;42:1135–43.
71. Klöpfer-Krämer I, Brand A, Kröger I, et al. Development of the center of pressure velocity in the healing process after intra-articular calcaneus fractures. Gait Posture. 2022;95:135–40. https://doi.org/10.1016/j.gaitpost.2022.04.015.

General Examination of the Ankle

6

Erdem Şahin and Altuğ Tanrıöver

6.1 General Principles

Physical examination of ankle is the most important component to diagnose and manage foot and ankle problems. With a systemic foot and ankle examination, precise diagnosis can be established as most of the structures are palpable, and specific joint and muscle functions can be evaluated. Having a routine order of palpation and manipulation of ankle joint and application of the routine sequence for all patients facilitates physical examination and documentation of findings. Through understanding of topographical and functional anatomy helps performing a diagnostic physical examination.

For a proper physical evaluation, examination room should be comfortable, well illuminated, and wide enough to assess standing and walking patterns. Also providing a regular step to simulate stair climbing and height blocks to determine extremity discrepancy is helpful (Fig. 6.1).

After having detailed history of the complaint with the patient's medical background, both lower extremities should be visualized completely for general alignment. Pelvic asymmetry, length discrepancy, hip and knee alignment disorders should be noted. For foot and ankle examination, the patient should sit comfortably with both legs dangled, with the physician sitting in front of the patient. This position allows physician to be oriented to the anatomy better (Fig. 6.2). Systemic physical examination requires a sequential approach, starting from general visual appearance to provocative tests. If painful tests are performed before simple palpation, it may interfere determination of tender points and disturb the patient's compliance to further examination. Proper sequence of physical examination is as follows:

E. Şahin (✉) · A. Tanrıöver
Cankaya Orthopedics, Ankara, Turkey

S. Sharma et al. (eds.), *Clinical and Radiological Examination of the Foot and Ankle*, https://doi.org/10.1007/978-981-97-4202-8_6

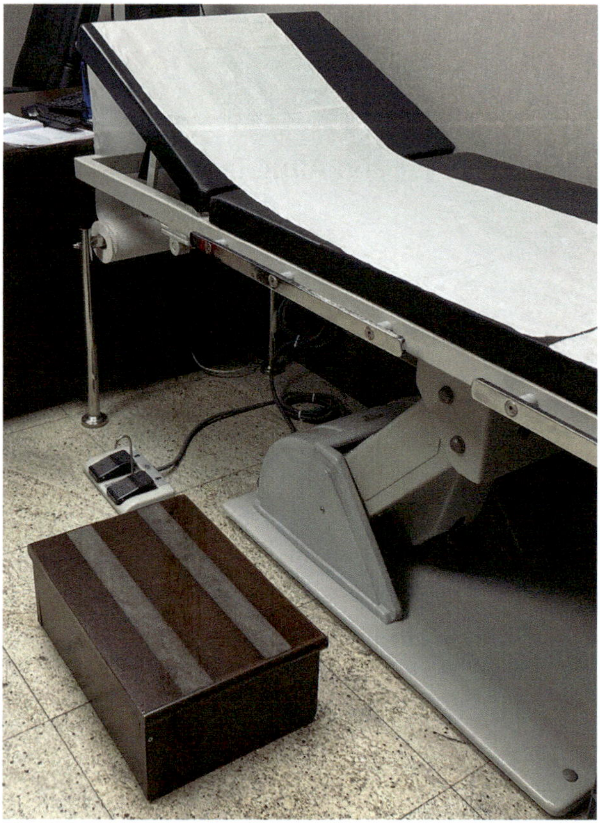

Fig. 6.1 A comfortable examination bed and a regular size step are required for a proper foot and ankle examination

1. Inspection of general anatomy from anterior, lateral, and medial sides and posterior.
2. Sitting examination.
   (a) Palpation, evaluation of range of motion.
   (b) Muscle function.
   (c) Ligament examination.
   (d) Neurological examination.
3. Standing examination.
   (a) Alignment.
   (b) Instability.
4. Gait examination.
   (a) Muscle balance and dynamic deformity analysis.
   (b) Neurologic conditions.
   (c) Shoe examination.

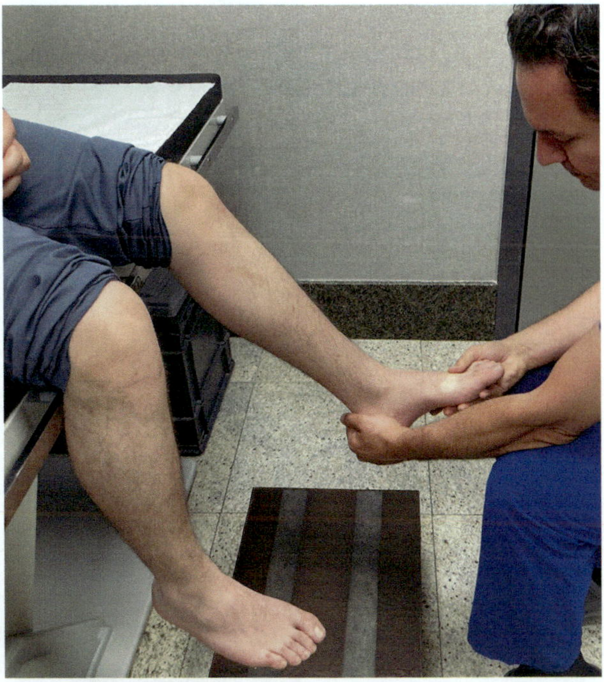

Fig. 6.2 The patient should sit comfortably with both legs dangled, with the physician sitting in front of the patient

6.2 Inspection of the Ankle

Surface anatomy of the ankle joint gives substantial number of clues to diagnosis and general medical status of the patient. Ankle should be visualized from anterior and posterior. Skin condition, hair pattern, discoloration, oedema are important to note. Both lower extremities should be inspected for any asymmetry. With single extremity inspection atrophy of muscles, single sided neurovascular problems may be missed. Especially Achilles tendon thickness relative to normal side is important to note for diagnosis and management of tendinosis, and tendon thickness is also noted for follow up after surgical treatment of tendon rupture or tendinosis.

Topographic inspection depicts possible tendon and synovia problems as most structures are visible (Fig. 6.3). Before palpation and manipulation of the ankle joint, knowing painful regions helps establishing preliminary diagnosis. Avoiding tender points in the beginning helps for a less painful further physical examination.

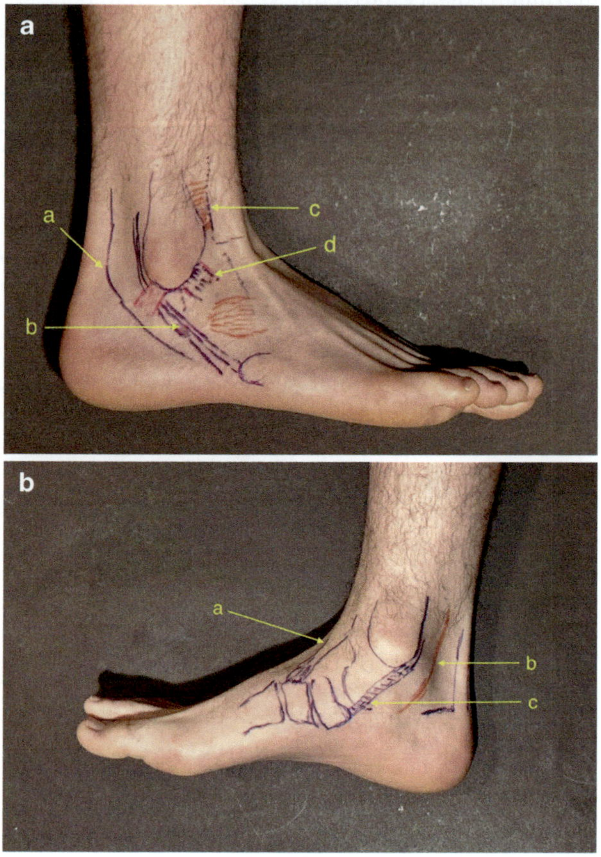

Fig. 6.3 (**a**, **b**) Most structures are visible and palpable in ankle examination A: Sural nerve (**a**), peroneus longus and brevis (**b**), anterior inferior tibiofibular ligament (**c**), anterior talofibular ligament (**d**); B: Tibialis anterior tendon (**a**), posterior tibial artery (**b**), tibialis posterior tendon (**c**)

6.3 Sitting Examination

6.3.1 Palpation and Manipulation of the Ankle

The patient is positioned sitting with both legs dangled freely and visible. Physician sits in front of the patient (Fig. 6.2). After gathering history, possible tender points are taken into consideration. Palpation of tendons and joints should start from non-painful and non-tender points. Before palpation, ask the patient to point out the most painful point or region with single finger helps determination of the pathology. Tendo calcaneus, tibialis anterior, tibialis posterior, and peroneal tendons are easily palpable. Lateral and medial malleoli and bony structures should be palpated for tenderness. Swelling, warmth, and skin changes should be noted. To start, most vital thing to note is vascular status. Tibialis posterior and dorsalis pedis arterial pulses are palpated in both extremities.

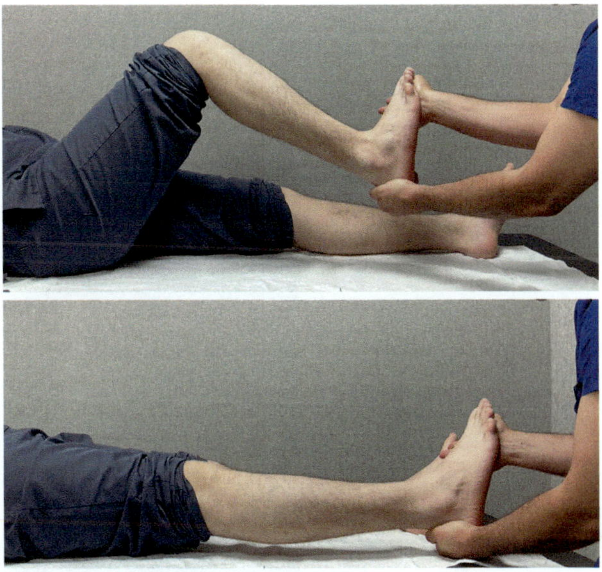

Fig. 6.4 Silfverskiold test. A positive test is more than 10° increased equinus deformity as the knee is extended from flexed position

A proper sequence of orthopedic examination may start with determining major deformities and alignment disorders. Tibiotalar joint range of motion is tested as subtalar joint is held stable with one hand holding calcaneus. To diagnose and specify equinus deformity, ankle is dorsiflexed both with knee flexed and extended (Silfverskiold test) (Fig. 6.4). A positive test is increased equinus deformity as the knee is extended from flexed position more than 10°. Positive test shows an isolated gastrocnemius contracture, which is shown to be an etiologic factor for intractable metatarsalgia and plantar fasciitis ([1].

6.3.2 Ligament Examination

6.3.2.1 Anterior Drawer Test

The most common orthopedic injury is ankle sprain. Usually, anterior talofibular ligament with or without calcaneofibular ligament is torn. In acute injury setting, lateral sided swelling and tenderness is present. For chronic injuries and lateral ankle instability inspection may be normal. Anterior drawer test with comparison of both ankles is a valuable tool to determine lateral instability. The patient sits with knees at 90° flexion. As calcaneus is pulled anteriorly, counter force is applied to tibia. Anterior translation of foot under tibia is considered as positive. Both feet should be tested as asymmetric translation is more valuable. A dimple formation anterior to lateral malleolus during anterior drawer test implies absence of a functioning anterior talofibular ligament (suction sign) (Fig. 6.5).

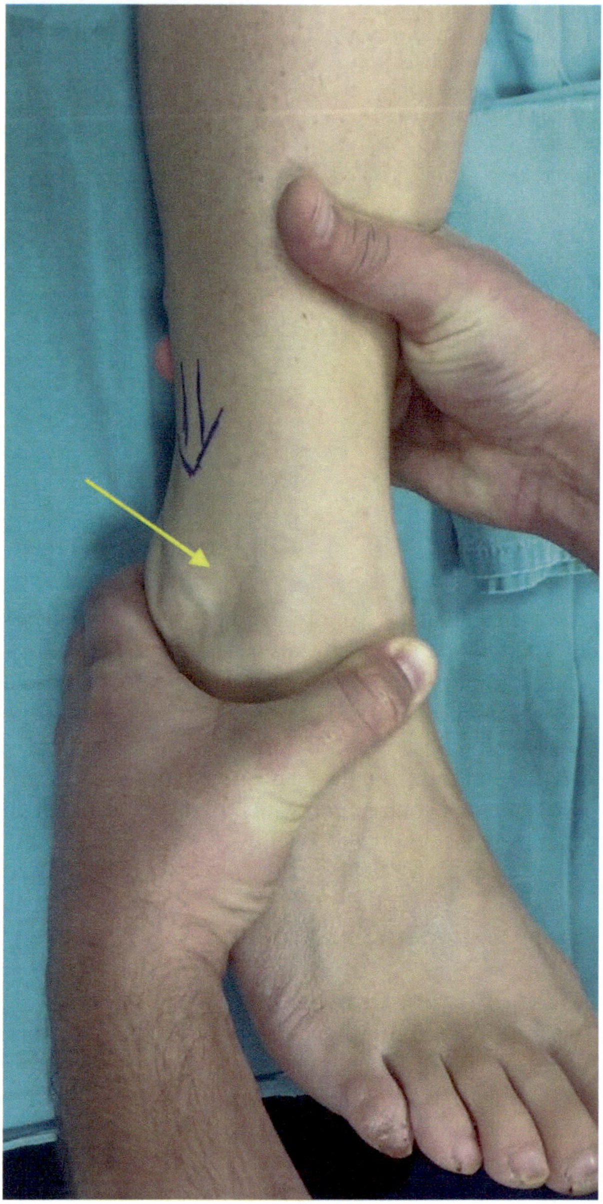

Fig. 6.5 A dimple formation anterior to lateral malleolus during anterior drawer test implies absence of a functioning anterior talofibular ligament (suction sign)

6.3.2.2 Varus Stress Test (Talar Tilt)

Similar to anterior drawer test, calcaneus is held, and subtalar joint is stabilized at neutral dorsiflexion. Ankle joint is forced to varus alignment. Like anterior drawer test, asymmetric varus angulation of ankle joint shows lateral instability, especially calcaneofibular ligament insufficiency (Fig. 6.6). As physician applies varus force,

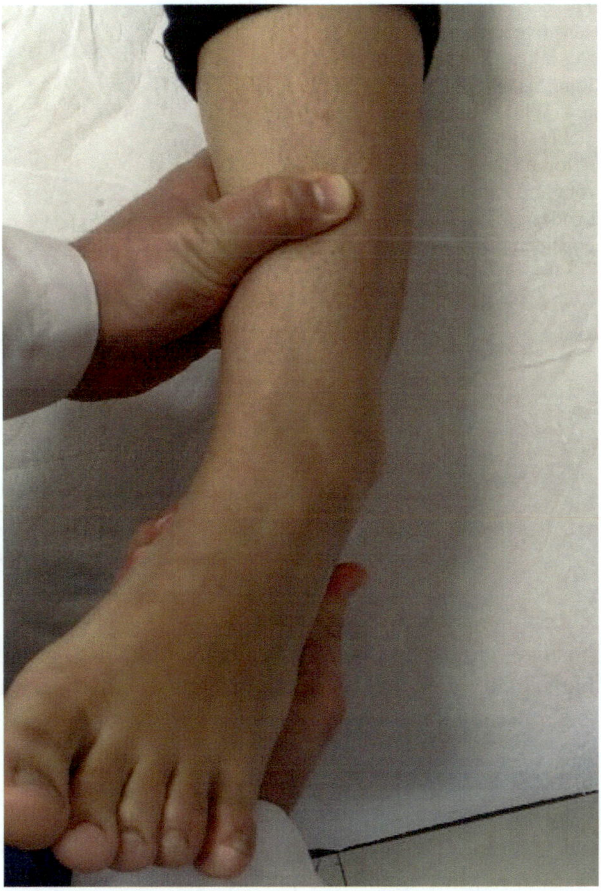

Fig. 6.6 A varus force is applied to calcaneus as tibia is stabilized. Asymmetric varus angulation shows calcaneofibular ligament insufficiency

firm end point for calcaneal medial displacement may indicate a partial lesion however, a soft end point indicates a complete calcaneofibular ligament rupture or major instability.

6.3.2.3 Syndesmosis Examination

Distal tibiofibular syndesmosis injuries are common and usually accompanied by other lesions, most frequently lateral malleolus fractures and deltoid ligament lesions. Unfortunately, these injuries may be missed due to lack of highly sensitive and specific tests or radiologic findings. Suspicion of a possible injury is important for the management as missed injuries frequently have worse prognosis, resulting arthritis [2].

6.3.2.4 Squeeze Test

The patient sits comfortably on the table, with both legs dangled and at 90° flexion. Compression of fibula to the tibia above midpoint of calf provokes pain over syndesmosis area in patients having syndesmosis injury. Provocation of pain is noted

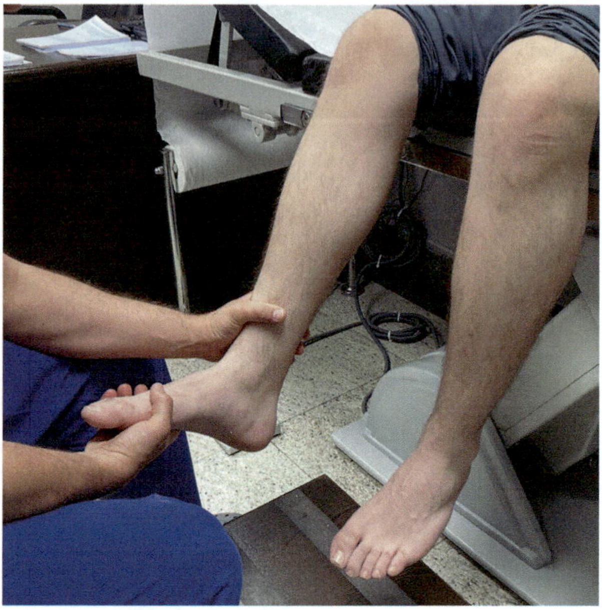

Fig. 6.7 One of the most sensitive and specific tests to evaluate syndesmosis injury is external rotation test. While tibia is held stable with one hand, foot is forced to external rotation. Pain on syndesmotic ligament is considered as ligament tear

positive, indicating injury to the ligament. Pain in proximal fibula is consistent with high ankle sprain and fibula fracture. Proximal fibula fracture after high ankle sprain is commonly known as Maisonneuve fracture.

6.3.2.5 External Rotation Test

Like the squeeze test, the patient is positioned over the side of table. Passive external rotation force is applied with ankle in neutral position. Pain over the syndesmosis area implies ligament rupture (Fig. 6.7).

6.3.2.6 Deltoid Ligament Examination (Medial Ankle Instability)

Medial ankle instability is relatively rare. Chronic instability is usually accompanied by other lesions such as alignment disorders, chondral lesions, or neurologic problems. Acute injuries are easily detected by ecchymosis on medial malleolus, swelling, and tenderness. To test ligament insufficiency, similar to talar tilt test for lateral instability, ankle is held in neutral position. One hand stabilizes tibia while physician applies valgus stress force holding calcaneus. Valgus displacement shows a deltoid ligament lesion. Contralateral side should be checked beforehand. Concomitant syndesmosis lesion should also be kept in mind.

6.3.3 Muscle Examination

Four major muscle contractions and forces are noted: gastrosoleus, tibialis anterior, tibialis posterior, and peroneal muscles. Gastrosoleus contraction is palpable,

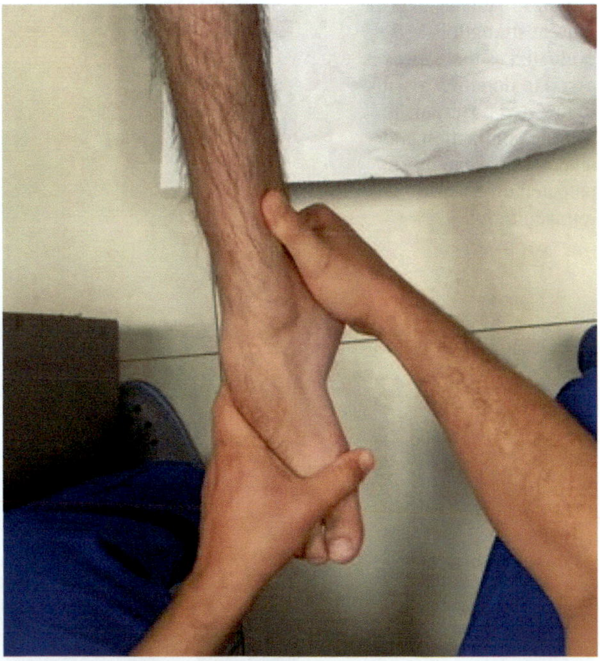

Fig. 6.8 Tibialis posterior muscle strength can be tested as the patient is asked to invert his foot in flexed ankle

strength is assessed with double and single leg raise tests. Tibialis posterior strength can be tested when the patient is sitting, ankle in flexed position, and the patient is asked to do forced inversion (Fig. 6.8). Tibialis anterior contraction and force may be tested as ankle is flexed and inverted; the patient is asked to dorsiflex his ankle (Fig. 6.9). Tibialis anterior muscle contraction and tenderness over tendon and extensor retinaculum is noted. Peroneal muscles are tested with forced eversion of foot. Peroneal tendon subluxation or dislocation is best diagnosed by physical examination. Provocation of subluxation or dislocation is possible as ankle is dorsiflexed and forced to evert (Fig. 6.10).

A regular muscle examination also requires testing long and short flexor and extensor muscles of toes. Toe flexors play important role for ankle function and extensor muscles provide balance and coordination to ankle motions. Intrinsic foot muscle atrophy is important to note, especially for neurologic problems and follow up of diabetic patients as imbalance of these muscles results to severe foot deformities.

6.3.4 Neurologic Examination

Muscle strength tests provide information about motor neuron status. Sensorineural examination is a crucial part of neurological examination. Hypoesthesia, paresthesia, and hyperesthesia of foot and ankle should be checked symmetrically. Subtle hypoesthesia is a common finding for diabetes and peripheric neuropathies.

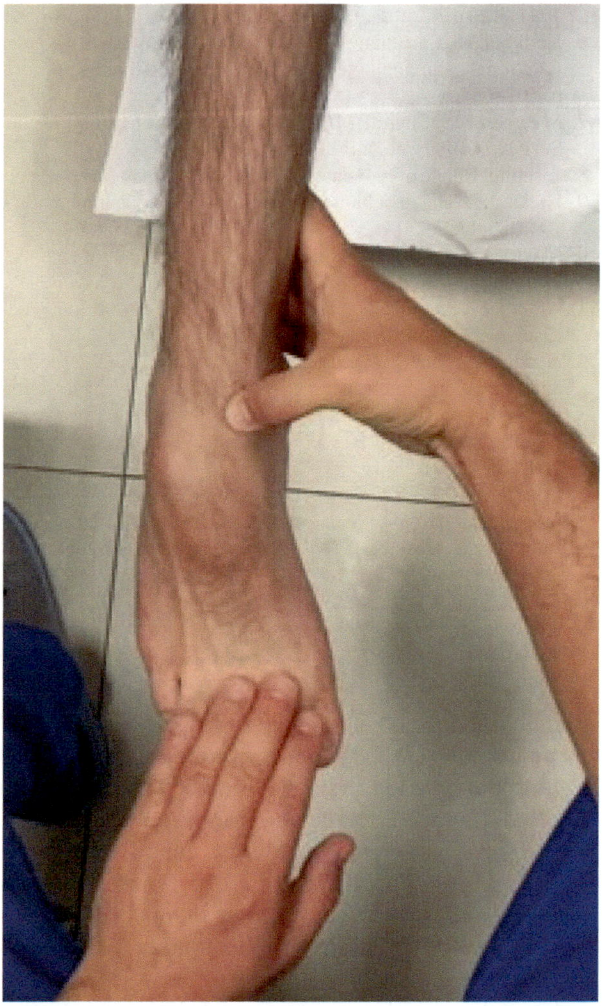

Fig. 6.9 Tibialis anterior muscle strength is evaluated as the patient is asked to dorsiflex his ankle in inverted position. Muscle contraction and tendon tonus can be palpated on extensor retinaculum

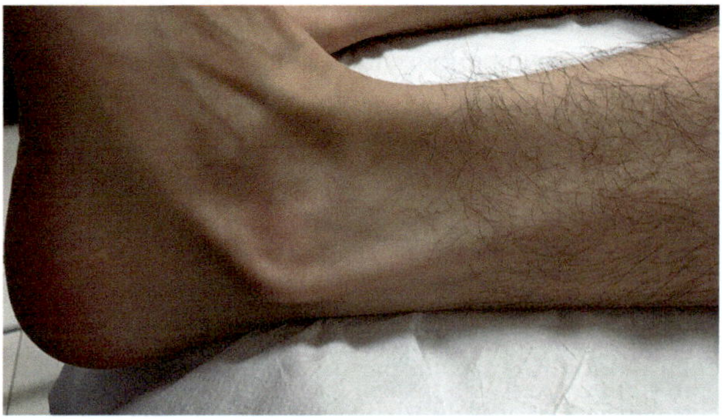

Fig. 6.10 Peroneal tendon subluxation can be diagnosed with physical examination. Provocation of subluxation or dislocation is possible with forced ankle dorsiflexion and eversion

Tarsal tunnel is located posterior to medial malleolus and extends distally. Through the tunnel, posterior tibial nerve gives branches. Tenderness with percussion (Tinnel test) over the tunnel implies a possible nerve entrapment.

Superficial peroneal nerve runs through lateral cruris compartment. It pierces fascia and becomes superficial about 5 cm above lateral malleolus. Paresthesia and tenderness in this region are compatible with nerve entrapment or possible neuroma. Superficial peroneal nerve may be injured in ankle arthroscopy cases as it is directly under skin in anterolateral portal.

Sural and saphenous nerves are vulnerable within commonly performed surgeries such as Achilles tendon repair or medial malleolus fixation.

6.4 Standing Examination

6.4.1 Alignment

Observing the patient in standing position is required to evaluate deformities. Both legs should be visualized from anterior and posterior. Ankle varus or valgus deformities are usually more evident when the patient is weight-bearing (Fig. 6.11). Contractures, equinus, or pes calcaneus is noted both sitting and standing.

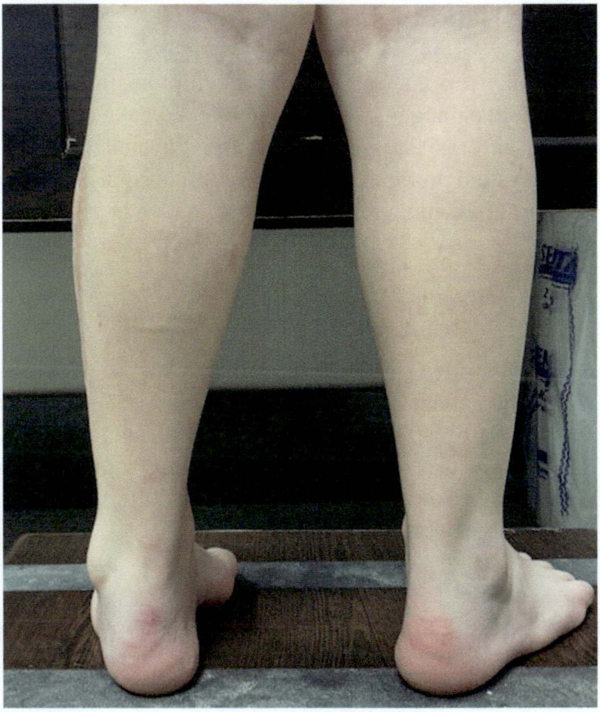

Fig. 6.11 Both lower extremities should be visualized anterior and posteriorly weight-bearing. Calcaneal varus deformity is seen on left side of the patient

6.4.2 Instability

Ankle instability is evaluated with ligament examination. Weight-bearing examination of ankle determine proprioception of the patient. The patient is asked to do single leg raise both with support and without support. Double and single leg raise tests confirms instability.

6.5 Gait Examination

6.5.1 Muscle Balance and Dynamic Deformity Analysis

The patient should be observed as walking barefoot on a straight line, at least 5 or 6 steps, if possible, longer distances. There should not be any distractions to interfere with the patient's stability. Muscle strength or ligaments may seem normal when tested consecutively but imbalance or discoordination becomes apparent when walking. The patient is asked to do tiptoe and heel walking. In addition, hip and knee alignment and range of motion disorders may be predisposing foot and ankle pain, which is best evaluated by gait examination.

6.5.2 Neurologic Conditions

Especially for older patients with symptoms relating to inability to walk, climbing stairs, and pain, gait examination may reveal a diagnosis of an early neurodegenerative disease as Parkinson's disease. Physician should be suspicious of a central nerve system problem in a patient with normal alignment and ligament examination but obscure complaints and symptoms.

6.5.3 Shoe Examination

It is a good idea to check the patient's shoes. Wearing characteristics of the shoes, which the patient has been using for a while, gives a better understanding of ongoing pedobarographic abnormalities. If the lateral side of sole is worn out, the patient may have a varus deformity or dynamic supination, causing lateral overpressure (Fig. 6.12). Likewise, asymmetric wear pattern may imply degenerative deformities or neurologic abnormalities.

Key Points

- With a systemic foot and ankle examination, precise diagnosis can be established as most of the structures are palpable, and specific joint and muscle functions can be evaluated.
- For a proper physical evaluation, examination room should be comfortable, well illuminated, and wide enough to assess standing and walking patterns.

Fig. 6.12 Patient's shoes give information about dynamic gait properties. Lateral of the sole is worn out more than medial side, showing imbalance in weight-bearing

- For foot and ankle examination, the patient should sit comfortably with both legs dangled, with the physician sitting in front of the patient.
- Palpation of tendons and joints should start from non-painful and non-tender points.

Author Contribution Statement E.S. performed the literature search and contributed to the writing of the paper; A.T. contributed to the writing and critically reviewed the article. All authors read and approved the final version.

Funding This work did not receive any specific grant from any funding agency in the public, commercial or not-for-profit sector.

ICMJE Conflict of Interest Statement The authors have nothing to disclosure.

References

1. DiGiovanni CW, Kuo R, Tejwani N, Price R, Hansen ST Jr, Cziernecki J, Sangeorzan BJ. Isolated gastrocnemius tightness. J Bone Joint Surg Am. 2002;84(6):962–70. https://doi.org/10.2106/00004623-200206000-00010. PMID: 12063330.
2. Baltes TPA, Al Sayrafi O, Arnáiz J, Al-Naimi MR, Geertsema C, Geertsema L, Holtzhausen L, D'Hooghe P, Kerkhoffs GMMJ, Tol JL. Acute clinical evaluation for syndesmosis injury has high diagnostic value. Knee Surg Sports Traumatol Arthrosc. 2022;30(11):3871–80. https://doi.org/10.1007/s00167-022-06989-2. Epub 2022 May 4. PMID: 35508553; PMCID: PMC9568458.

General Examination of the Foot

7

Vishnu Baburaj, Prasoon Kumar, and Sandeep Patel

7.1 Introduction

Physical examination remains a critical diagnostic tool for assessing the foot and ankle. This chapter aims to provide a systematic approach to performing a clinical examination of the foot. A thorough physical examination should first look for any abnormalities by a thorough inspection, examine structural integrity through palpation, mobility, and strength testing while also considering epidemiologic factors such as age, gender, employment, and activities. With this information in hand, physicians can individualize the history, physical examination, and treatment of patients with foot and ankle problems.

7.2 Preparing for the Examination

Before delving into the intricacies of the foot examination, it is imperative to set the stage, creating an environment conducive to both accurate assessment and patient comfort. This preparatory phase is instrumental in ensuring that the examination proceeds smoothly and yields valuable insights.

Make sure the examination room is well-lit. Good lighting helps us see details clearly, ensuring nothing is missed during the examination. Ensure the patient is comfortable on the examination table or chair. This helps them feel relaxed and less anxious, especially if they have foot or ankle pain. Talk to the patient openly. Explain what you will do during the examination, listen to their questions and concerns, and ask about their medical history. This builds trust and helps them feel involved in

V. Baburaj · P. Kumar · S. Patel (✉)
Foot and Ankle Biomechanics, Experimentation & Research Laboratory, Department of Orthopedics, Postgraduate Institute of Medical Education and Research, Chandigarh, India

S. Sharma et al. (eds.), *Clinical and Radiological Examination of the Foot and Ankle*, https://doi.org/10.1007/978-981-97-4202-8_7

their care. Before starting, get the patient's permission for the examination. Tell them about any potential discomfort or risks involved. This ensures they understand and agree to the process. Keep talking to the patient during the examination. Let them know what you are doing and why. If they feel any discomfort, encourage them to tell you so you can adjust as needed.

7.3 Physical Examination

The examination should begin the moment the patient enters into the clinic. Look for any gait abnormalities that could potentially point toward foot pathologies. Also take a note of any walking aids the patient might be using such as canes or crutches. To facilitate a comprehensive examination, it is ideal for patients to be adequately exposed, allowing for a thorough evaluation of the lower extremities. Patients should ideally wear shorts, enabling easy access to the lower limbs while barefoot.

For ease of understanding, we divide the foot examination into three regions: forefoot, midfoot, and hindfoot.

7.3.1 Inspection

7.3.1.1 Forefoot

Deformities: Begin by carefully examining the forefoot for any deformities. Look for common deformities such as hallux valgus (bunion) or hallux varus, as well as metatarsus adductus. These deformities can significantly impact gait and overall foot function. Look out for hammer toes and mallet toe deformities.

Pay close attention to the spaces between the toes. Look for any skin lesions, growths, or other abnormalities. In particular, keep an eye out for signs of fungal infections or the presence of ulcers that may require immediate attention. Look for hair abnormalities that suggest impaired circulation. Inspect closely for any nail changes.

Toe clawing: Check for clawing of the toes, which can be indicative of inflammatory arthropathies [1]. Clawing can lead to subsequent ulcerations over the tips of the toes, making it a crucial observation in the examination.

7.3.1.2 Midfoot

Assess the midfoot for any loss of arch or an exaggerated arch (pes planus or pes cavus, respectively). Changes in the arch can significantly impact the foot's biomechanics and may contribute to various musculoskeletal issues [2].

7.3.1.3 Hindfoot

Heel alignment: Examine the alignment of the heel, making a note of heel varus (inward tilt) or valgus (outward tilt), which can be associated with various conditions affecting gait and weight distribution such as a cavovarus or a planovalgus foot. Assess for "peekaboo sign" and "too many toes sign" [3, 4]. Normally, on inspection from the front of the foot, the medial aspect of the heel should not be visible. If the medial heel pad is seen, it is referred to as "peekaboo sign," indicating a cavovarus foot. Conversely, on inspecting from the back of the foot, only the fourth and fifth toe should be visible laterally. If any more toes are seen, it is called a "too many toes sign," typical of a planovalgus foot.

Pay attention to the width of the heels. Broadening of the heels may be indicative of a malunited calcaneus fracture, which can lead to persistent discomfort and altered gait. Look for any areas of swelling of the hindfoot. Swelling can signal various issues, such as an inflamed tendon sheath in the peroneal tendons or tibialis posterior. Additionally, observe for swelling over the calcaneal spur, which may indicate insertional Achilles tendinitis [5].

Inspection of the sole: Examine the sole of the foot for the presence of callosities, which can indicate areas of excessive pressure or friction. Additionally, scrutinize the sole for any ulcerations, a particularly concerning finding in patients with diabetes or compromised circulation [6].

7.3.2 Palpation

Each and every bone and joint should be palpated for abnormalities. Before starting this examination, enquire from the patient about potential sites of pain and tenderness so as to avoid excessive pain while palpating those areas. Use the dorsum of the fingers to assess skin temperature and compare it with the opposite side. Next, adhere to a structured approach to palpating each important structure so that no part is overlooked. Commence the examination by palpating the metatarsophalangeal (MTP) joints. Tenderness over the first MTP joint here can have a range of differential diagnoses, including hallux rigidus and gouty arthritis [7]. Tenderness over the plantar surface could indicate plantar plate rupture and sesamoid bone pathologies.

Proceed to assess the region over the Lisfranc and Chopart joints. Tenderness in these areas can raise suspicions of inflammatory arthritis, necessitating further evaluation. A tender Lisfranc joint with a history of trauma could indicate a chronic Lisfranc injury and secondary arthritis. The base of the fifth metatarsal should be specifically palpated to look for signs of injury. Palpate the navicular bone for any tenderness. The presence of tenderness in this area may raise suspicion of a stress fracture, warranting further investigation. Check for tenderness just beneath the fibula. Subfibular tenderness in a horseshoe pattern may indicate subtalar arthritis, necessitating a comprehensive assessment. Malunited calcaneal fractures with lateral wall exostoses could also lead to subfibular impingement and tenderness.

Gently palpate over the spring ligament, checking for areas of tenderness. Tenderness in this region could signify ligamentous issues, potentially contributing to arch abnormalities. Also palpate over ligaments such as anterior and posterior talofibular ligament and calcaneofibular ligament. Examine the tibialis posterior tendon for any tenderness. Tenderness in this area may be associated with tendonitis or dysfunction, indicating an early stage of adult acquired flat foot [8].

Tenderness in the heel region can be associated with conditions such as plantar fasciitis or insertional Achilles tendinitis. Palpate the Achilles tendon along its length, scrutinizing for tenderness, gaps, or thickening.

7.3.3 Range of Motion and Muscle Strength

Assess for active and passive range of motion (ROM) in each individual joint. Instruct the patient to carry out plantar and dorsiflexion of the ankle, as well as pronation and supination of the foot comparing with the opposite side. Next, proceed to assess passive movements. Typically, the joints should demonstrate smooth pain-free motion and a firm end-point. If any deformity is identified, check if it flexible (correctable) or rigid.

Abnormal hypermobility may be seen in the first MTP joint in hallux valgus. The normal ROM within the first MTP joint is usually restricted to sagittal movements. In hallux valgus, abnormal motion such as abduction and eversion may be observed here [9]. One hand should stabilize the midfoot while assessing for ROM in the hallux. Impaired motion within the first MTP joint is seen in hallux rigidus. It leads to an atypical gait pattern marked by foot supination, walking on the outer border of the foot, and a compensatory hyperextension at the hallux interphalangeal joints; ultimately leading to degenerative arthritis.

Assess adduction and abduction at the level of midfoot joints. Make sure that one hand is used to stabilize the talus and calcaneus before looking at midfoot movements. Subtalar movements are assessed with the patient lying down in a prone position with the knee flexed to 90°. The ankle is kept in maximal dorsiflexion so as to limit side-to-side ankle motion, and the midfoot joints are locked by keeping the foot inverted (Fig. 7.1). The heel is then turned inward and outward to assess subtalar motion. Ankle ROM is assessed in both knee flexion and extension to assess for the gastrocnemius tightness, again with midfoot joints locked in inversion (Silfverskiöld test, Fig. 7.2) [10]. Various pathologies such as bone spurs, accessory bones, and arthritis tend to cause limitations in ankle ROM.

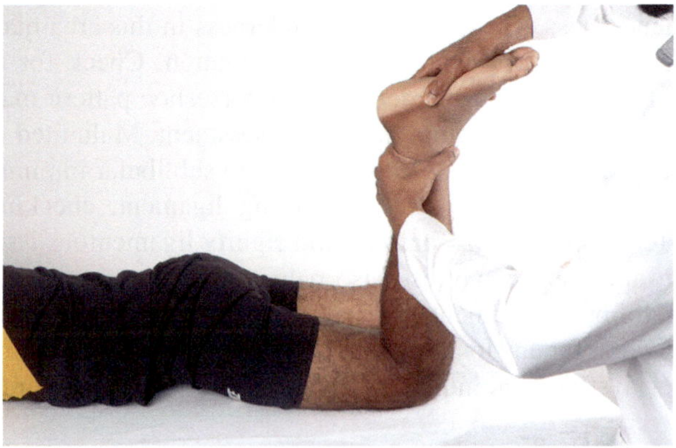

Fig. 7.1 Examination to assess subtalar movements

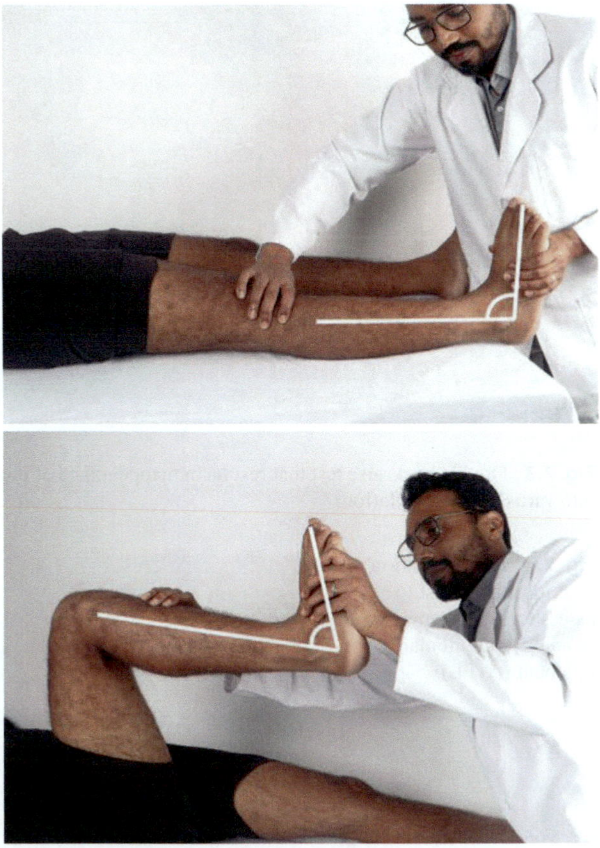

Fig. 7.2 Demonstration of Silfverskiöld test to assess for gastrocnemius tightness

Muscle strength is assessed by asking the patient to walk with the foot held in dorsiflexion, with the heel bearing the entire weight, followed by walking tip-toed on a plantar flexed foot. They are also asked to walk with supinated and pronated feet, bearing weight on the lateral and medial borders of the foot, respectively.

7.3.3.1 Special Tests

Heel Raise Test

The patient is made to stand on his or her toes, while both heels are lifted up. This should result in the hindfoot going into varus, suggesting normal motion of the subtalar joint (Fig. 7.3). If the medial arch reappears on doing this maneuver in a patient with planovalgus feet, it indicates flexible pes planus [11].

Anterior Drawer Test

With the patient in sitting position and feet hanging, one hand is used to stabilize the patient's leg just above the ankle. The other hand is used to pull the foot forward while maintaining the ankle in plantarflexion. The anterolateral aspect of the talus is

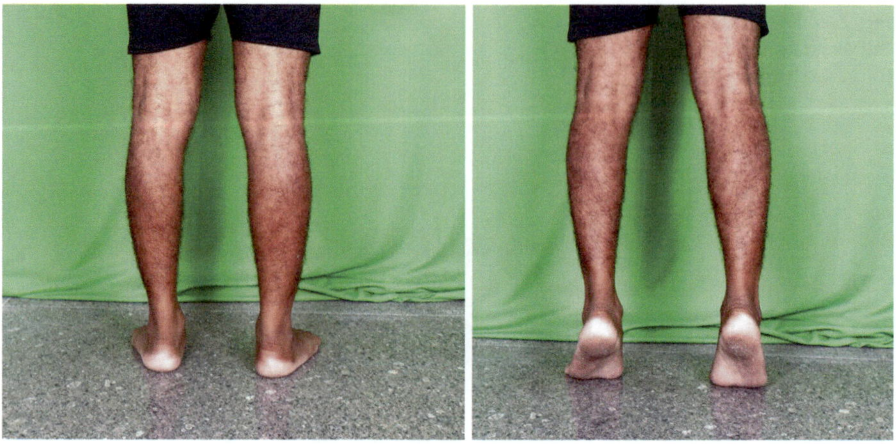

Fig. 7.3 Double heel raise test that results in reappearance of the medial arche and hindfoot going into varus in flexible flatfoot

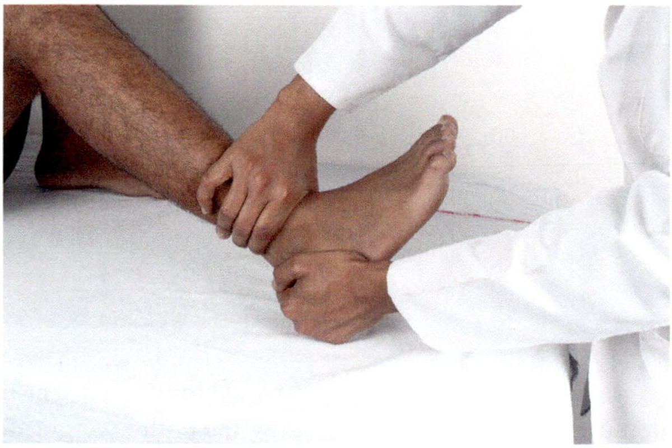

Fig. 7.4 Demonstration of anterior drawer test to assess for the competency of anterior talofibular ligament (ATFL)

observed for excessive anterior translation (Fig. 7.4). A positive test indicates laxity of the lateral ligament complex including anterior talofibular ligament (ATFL) insufficiency [12].

Coleman Block Test

In a cavovarus foot, this test is used to determine if the hindfoot varus is forefoot-driven or not. The patient is asked to stand on a cuboid block such that the lateral border is kept on the block and the medial border is freely hanging. If the hindfoot varus is flexible (forefoot-driven), it will correct and the heel will turn into valgus. In a fixed deformity (hindfoot-driven), the heel will continue to remain in varus (Fig. 7.5) [13].

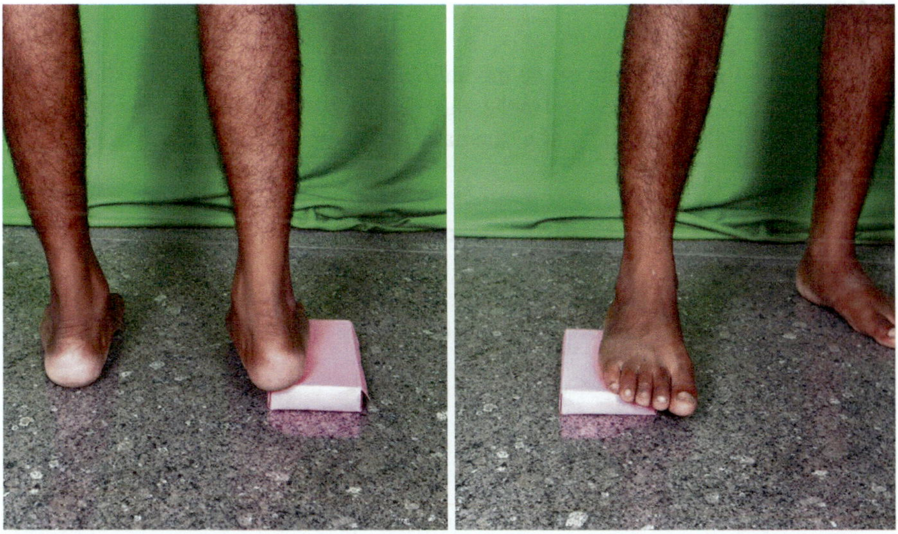

Fig. 7.5 Demonstration of Coleman block test to check for forefoot-driven hindfoot varus

Silfverskiöld Test

With the patient lying down and with the knee extended, maximally dorsiflex the ankle. Now repeat the same maneuver with the knees flexed to 90°. If there is an improvement in ankle dorsiflexion with the knee flexed, it indicates the presence of gastrocnemius tightness (Fig. 7.2).

Tests for Morton's Neuroma

The thumb index finger squeeze test: Apply pressure over the intermetatarsal space by keeping the thumb dorsally and the index finger in the plantar aspect, eliciting pain in a positive test.

Mulder's click test: The foot is dorsiflexed and the metatarsals are squeezed. The presence of a Morton's neuroma is associated with an audible click [14].

7.4 Examination for Nerve Entrapment Syndromes

7.4.1 Tarsal Tunnel Syndrome

It is caused by compression of the tibial nerve in the flexor retinaculum. A Tinnel's sign can be elicited by gently tapping just inferior and posterior to the medial malleolus.

7.4.2 Jogger's Foot

It is because of compression of the medial plantar nerve between the abductor hallucis and the knot of Henry. A Tinel's sign can be produced by tapping about two finger breadths distal and plantar to the navicular tuberosity.

7.4.3 Anterior Tarsal Tunnel Syndrome

It results from compression of the deep peroneal nerve beneath the inferior extensor retinaculum. Tinnel's sign can be elicited by tapping just lateral to the dorsalis pedis arterial pulse at the level of the talonavicular joint.

7.4.4 Baxter's Neuropathy

It occurs because of inferior calcaneal nerve entrapment between the abductor hallucis and quadratus plantae. There would be pain over the medial calcaneus and tenderness over the origin of the abductor hallucis, which on percussion may lead to laterally radiating paresthesia [15].

7.5 Neurovascular Examination

Palpate the anterior tibial, dorsalis pedis, and posterior tibial pulse. Observe the volume and character of the pulse and compare it with the radial artery pulse. Diminished pulse volume can be seen in peripheral vascular disease. A neurologic examination should include assessment of sensory and motor system, along with balance and proprioception.

7.6 Inspection of Footwear

Foot examination is not complete unless a thorough inspection of the patient's footwear is carried out. The condition and design of the patient's footwear can provide significant diagnostic clues. Begin by examining the patient's shoes, differentiating between commercial and surgical footwear. Pay particular attention to the wear patterns present on the soles of the shoes, as these patterns often reveal information about how the foot interacts with the ground. Examine the distribution of wear on the shoe soles. Commonly, wear is most pronounced on the outside of the heel, a sign of typical gait. However, different patterns of wear may indicate abnormalities in foot alignment. Early lateral, proximal, and mid-shoe wear may suggest a supination deformity, while wear on the medial border indicates a pronation deformity. Inquire about the use of any orthotic inserts. Inspect any insoles present within the shoes, and ask the patient about their comfort levels with various insole types gaining insight into which type of insole provides comfort and which may cause discomfort or pain.

7.7 Examination of Other Systems

For any case with a suspected foot pathology, it is essential to carry out a spine examination, as well as hip and knee joint examination. Also, make a note of any limb length discrepancy or any gross spinal deformity.

7.8 Conclusion

A meticulous examination of the foot is crucial for accurate diagnosis and personalized treatment. By systematically inspecting, palpating, and assessing the foot's mobility and strength, clinicians can uncover valuable insights into various pathologies. Additionally, special tests and neurovascular examinations further enhance diagnostic precision. Lastly, attention to footwear and consideration of other systemic factors ensure a comprehensive approach to foot health, ultimately improving patient outcomes.

References

1. Bal A, Aydog E, Aydog ST, Cakci A. Foot deformities in rheumatoid arthritis and relevance of foot function index. Clin Rheumatol. 2006;25:671–5.
2. Rodgers MM. Dynamic foot biomechanics. J Orthop Sports Phys Ther. 1995;21(6):306–16.
3. Shim DW, Suh JW, Park KH, Lee JW, Byun J, Han SH. Diagnosis and operation results for chronic lateral ankle instability with subtle cavovarus deformity and a peek-a-boo heel sign. Yonsei Med J. 2020;61(7):635.
4. Geideman WM, Johnson JE. Posterior tibial tendon dysfunction. J Orthop Sports Phys Ther. 2000;30(2):68–77.
5. Sayana MK, Maffulli N. Insertional Achilles tendinopathy. Foot Ankle Clin. 2005;10(2):309–20.
6. Jeffcoate WJ, Harding KG. Diabetic foot ulcers. Lancet. 2003;361(9368):1545–51.
7. Keen HI, Redmond A, Wakefield RJ, Freeston J, Grainger AJ, Hensor EM, et al. An ultrasonographic study of metatarsophalangeal joint pain: synovitis, structural pathology and their relationship to symptoms and function. Ann Rheum Dis. 2011;70(12):2140–3.
8. Kohls-Gatzoulis J, Angel JC, Singh D, Haddad F, Livingstone J, Berry G. Tibialis posterior dysfunction: a common and treatable cause of adult acquired flatfoot. BMJ. 2004;329(7478):1328–33.
9. Doty JF, Coughlin MJ. Hallux valgus and hypermobility of the first ray: facts and fiction. Int Orthop. 2013;37:1655–60.
10. Singh D. Nils Silfverskiöld (1888–1957) and gastrocnemius contracture. Foot Ankle Surg. 2013;19(2):135–8.
11. Erol K, Karahan AY, Kerimoğlu Ü, Ordahan B, Tekin L, Şahin M, et al. An important cause of pes planus: the posterior tibial tendon dysfunction. Clin Pract. 2015;5(1):699.

12. Tohyama H, Beynnon BD, Renström PA, Theis MJ, Fleming BC, Pope MH. Biomechanical analysis of the ankle anterior drawer test for anterior talofibular ligament injuries. J Orthop Res. 1995;13(4):609–14.
13. Younger AS, Hansen ST Jr. Adult cavovarus foot. J Am Acad Orthop Surg. 2005;13(5):302–15.
14. Mahadevan D, Venkatesan M, Bhatt R, Bhatia M. Diagnostic accuracy of clinical tests for Morton's neuroma compared with ultrasonography. J Foot Ankle Surg. 2015;54(4):549–53.
15. Beltran LS, Bencardino J, Ghazikhanian V, Beltran J. Entrapment neuropathies iii: lower limb. Semin Musculoskelet Radiol. 2010;14:501–11.

Ankle Instability

8

João Caetano, Francisco Guerra-Pinto,
and Nuno Côrte-Real

8.1 Introduction

Ankle sprains are the most common musculoskeletal injury, with up to 70% of the general population reporting having incurred an ankle injury during their lifetime [1]. The incidence of an ankle sprain is 3.29 per 1000 person per year, resulting in a tremendous healthcare burden and high societal economic costs [2]. There is a high propensity for recurrence (as high as 80%) and the development of permanent symptoms [3]. Approximately 30% of the patients will develop a condition called chronic ankle instability (CAI), which is characterized by a history of one severe lateral ankle sprain (LAS) followed by recurrent sprains, feeling of ankle "giving away," persistent pain or swelling, or self-reported deficits in ankle function for more than 1 year following the initial episode [3, 4].

The majority of ankle sprains occurs with an inversion–supination trauma with a plantarflexed foot, stressing the lateral ligament complex. There is likely an injury pattern that starts in the anterior talofibular ligament (ATFL) and progresses both anteriorly, trough the anterior capsule and anterior deltoid ligament bundles, and posteriorly, trough the calcaneofibular ligament (CFL), posterior talofibular ligament (PTFL), and peroneal tendons. An isolated rupture or tear of the ATFL, the

J. Caetano (✉) · N. Côrte-Real
Department of Orthopaedics, Hospital de Cascais, Dr. José de Almeida, Alcabideche, Portugal

F. Guerra-Pinto
Department of Orthopaedics, Hospital de Sant'Ana, Parede, Portugal

Department of Orthopaedics, Hospital da Cruz Vermelha, Lisboa, Portugal

Department d'Anatomia i Embriologia Humana, Universitat de Barcelona, Barcelona, Spain

Nova Medical School, Lisbon, Portugal

S. Sharma et al. (eds.), *Clinical and Radiological Examination of the Foot and Ankle*, https://doi.org/10.1007/978-981-97-4202-8_8

most fragile, can be found in up to 75% of the LAS [5]. In around 20–30% of cases, there is a combined rupture of the ATFL and CFL, with data suggesting worse clinical outcome and a delayed or even decreased return to full sports activities for these patients [6].

The clinical examination of an ankle trauma is extremely important, as many ankle problems can be misdiagnosed as simple ankle sprains. The lateral ligamentar complex is not the only structure at risk. Therefore, it is crucial to have a good differential diagnosis system for every acute ankle sprain injury and possible concomitant lesions (osteochondral lesions, bone bruise/contusion, bony avulsions, or other tendon lesions).

An associated syndesmotic lesion can be present in up to 20% of patients, particularly when an external rotation force is applied in a dorsiflexed ankle [7, 8]. These injuries, referred to as "high ankle sprains," are associated with important morbidity and delayed return to sports [9]. The incidence is higher in contact sports and sports where the foot is locked in a boot such as skiing or ice hockey. They are frequently overlooked in the emergency room, with 20% being missed during the initial examination. A combined complete rupture of two of the three syndesmosis components represents a loss of more than 50% of resistance to tibiofibular diastasis and may result in instability [10].

The deltoid ligament complex plays a crucial role in stabilizing the medial ankle joint. The superficial component primarily limits external rotation and resists valgus stress of the ankle, whereas the deep component mainly resists ankle eversion and lateral migration of the talus. Isolated deltoid ligament injuries are rare, accounting for approximately 3–4% of all ankle ligament injuries [11]. They usually result from a combined eversion/external rotation trauma and are frequently associated with complex ligament lesions or fractures on the lateral aspect of the ankle joint. This injury mechanism starts with a deltoid ligament stress, progresses through the syndesmosis, and ends at the proximal fibula, resulting in a Maisonneuve fracture.

8.2 Clinical Examination

8.2.1 Common Clinical Findings and Their Significance

The examination of a patient with an ankle sprain begins in the exact moment when the patient enters the emergency room (ER). The clinician should acknowledge if the patient is able to bear weight and if there is a limp, suggesting a painful gait. Before examining the ankle, the mechanism of injury should be determined to provide clues for injured structures.

One should progress with inspection, noting the degree and location of swelling, and if any ecchymosis is present. A complete rupture is likely if there is edema and hematoma associated with local tenderness and a positive stress test as discussed earlier.

Anterolateral edema has 89% sensitivity and 35% specificity for complete ligamentar rupture [12]. The local palpation brings very relevant information regarding the location of the injured structures after a sprain. Palpation on the anterior aspect of the distal fibula is useful for ATFL evaluation: if there is no pain on this location, a rupture is very unlikely (Se = 96%; Sp = 7%) [13].

Edema and pain proximal to the tibiotalar joint may suggest a syndesmosis lesion. Pain on palpation over the anterior syndesmosis shows good sensitivity (92%) and specificity (79%) for anterior inferior tibiofibular ligament (AITFL) injury [14]. There are several stress tests to evaluate the syndesmosis integrity (see next section), which, when combined with palpation, enhances sensitivity and specificity of our investigation. However, these clinical signs are frequently insufficient to confirm the diagnosis, so complementary imaging is thus essential.

An associated deltoid involvement is indicated by pain, tenderness, local swelling, or ecchymosis on the anteromedial part of the ankle.

Harmon [15] presented a five-step systematic approach to avoid missing potentially serious injuries: (1) palpation of bony structures, (2) palpation of ligamentous structures, (3) assessment of range of motion. (4) testing ankle muscles, and (5) special tests (see next section) (Fig. 8.1).

In the first few days following an ankle supination trauma, swelling and tenderness often limit the patient's ability to tolerate a stress test, preventing the clinician to assess the ligament integrity. Some days after the sprain (usually 5–7 days), the clinical examination will be tolerable and will provide the clinician with relevant information, improving the diagnostic accuracy [16]. This is designated "delayed clinical examination" and it is determinant to the prognosis and therapeutic plan for the patient.

It is also important to determine if there were previous episodes of LAS, as this may represent that the patient already developed chronicity. In patients with CAI, every new ankle sprain does not necessarily result in ligament rupture, as this tipically occurs during the initial episode. Therefore, subsequent episodes may exhibit less pain, swelling, and tenderness compared to the initial episode. The most relevant finding will be a positive stress test, meaning increased ligamentar insufficiency, which will be unpainful and easy to assess.

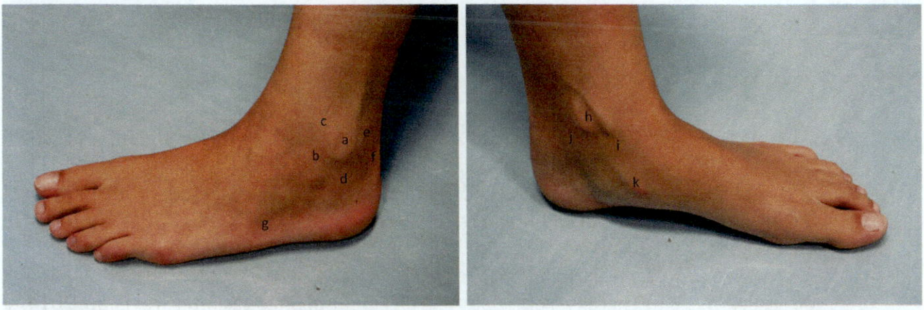

Fig. 8.1 Sequential structures that should not be missed on palpation: (**a**) lateral malleolus, (**b**) ATFL, (**c**) anterior syndesmosis, (**d**) CFL, (**e**) peroneal tendons, (**f**) Achilles tendon, (**g**) base of the fifth metatarsal, (**h**) medial malleolus, (**i**) deltoid ligament, (**j**) posterior tibial tendon (PTT), (**k**) navicular insertion of the PTT

The International Ankle Consortium reached a consensus on the clinical evaluation of lateral ankle sprains (LAS) [17]. The consensus established a minimum standard for clinical diagnosis assessment, emphasizing the importance of identifying the mechanism of injury and evaluating the bones and ligaments of the ankle joint. Additionally, the expert panel developed the Rehabilitation-Oriented Assessment of Sensorimotor Impairments and Mechanical Dysfunction (ROAST) under the International Ankle Consortium. This tool helps clinicians in identifying impairments associated with Chronic Ankle Instability (CAI) [17].

8.2.2 Techniques for Clinical Examination and Their Significance

8.2.2.1 Lateral Ankle Ligamentar Complex Tests

The **anterior drawer test** (Fig. 8.2) classically assesses the integrity of the ATFL. It is performed with the knee bent and the foot in a slightly plantar flexed position. The tibia and fibula are stabilized with one hand while the other translates the foot forward, bringing the talus anteriorly. Some prefer to pull the foot by grabbing the heel. The translation grade should be compared to the contralateral side due to differences in laxity between subjects. The test has a sensitivity of 76% (62–86%) and a specificity of 79% (59–91%) for ATFL rupture [13].

An ATFL rupture might be better evaluated by a rotational vector owing to the intact and competent medial ankle ligaments that will block any pure anterior translation of the talus underneath the tibia (Fig. 8.3). Guerra-Pinto et al. [18] found that the ATFL integrity can be tested using a pure rotational force, a **pivot test** (Fig. 8.4).

The technique for the ankle pivot test uses the same patient positioning as the anterior drawer: with the ankle relaxed, and the limb hanging free, the examiner will

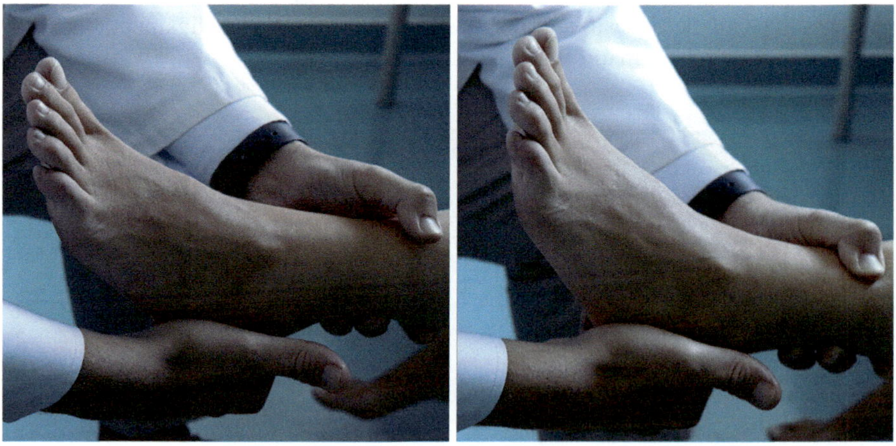

Fig. 8.2 Anterior drawer test—with the foot hanging and the knee slightly bent; the left hand will block the talofibular joint and the right hand holds the patient's foot around the heel and then applies an anterior force. The anterior translation of the foot is compared with the contralateral side

Fig. 8.3 Axial cut showing the position of the deltoid ligament (red line) and its role while blocking anterior talus translation. The intact medial ligaments maintain the strict posteroanterior stability while working has a fulcrum for this rotational movement

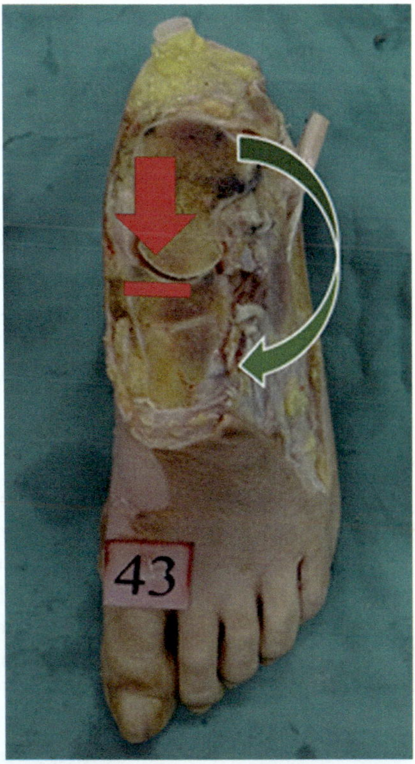

Fig. 8.4 Pivot test—the right thumb fingertip will appreciate the magnitude of the talar rotation gliding while the right hand (second, third, and fourth fingers) forces an internal rotation of the foot over a blocked leg (left hand). The rotation laxity should be compared to the contralateral side

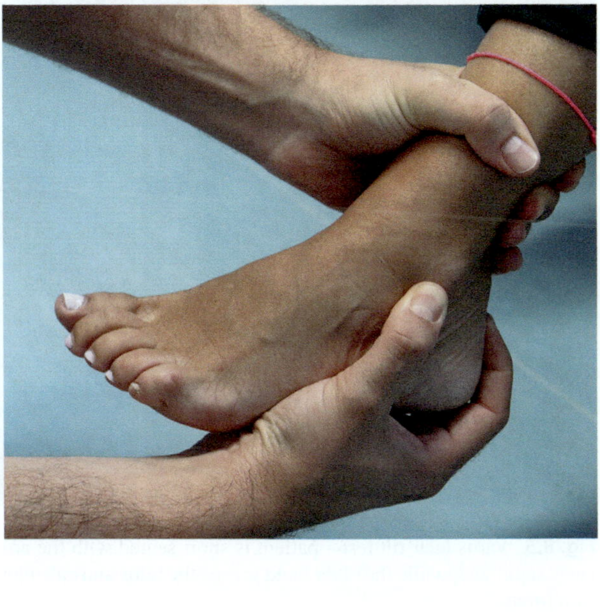

hold the plantar aspect of the patient's right foot in the left palm of his hand; the thumb will embrace the lateral aspect of the rearfoot and reach, with its fingertip, the anterior aspect of the lateral malleolus; the second, third, and fourth fingers will surround the calcaneus on its posterior aspect. The thumb and the second, third, and fourth fingers will thus assume a claw-like positioning; the next movement will be an internal rotational twist with the left hand, around an imaginary fulcrum on the medial side of the ankle; the thumb fingertip, which is feeling the anterior aspect of the lateral malleolus and the lateral aspect of the talus, will appreciate the magnitude of the **ankle pivot test**. A healthy ligament typically allows only 2 or 3 mm of separation between these two bony references. It also provides a firm stop to the movements, whereas a torn ligament to restrict internal rotation of the talus. Significant asymmetry between healthy and injured ankles will suggest a rupture of the ATFL.

The stability of the CFL can be assessed by the **varus talar tilt test** (Fig. 8.5), which consists of applying inversion of the rearfoot on a slightly dorsiflexed ankle, comparing the lateral opening up (lateral talar laxity) with the contralateral side. A difference of more than 10° is considered positive.

The authors suggest that the talar tilt test may be more accurate in diagnosing lateral ankle ligament insufficiency when performed with the foot prepositioned in internal rotation (Fig. 8.6) [19]. This positioning creates an oblique vector that mimics the supination and adduction trauma responsible for lateral ankle ligament rupture, aligning between the pivot test and the traditional talar tilt test.

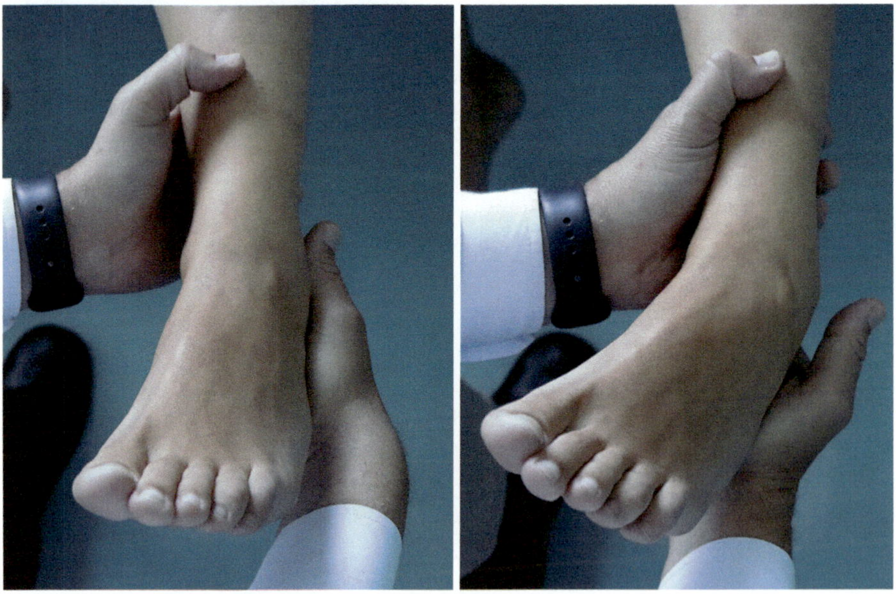

Fig. 8.5 Varus talar tilt test—patient is short seated with the ankle in neutral position. The distal leg is stabilized, while the other hand grasps the talus and calcaneus as a unit and provides an inversion force

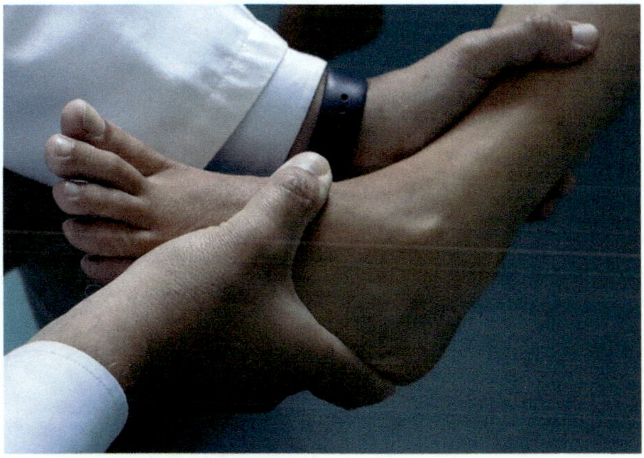

Fig. 8.6 Modified varus talar tilt test—inversion of the rear foot after prepositioning the foot in internal rotation

Table 8.1 Interpretation of the clinical examination in a stepwise approach using the ankle pivot test and the modified varus talar tilt test

	No rupture	Partial ATFL rupture	Complete ATFL rupture	Double (ATFL + CFL) rupture
Ankle pivot test	No displacement Firm end-point	Minimal laxity Firm end-point	Positive pivot test No end-point	Positive pivot test No end-point
Varus talar tilt (w/internal rotation)	–	–	Firm end-point	Positive varus talar tilt (IR) No end-point

A stepwise approach beginning with the **ankle pivot test, followed** by the **modified talar tilt test** (with the foot in internal rotation), can assist in assessing the grade (partial/complete) and extension (single/double) of the ligamentous rupture as shown in Table 8.1.

8.2.2.2 Syndesmosis Tests

The syndesmosis tests are considered positive if pain is elicited. The **external rotation stress test (Frick test)** (Fig. 8.7) is performed by applying an external rotation to the foot with the knees and ankle at 90% position, reproducing the mechanism of injury. It shows good interrater reliability compared to others [20]. The **squeeze test** consists of compressing the tibia against the fibula at the mid-point of the calf, promoting distal tibiofibular distraction. In the **cotton test**, medial and lateral forces are applied to the talus. In the **crossed-leg test**, the patient is seated with the injured leg crossed over healthy leg on the mid-fibula site; then, pressure is applied on the knee toward the floor. Fibular anteroposterior motion can be augmented in case of a ligament tear and will produce a positive **fibular translation test**. Clinical improvement by tightening the ankle with tape at the syndesmosis level (Fig. 8.8) suggests syndesmosis lesion (acts like an external fixator, stabilizing the syndesmosis).

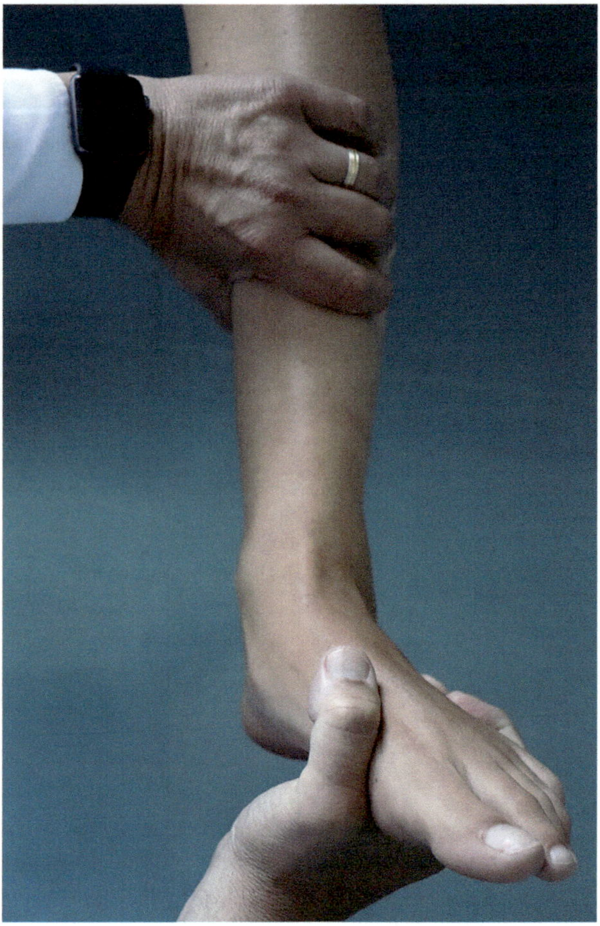

Fig. 8.7 The Frick test (external rotation stress test), to test the syndesmosis, is performed by applying an external rotation to the dorsiflexed foot, with the knee flexed 90°, replicating the mechanism of injury

The syndesmosis stress tests show low sensitivity and high specificity [20], meaning that patients are very likely to actually have a syndesmotic injury when positive. However, the low sensitivity may be responsible for underdiagnosis in urgent care.

8.2.2.3 Deltoid Ligament Tests

The superficial deltoid ligament can be assessed on a relaxed sitting position with hanging leg, using the **external rotation test**. The deep deltoid can be evaluated by the **eversion stress test**, where valgus tilt stress is applied to the heel, while grabbing the tibia with the other hand. An insufficiency of the deltoid ligament complex will also originate a positive **anterior drawer test**, especially if combined with a lateral injury. Applying the same rationale as in the pivot test for the lateral instability evaluation, if we combine the anterior drawer test with an external rotation

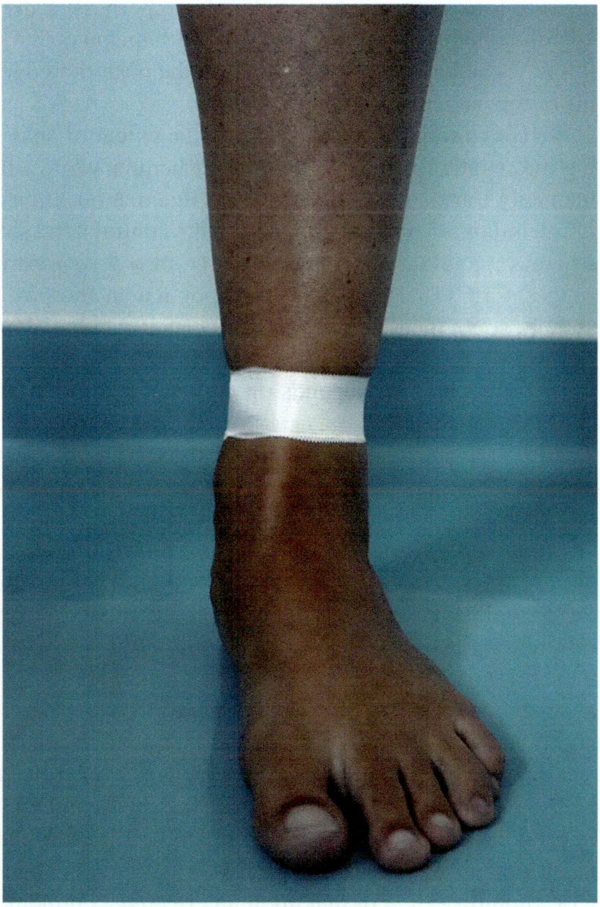

Fig. 8.8 Taping the ankle at the syndesmosis level may improve pain symptoms while walking or jumping (act as an external brace)

component, we might isolate the deltoid for evaluation, forcing an anterolateral migration of the talus.

Whenever an ankle eversion trauma causes pain and instability over the deltoid area, the syndesmosis and proximal fibular section should be palpated to exclude a Maisonneuve fracture or syndesmosis instability.

8.3 Radiological Examination

8.3.1 Common Radiological Findings and Their Significance

8.3.1.1 Lateral Ankle Ligamentar Complex

The Ottawa Ankle Rules (OAR) determine the need for radiographs in LAS in order to avoid unnecessary imaging to exclude an ankle fracture (Se > 97% [21]). The OAR recommends an ankle X-ray series for patients who are unable to bear weight or walk four steps and have palpable tenderness at the posterior edge or tip of the lateral malleolus or at the base of the fifth metatarsal. A foot X-ray series is required if there is palpable bone tenderness at the posterior edge or tip of the medial malleolus or the navicular bone.

Stress radiographs can quantify the extent of ankle-joint laxity by demonstrating the separation of the bony joint structures while a force is applied. The ankle is manually translated in the suspected ligamentous injury: a total anterior translation of >9 mm or > 5 mm compared with the contralateral side may suggest ATFL lesion, whereas a total talar tilt angle of >10° or > 5° compared to the contralateral ankle suggests a CFL injury [22]. Very popular in the past, its use is declining. Some authors doubt their benefit for the decision-making algorithm and the results can be misleading because of low sensitivity, high levels of variability, and inconsistency [23]. The result also depends on patient cooperation, which can be difficult, especially in acute lesions.

Alternatively, to avoid radiation exposure, **arthrometers** provide a cost-effective method for dynamic assessment of ligamentous integrity. These ankle devices stress the lateral ankle ligaments, replicating manual stress tests such as the anterior drawer or talar tilt test, measuring the linear or angular translation in a controlled system. A systematic review [24] found high heterogeneity on the laxity outcomes due to wide range of arthrometers used, their different mechanism of action, and the different assessment procedures. Ankle arthrometry may provide useful objective data for investigation purpose, yet further clinical validation of this technique against other diagnostic tools is needed [23], as well as standardization of the measuring methods [24].

Diagnostic **ultrasound** is an inexpensive, versatile, and portable tool, which can be used in acute phase without stressing the ligaments. Ultrasonography is reliable in differentiating the normal ATFL and CFL from the torn or sprained ligament. It has a sensitivity of 94–100% for ATFL ruptures and a sensitivity of 92–93% for CFL ruptures [23, 25], making it an effective complementary tool for primary evaluation and early diagnosis. It also shows good correlation with intraoperative findings [26]. Common findings include discontinuity, hypoechogenicity or lack of visualization. It may be also used for dynamic assessment of the translation between the articular surfaces and ligamentar behavior under stress. A cadaveric study found good correlation between ultrasonographic and fluoroscopic values measured during simulated anterior drawer and talar tilt test [27]. The main disadvantage is that it does depend on the examiner and the way it is performed.

Magnetic resonance imaging (MRI) is not routinely performed following an acute ankle sprain and is typically reserved for cases involving high-energy trauma or suspected associated lesions, such as osteochondral lesions. MRI plays a crucial role in the subacute phase and in assessing chronic ankle instability (CAI). One of its key advantages is its ability to provide detailed visualization of ankle anatomy, ligament injuries, and any concurrent pathologic conditions resulting from ankle trauma.

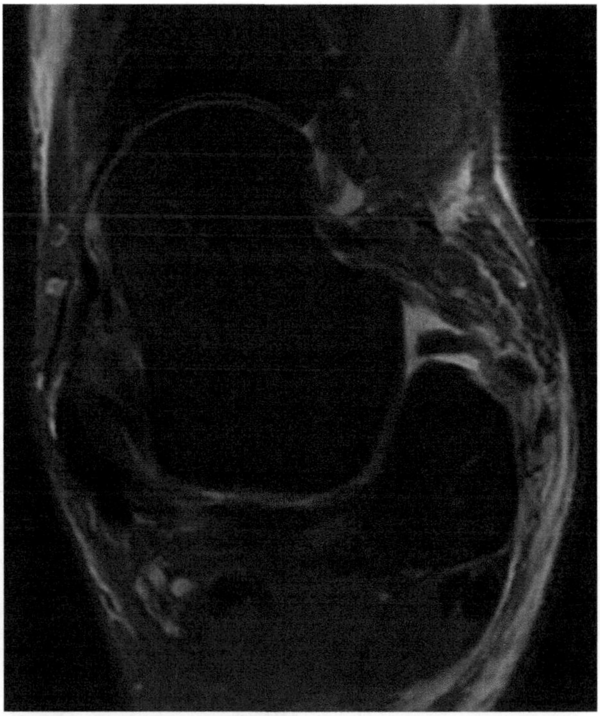

Fig. 8.9 The ATLF is best visualized on oblique axial images. The image shows an acute complete rupture of the ATFL with avulsed component and local effusion in an MRI axial oblique view of the left ankle (proton density [PD]; fast spin-echo [FSE]; fat suppressed [FS] imaging)

In acute ligamentous injuries, MRI reveals discontinuity, detachment (refer to Fig. 8.9), contour irregularity, thinning, and/or increased intraligamentous signal on T2-weighted images (indicative of edema or hemorrhage), while subacute or chronic tears present as ligament heterogeneity, waviness, thickening (refer to Fig. 8.10), thinning, elongation, or absence [28]. The altered morphology of the ligaments is more clearly visualized during the remodeling phase (typically 15–28 days post-injury) [28]. When compared to operative findings, MRI demonstrates a sensitivity and specificity of 100% and 50% for diagnosing rupture of the anterior talofibular ligament (ATFL) and 92% and 100% for the calcaneofibular ligament (CFL), respectively [29].

The MRI is also very useful for evaluation of concomitant injuries, mainly bone contusions (44%), other ligaments involvement, or osteochondral acute lesions (8% of an ALS) [8]. It is also an essential tool for surgery decision or planning. The main limitations of this method are the cost, availability, time, and motion artifacts [30].

8.3.1.2 Syndesmosis

The syndesmosis radiographic study comprises comparative weight-bearing X-rays that includes three projections—anteroposterior, mortise (20° of foot internal rotation), and lateral views—to evaluate three radiographic parameters: the tibiofibular

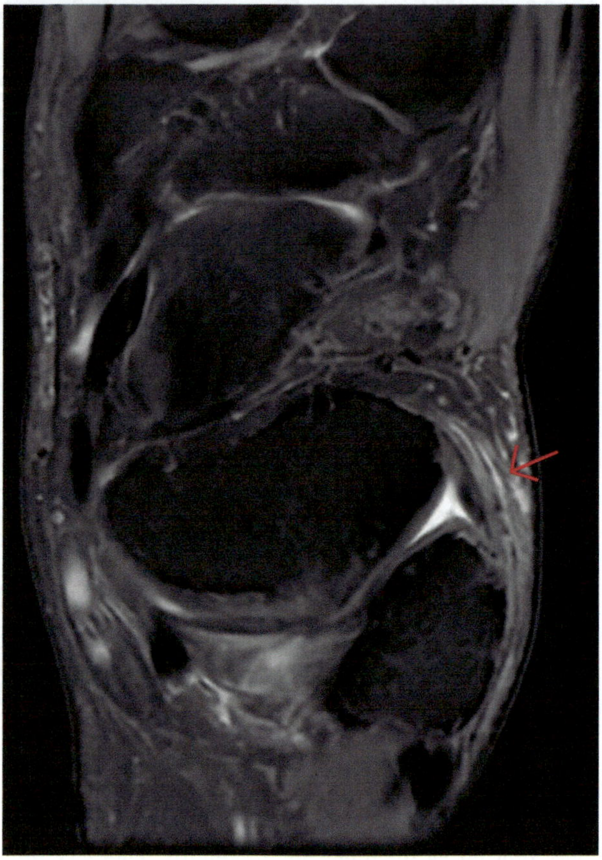

Fig. 8.10 Chronic ATFL partial rupture, with ligament heterogeneity, thickening, and residual edema (PD FSE FS oblique axial, left ankle MRI cut)

clear space (TFCS), the tibiofibular overlap (TFO), and the medial clear space (MCS) (Fig. 8.11 and Table 8.2).

The tibiofibular clear space (TFCS) is defined as the distance between the lateral border of the posterior tibial malleolus and the medial edge of the distal fibular epiphysis, measured approximately 10 mm above the tibial plafond in both the anteroposterior (AP) projection and the mortise view. It is considered pathological if it exceeds 6 mm. The tibiofibular overlap (TFO) measures the distance between the lateral edge of the anterior tibial tubercle and the medial edge of the distal fibular epiphysis, which should be greater than 6 mm in the AP projection and greater than 1 mm in the mortise view. The TFCS may offer greater reliability compared to the TFO because it is not influenced by ankle or foot rotation [31].

The medial clear space (MCS) is defined as the widest distance between the lateral edge of the medial tibial malleolus and the medial edge of the talus, which should be equal to or less than 4 mm, or equal to or less than the space between the tibial plafond and the talar dome [32]. If radiographs are not properly calibrated, comparing measurements with those of a healthy ankle can be beneficial.

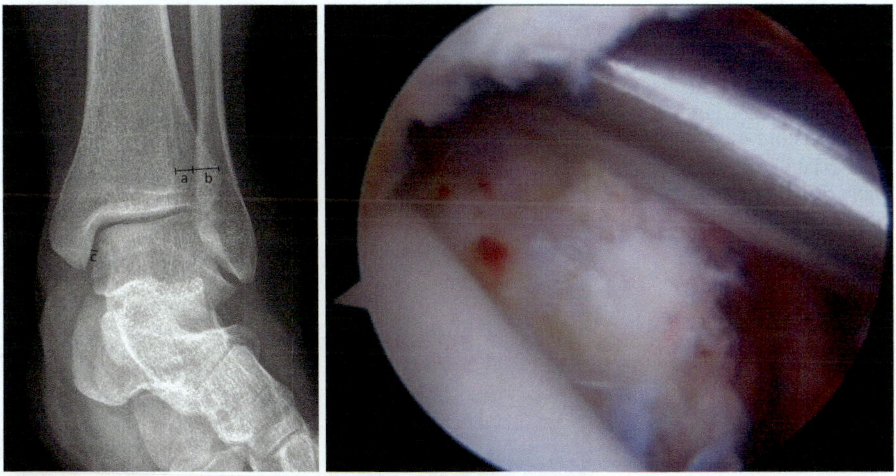

Fig. 8.11 Stress radiograph (forced external rotation) of the left ankle, showing the three measurements that should be made for syndesmosis injury diagnosis: (**a**) tibiofibular clear space (TFCS) and (**b**) tibiofibular overlap (TFO); the measurements should be made 1 cm above the tibiotalar articular line. The TFCS for this patient was 6.9 mm and a syndesmotic instability was confirmed by ankle arthroscopy, with the shaver instrument passing through the anterior syndesmosis without any resistance

Table 8.2 Normal radiographic values for each X-ray projection at 1 cm above the tibiotalar line

	Normal radiographic values	
	Anteroposterior	Mortise
Tibiofibular clear space (TFCS)	≤6 mm	≤6 mm
Tibiofibular overlap (TFO)	≥6 mm	≥1 mm
Medial clear space	≤4 mm	

However, these radiographic measurements demonstrate poor reproducibility, with a sensitivity of 82% and specificity of 75% for TFCS, and 36% and 78% for TFO, respectively. Additionally, these results do not consistently correlate with findings from MRI scans [32]. Even with a syndesmosis ligament rupture, the tibiofibular distance can remain unchanged. Stress radiographs using the Télos apparatus, with forced external rotation or gravity, may detect latent diastasis in the case of a complete syndesmosis tear but exhibit low reliability [33].

Computerized tomography (CT) scans are useful for assessing tibiofibular diastasis, fibular rotation, and joint asymmetry (including peroneal groove anatomy) but have low sensitivity [34]. Some reports discuss the use of weight-bearing CT scans, which can increase diastasis and alter the fibular angulation, but their clinical role remains unclear [35].

Ultrasound can be beneficial for detecting anterior inferior tibiofibular ligament (AITFL) lesions but is less effective in identifying interosseous membrane and posterior inferior tibiofibular ligament (PITFL) lesions, making it less suitable for

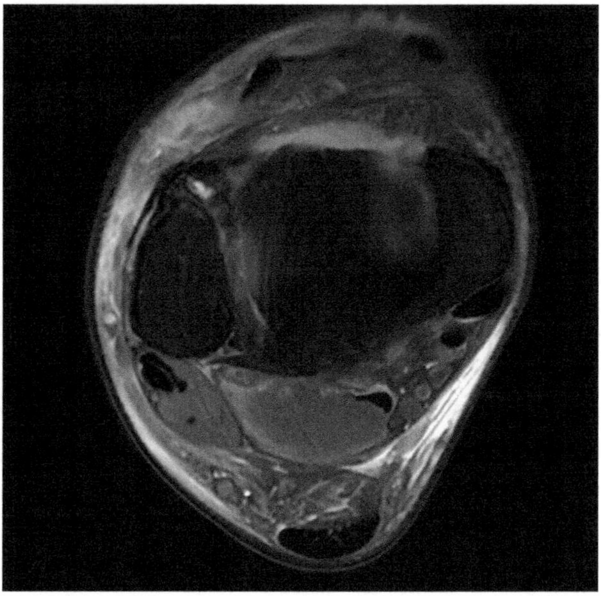

Fig. 8.12 The AITLF is best visualized on axial images. The image shows an acute partial AITFL rupture surrounded by extensive fluid signal in an MRI axial view of the right ankle (PD FSE FS axial, right ankle MRI cut)

syndesmosis assessment. Dynamic ultrasound, although promising, is highly dependent on the examiner, has a steep learning curve, and demonstrates low reproducibility.

The next imaging option is MRI. While ankle arthroscopy is considered the gold standard for diagnosing syndesmotic ligament ruptures, recent findings suggest that MRI offers comparable accuracy [34] without the invasiveness associated with surgical procedures. Takao et al. [36] compared MRI with arthroscopic results, demonstrating MRI's sensitivity and specificity of 100% and 93%, respectively, for AITFL injuries, and 100% sensitivity and specificity for PITFL injuries (Fig. 8.12).

While MRI provides precise information about ligament integrity, it does not stress the ankle or assess tibiofibular stability. Syndesmotic instability can be definitively confirmed through ankle arthroscopy, which allows dynamic evaluation and detection of subtle displacements in multiple planes. Furthermore, this procedure guides surgeons in achieving anatomically precise reduction.

8.3.1.3 Deltoid Ligament

On radiographic evaluation of the medial ligamentar complex, Gibson et al. demonstrated that if there is more than 4 mm of medial clear space after an external rotation stress test, it is considered a medial instability (Fig. 8.13) [37]. If an eversion trauma caused the deltoid lesion, a knee radiograph should be requested to exclude a Maisonneuve fracture (Fig. 8.14).

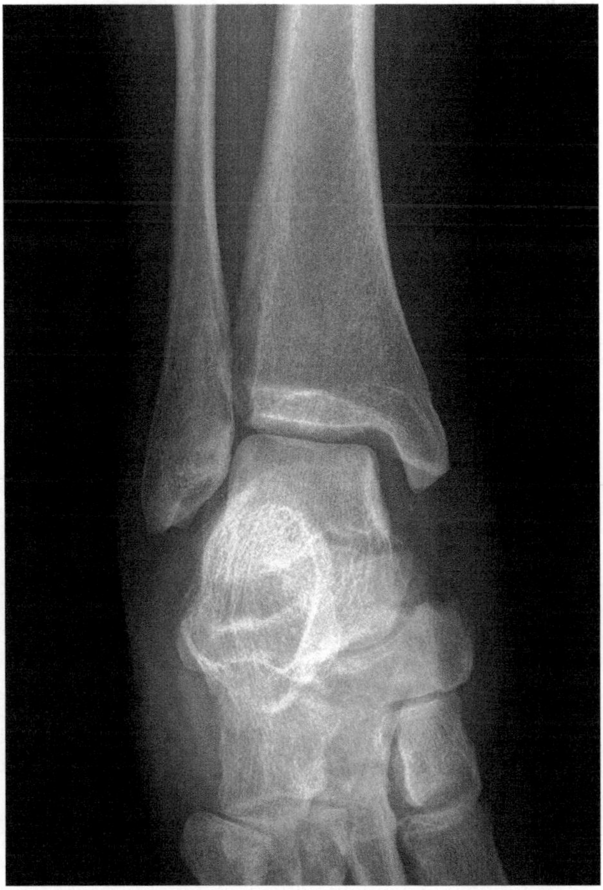

Fig. 8.13 The AP ankle radiograph shows a Deltoid insufficiency after an external rotation ankle trauma, with evident medial diastasis

In cases of chronic medial ankle instability, standard weight-bearing radiographs can reveal segmental deformities and associated lesions, such as osteochondral lesions (OCL), bony impingements, or osteoarthritic changes. Comparing with the contralateral ankle may be necessary to identify more subtle injuries.

Ultrasound is a valuable tool for initial screening of suspected acute deltoid ligament rupture. Both in cases of acute and chronic medial ankle instability, MRI is crucial for characterizing the extent and location of the rupture, and has demonstrated reliability [38]. MRI helps in preoperative planning and can influence the treatment approach.

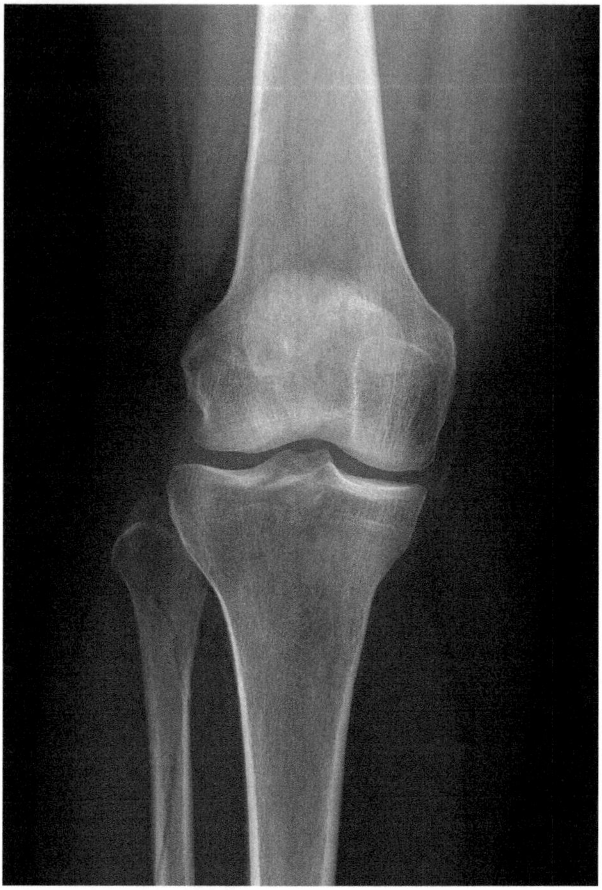

Fig. 8.14 On the same patient, the knee AP radiograph shows a proximal fibular fracture; this combination is the definition of a Maisonneuve fracture

Key Points

- Lateral ankle sprain injury is the most common musculoskeletal injury and is often an inversion–supination trauma. The clinical examination of an ankle sprain is of extreme importance in order to avoid missing serious injuries and prevent the evolution to chronic ankle instability.
- Determining the mechanism of injury is essential to guide our investigation. A delayed clinical examination, 5–7 days after the trauma, will be more tolerable regarding stress tests, improving its diagnostic accuracy.
- The anterior drawer and talar tilt tests are classic stress tests to test the ATFL and CFL, respectively. Due to their low sensitivity, the authors consider the pivot test and the modified talar tilt test a better option to test these ligaments. Nevertheless, their applicability still lacks scientific validation.
- Ultrasound is an inexpensive and accessible tool with good reliability for ATFL and CFL evaluation. MRI offers 100% sensibility for ATFL ruptures and is very useful to detect concomitant injuries.

- Syndesmosis injuries result from external rotation forces and are frequently overlooked in the emergency room (ER). They are associated with persistent symptoms and delayed return to sports.
- The syndesmosis stress tests show low sensitivity for ligament rupture diagnosis, thus complementary imaging is essential to confirm the clinical suspicion.
- Stress radiographs exhibit poor reproducibility for syndesmotic evaluation and do not correlate well with MRI findings. MRI appears to have similar accuracy as an ankle arthroscopy in confirming ligament ruptures.
- Isolated deltoid ligament lesions are uncommon and other concomitant ligamentar or osseous lesions should be looked. Ultrasound is an important tool for initial screening and an MRI will confirm the diagnosis.

References

1. Hiller CE, Nightingale EJ, Raymond J, et al. Prevalence and impact of chronic musculoskeletal ankle disorders in the community. Arch Phys Med Rehabil. 2012;93:1801–7. https://doi.org/10.1016/j.apmr.2012.04.023.
2. Cooke MW, Lamb SE, Marsh J, Dale J. A survey of current consultant practice of treatment of severe ankle sprains in emergency departments in the United Kingdom. Emerg Med J. 2003;20:505–7. https://doi.org/10.1136/emj.20.6.505.
3. Gribble PA, Delahunt E, Bleakley CM, et al. Selection criteria for patients with chronic ankle instability in controlled research: a position statement of the international ankle consortium. J Athl Train. 2014;49:121–7. https://doi.org/10.4085/1062-6050-49.1.14.
4. Lin C-I, Houtenbos S, Lu Y-H, et al. The epidemiology of chronic ankle instability with perceived ankle instability—a systematic review. J Foot Ankle Res. 2021;14:41. https://doi.org/10.1186/s13047-021-00480-w.
5. Fong DT, Chan Y-Y, Mok K-M, et al. Understanding acute ankle ligamentous sprain injury in sports. Sports Med Arthrosc Rehabil Ther Technol. 2009;1:14. https://doi.org/10.1186/1758-2555-1-14.
6. Pacheco J, Guerra-Pinto F, Araújo L, et al. Chronic ankle instability has no correlation with the number of ruptured ligaments in severe anterolateral sprain: a systematic review and meta-analysis. Knee Surg Sports Traumatol Arthrosc. 2021;29:3512–24. https://doi.org/10.1007/s00167-021-06610-y.
7. McCollum GA, van den Bekerom MPJ, Kerkhoffs GMMJ, et al. Syndesmosis and deltoid ligament injuries in the athlete. Knee Surg Sports Traumatol Arthrosc. 2013;21:1328–37. https://doi.org/10.1007/s00167-012-2205-1.
8. Roemer FW, Jomaah N, Niu J, et al. Ligamentous injuries and the risk of associated tissue damage in acute ankle sprains in athletes: a cross-sectional MRI study. Am J Sports Med. 2014;42:1549–57. https://doi.org/10.1177/0363546514529643.
9. Lubberts B, D'Hooghe P, Bengtsson H, et al. Epidemiology and return to play following isolated syndesmotic injuries of the ankle: a prospective cohort study of 3677 male professional footballers in the UEFA elite Club injury study. Br J Sports Med. 2019;53:959–64. https://doi.org/10.1136/bjsports-2017-097710.
10. Ogilvie-Harris DJ, Reed SC, Hedman TP. Disruption of the ankle syndesmosis: biomechanical study of the ligamentous restraints. Arthrosc J Arthrosc Relat Surg. 1994;10:558–60. https://doi.org/10.1016/s0749-8063(05)80014-3.
11. Ribbans WJ, Garde A. Tibialis posterior tendon and deltoid and spring ligament injuries in the elite athlete. Foot Ankle Clin. 2013;18:255–91. https://doi.org/10.1016/j.fcl.2013.02.006.

12. Nilsson S. Sprains of the lateral ankle ligaments. An epidemiological and clinical study with special reference to different forms of conservative treatment. Part I. Epidemiological and clinical considerations. J Oslo City Hosp. 1982;32:3–29.
13. Pijnenburg ACM. Acute ankle injuries: diagnostic and therapeutic strategies on evidence-based grounds. 2006.
14. Rammelt S, Obruba P. An update on the evaluation and treatment of syndesmotic injuries. Eur J trauma Emerg Surg. 2015;41:601–14. https://doi.org/10.1007/s00068-014-0466-8.
15. Harmon KG. The ankle examination. Prim Care. 2004;31(1025–37):x. https://doi.org/10.1016/j.pop.2004.07.008.
16. van Dijk CN, Lim LS, Bossuyt PM, Marti RK. Physical examination is sufficient for the diagnosis of sprained ankles. J Bone Joint Surg Br. 1996;78:958–62. https://doi.org/10.1302/0301-620x78b6.1283.
17. Delahunt E, Bleakley CM, Bossard DS, et al. Clinical assessment of acute lateral ankle sprain injuries (ROAST): 2019 consensus statement and recommendations of the international ankle consortium. Br J Sports Med. 2018;52:1304–10. https://doi.org/10.1136/bjsports-2017-098885.
18. Guerra-Pinto F, Côrte-Real N, Mota Gomes T, et al. Rotational instability after anterior talofibular and calcaneofibular ligament section: the experimental basis for the ankle pivot test. J foot ankle Surg. 2018;57:1087–91. https://doi.org/10.1053/j.jfas.2018.03.038.
19. Guerra-Pinto F, Côrte-Real N, Gomes TM, et al. Varus talar tilt combined with an internal rotation pivot stress assesses the supination instability vector in lateral ankle ligaments' injury—cadaver study. Foot ankle Surg. 2020;26:258–64. https://doi.org/10.1016/j.fas.2019.03.007.
20. de César PC, Avila EM, de Abreu MR. Comparison of magnetic resonance imaging to physical examination for syndesmotic injury after lateral ankle sprain. Foot Ankle Int. 2011;32:1110–4. https://doi.org/10.3113/FAI.2011.1110.
21. Bachmann LM, Kolb E, Koller MT, et al. Accuracy of Ottawa ankle rules to exclude fractures of the ankle and mid-foot: systematic review. BMJ. 2003;326:417. https://doi.org/10.1136/bmj.326.7386.417.
22. Shakked R, Sheskier S. Acute and chronic lateral ankle instability diagnosis, management, and new concepts. Bull Hosp Jt Dis. 2017;75:71–80.
23. Gribble PA. Evaluating and differentiating ankle instability. J Athl Train. 2019;54:617–27. https://doi.org/10.4085/1062-6050-484-17.
24. Guerra-Pinto F, Andrade R, Diniz P, et al. Lack of definition of chronic ankle instability with Arthrometer-assisted ankle joint stress testing: a systematic review of in vivo studies. J foot ankle Surg. 2021;60:1241–53. https://doi.org/10.1053/j.jfas.2020.04.026.
25. Hosseinian SHS, Aminzadeh B, Rezaeian A, et al. Diagnostic value of ultrasound in ankle sprain. J foot ankle Surg. 2022;61:305–9. https://doi.org/10.1053/j.jfas.2021.08.008.
26. Campbell DG, Menz A, Isaacs J. Dynamic ankle ultrasonography. A new imaging technique for acute ankle ligament injuries. Am J Sports Med. 1994;22:855–8. https://doi.org/10.1177/036354659402200620.
27. Saengsin J, Bhimani R, Sato G, et al. Use of por ultrasonography for the diagnosis of lateral ankle instability. J Orthop Res. 2022;40:2421–9. https://doi.org/10.1002/jor.25256.
28. Nazarenko A, Beltran LS, Bencardino JT. Imaging evaluation of traumatic ligamentous injuries of the ankle and foot. Radiol Clin North Am. 2013;51:455–78. https://doi.org/10.1016/j.rcl.2012.11.004.
29. Verhaven EF, Shahabpour M, Handelberg FW, et al. The accuracy of three-dimensional magnetic resonance imaging in the diagnosis of ruptures of the lateral ligaments of the ankle. Am J Sports Med. 1991;19:583–7. https://doi.org/10.1177/036354659101900605.
30. Griffith JF, Brockwell J. Diagnosis and imaging of ankle instability. Foot Ankle Clin. 2006;11:475–96. https://doi.org/10.1016/j.fcl.2006.07.001.
31. Pogliacomi F, De Filippo M, Casalini D, et al. Acute syndesmotic injuries in ankle fractures: from diagnosis to treatment and current concepts. World J Orthop. 2021;12:270–91. https://doi.org/10.5312/wjo.v12.i5.270.

32. Tourné Y, Molinier F, Andrieu M, et al. Diagnosis and treatment of tibiofibular syndesmosis lesions. Orthop Traumatol Surg Res. 2019;105:S275–86. https://doi.org/10.1016/j.otsr.2019.09.014.
33. Anand Prakash DA. Syndesmotic stability: is there a radiological normal?—a systematic review. Foot Ankle Surg. 2018;24:174–84. https://doi.org/10.1016/j.fas.2017.02.004.
34. Chun D-I, Cho J-H, Min T-H, et al. Diagnostic accuracy of radiologic methods for ankle syndesmosis injury: a systematic review and meta-analysis. J Clin Med. 2019;8:968. https://doi.org/10.3390/jcm8070968.
35. Conti MS, Ellis SJ. Weight-bearing CT scans in foot and ankle surgery. J Am Acad Orthop Surg. 2020;28:e595–603. https://doi.org/10.5435/JAAOS-D-19-00700.
36. Takao M, Ochi M, Oae K, et al. Diagnosis of a tear of the tibiofibular syndesmosis. The role of arthroscopy of the ankle. J Bone Joint Surg Br. 2003;85:324–9. https://doi.org/10.1302/0301-620x.85b3.13174.
37. Gibson PD, Ippolito JA, Hwang JS, et al. Physiologic widening of the medial clear space: what's normal? J Clin Orthop trauma. 2019;10:S62–4. https://doi.org/10.1016/j.jcot.2019.04.016.
38. Crim J. Medial-sided ankle pain: deltoid ligament and beyond. Magn Reson Imaging Clin N Am. 2017;25:63–77. https://doi.org/10.1016/j.mric.2016.08.003.

9 Peroneal Tendon Disorders

Dzenan Jahic, Jurij Stalc, Mohamed Abdelatif Djadoun, Ghislain N. Aminake, and Pieter D'Hooghe

9.1 Peroneal Tendon Pathology and Anatomical–Biomechanical Characteristics

The peroneus longus (PL) muscle and peroneus brevis (PB) muscle form the lateral compartment of the lower leg. Its tendons curl around posterior aspect of the tip of fibula going together through the peroneal tunnel formed by the retromalleolar groove of the fibula and the superior peroneal retinaculum (SPR), which arises from the posterior ridge of the lateral malleolus and inserts in the region of calcaneus, with existing several different anatomical variations of insertion. This part of peroneal tendons path is known as Zone A [1].

The tendons then pass by the peroneal tubercle of the calcaneus. In this part, they are covered by the inferior peroneal retinaculum—an extension of inferior extensor retinaculum. Its insertion to peroneal tubercle creates a septum between the tendons. This region is known as Zone B.

At the level of the cuboid bone, the peroneal tendons separate. The PL creates another turn around the cuboid bone and passes underneath the cuboid in a groove that is covered by plantar ligaments, forming a tunnel. It crosses toward the medial side of the foot inserting on the lateral plantar aspects of the base of the first metatarsal and medial cuneiform bone. The PB tendon continues straight inserting on the base of the fifth metatarsal. This part is known as Zone C (Fig. 9.1).

D. Jahic (✉)
General Hospital "Prim. dr. Abdulah Nakas", Sarajevo, Bosnia and Herzegovina

J. Stalc
Orthopedic Hospital Valdoltra, Ankaran, Slovenia
e-mail: jurij.stalc@ob-valdoltra.si

M. A. Djadoun · G. N. Aminake · P. D'Hooghe
Aspetar Orthopedic and Sports Medicine Hospital, Doha, Qatar

S. Sharma et al. (eds.), *Clinical and Radiological Examination of the Foot and Ankle*, https://doi.org/10.1007/978-981-97-4202-8_9

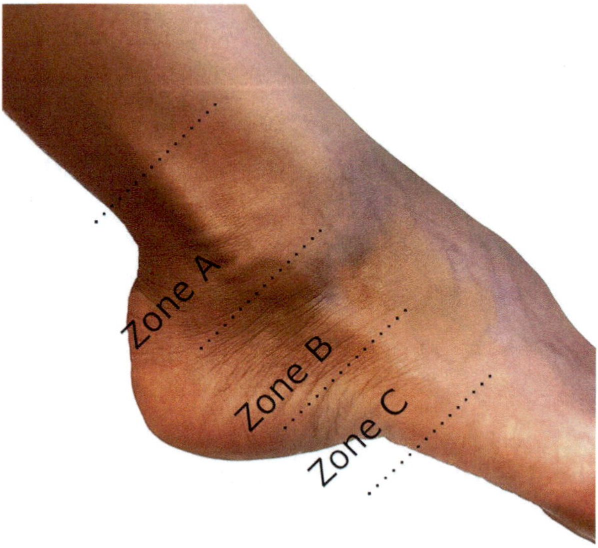

Fig. 9.1 Zones of peroneal tendons described by Brandes and Smith [1]. Three zones of tendon injury were described: Zone (**a**) retromalleolar region where tendons are covered by SPR; Zone (**b**) the region of peroneal tubercle where the tendons are covered by inferior retinaculum; Zone (**c**) the region of cuboid notch. Note the straight run of peroneus brevis in plantarflexed ankle. The evolution of human stance and gait and the new function of the first ray had dramatic effect on neutral position of the ankle. Peroneal tendons now need to turn around several bones to change direction, making them prone to injury

Peroneal muscles are primary ankle evertors and foot pronators. As the tendons are passing posteriorly to ankle axis, they also assist ankle plantarflexion. PL has plantarizing effect on the medial column. Peroneals are dynamic stabilizers of the lateral ankle as well as longitudinal and transverse arches of the foot [2–8].

The peroneal tendon region is prone to anatomical variability, which might have a significant role in etiopathogenesis of tendons disorders. A low-lying PB muscle, accessory peroneal muscles (peroneus quartus and peroneus quintus), shape of retromalleolar groove, shape of peroneal tubercle, and presence of os peroneum—sesamoid bone in PL at the level of cuboid bone turn—might all predispose the development of clinically important conditions [2–11].

Another important cofactor in the process of development of peroneal tendon disorders is cavovarus foot alignment. Cavovarus itself might be caused by improper peroneal muscle function (e.g., cavovarus in HMSN—hereditary motor and sensory neuropathy) but no matter the cause, this type of foot shape would exert additional strain on peroneal tendons predisposing the pathology.

Generally, three primary categories of peroneal tendon pathology exist: tendinopathy, tendon tears/ruptures, and tendon dislocation [12]. There are two zones where the peroneal tendons are more frequently affected—the hypovascular or avascular zones where tendons bend around the bones. The first one is behind the retromalleolar groove and the second one affecting the PL at the cuboid bone level. Those are the zones where peroneal tendinopathy would usually appear, followed by signs and symptoms of chronic inflammation [12–14].

Tendon tears and ruptures are more common in athletes than in nonathletes [14]. A tear is defined as a longitudinal split of tendon fibers, whereas rupture is a transverse loss of tendon integrity. The usual mechanism of injury is a forceful supination of contracted peroneals as happens in inversion sprain. Tendon degeneration, overuse tendon injury and anatomical abnormalities might preclude the tendon injury.

Peroneal tendon dislocation occurs as a result of a traumatic event. A forceful dorsiflexion of pronated foot is the usual mechanism of injury. The fibular origin of SPR is usually detached and peroneal tendons dislocated anteriorly over the lateral malleolus. Peroneal tendon dislocation might be acute or chronic, single event, or recurrent dislocations. Missing the diagnosis of peroneal tendon pathology at the time of first presentation is common—a delayed diagnosis happens in up to 40% of cases [12].

9.2 Clinical Examination

9.2.1 Common Clinical Findings and Their Significance

9.2.1.1 Medical History

The process of medical history taking in peroneal tendon disorders is no different as in other foot conditions. As pain is the predominant symptom, an engaged clinician would try to understand the pain pattern: quality of pain, duration, onset, triggers (what aggravates it, what relieves it), intensity, location, and radiation. A peroneal tendon pain would usually project to the lateral ankle. Other qualifiers might differ based on underlying pathology.

Pain associated with tendinopathy might be provoked by prolonged use of the tendons, e.g., walking or running on uneven terrain. It can be dull, burning, or throbbing sensation. It is usually alleviated with rest. The onset is rarely acute but patients might remember an event of overuse that can be associated with the start of pain.

Pain associated with acute rupture or tear is of sudden onset. Location of pain might differ among both tendons—with PB tear more frequently hurting in retromalleolar region and PL usually hurting at the level of cuboid tunnel. Pain associated with chronic tears is usually less specific and resembles the pattern of pain described for tendinopathy. Chronic inflammation is nonetheless a common denominator of both conditions.

Pain in acute dislocation is of sudden onset and of severe intensity. It might prevent the ankle motion and might mimic lateral malleolus fracture. In a case of chronic tendon instability, the pain can completely fade away and patient can sometimes demonstrate the dislocation maneuver without any pain.

Except from understanding the pain pattern, an experienced clinician would also try to understand the mechanism of injury, history of previous injury, the level of sports activity, changes in training activities, changes in shoe wear patterns, and so on.

As peroneal tendon injury is very common in long-lasting chronic lateral ankle instability, a detailed medical history of instability should be obtained. Many patients with chronic lateral ligament tear would not directly report instability as they might cope with it by means of dynamic joint stabilization, putting the peroneal tendons under prolonged increased stress. Lateral ankle instability has many faces: listen carefully for history of giving way, tripping on uneven terrains, and feeling of losing ankle power. Pain and swelling might be the symptoms of the injured lateral ligaments as well.

The general health aspect should not be forgotten. Ask for history of spinal problems, systemic illnesses (Lyme, gout, generalized arthritis), neurologic problems, congenital diseases, and so on. Extend the interview in case changed foot posture has been reported or observed. A cavovarus deformity has a very broad differential diagnosis range.

9.2.1.2 Inspection

Investigation is started by observing the patient's gait the moment he walks into the office. Some gait abnormalities as peroneal nerve palsy could be immediately diagnosed. Excessive supination during gait is also easily distinguished. This gait pattern is additionally confirmed by investigating the shoes and the shoe-wear pattern. Lateral heel wear is indication of excessive supination during gait.

The affected leg is then visualized from knees down. Any signs of swelling, discoloration, or deformity are searched for in comparison to unaffected ankle. This observation will usually reveal the location of pathology in a chronic setting when the chronic inflammation process produces an isolated and well-located swelling with or without associated skin redness (Fig. 9.2). An isolated swelling located in one of the zones of peroneal tendon injury is a common observation and a diagnostic one as well. A significant swelling over the lateral malleolus may indicate dislocated peroneal tendons.

In an acute setting, the global ankle inflammation would usually prevent the clinician to gather any information about the location of injury. In this scenario, a bluish discoloration of the skin might be more of help. Peroneal tendon injuries would usually produce a skin discoloration located over the Zone A or a bit more posterior over the triangle of Kager in comparison to lateral ligament tears, which would more often present with a discoloration over the lateral heel wall (Fig. 9.2).

9.2.1.3 Alignment Assessment

With patient standing in a relaxed weight-bearing stance, observe both legs from knees down from front and behind of the patient in order to search for any asymmetry or deformity. Signs of cavovarus should be sought for. A subtle cavovarus deformity is a frequent finding among those affected by peroneal tendon pathology.

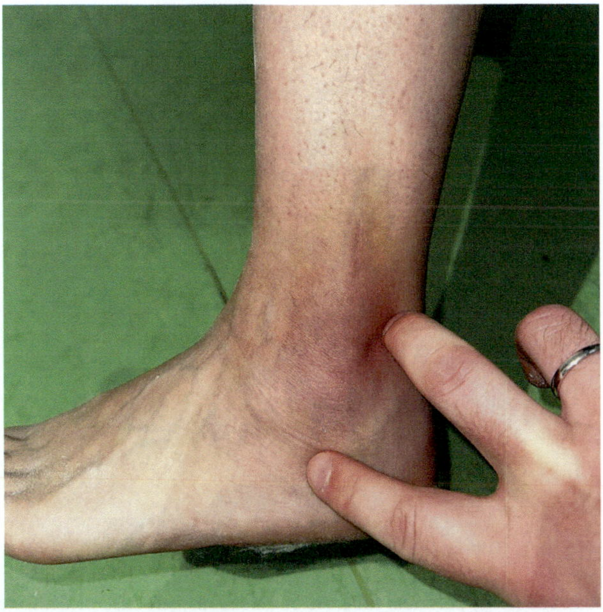

Fig. 9.2 Looking for the pain in this area (between the clinician's fingers). The patient was extremely painful in this region with very painful passive plantarflexion and supination—22 days after the "classic" ankle sprain

An increased foot arch, the absence of heel valgus, a "peek-a-boo" heel observable from front, or clearly visible big toe with the absence of lateral toes when seeing from behind (the medial toe sign) are all indicators of cavovarus deformity. In case of positive findings, perform Coleman block test to gain the understanding of deformity (forefoot driven vs. true hindfoot varus).

9.2.1.4 Palpation

Lateral ankle is very rarely covered with abundant soft tissues. In the majority of patients, the skeletal structures are covered only with a thin cutaneous layer, making palpation an ideal diagnostic tool. Comparing location of maximal tenderness upon palpation to anatomical structures helps narrowing the differential diagnosis of lateral ankle pain. The patient-reported pain is usually reproduced in this way. In many occasions the patient will report the point of maximal tenderness by performing the self-palpation. An extensive knowledge of topographic ankle anatomy is a must in order for palpation to be helpful (Fig. 9.3). Again, this diagnostic tool would be more of use in a subacute or chronic setting. In an acute trauma palpating the everywhere-hurting ankle would reveal very little information about the anatomic location of the injured tissue.

9.2.1.5 Muscle Strength

Function of peroneal muscles is then assessed. The patient is instructed to actively evert the ankle. The force of peroneal muscles is measured in a semiquantitative manner by resisting the movement. Both muscles are assessed as a functional unit.

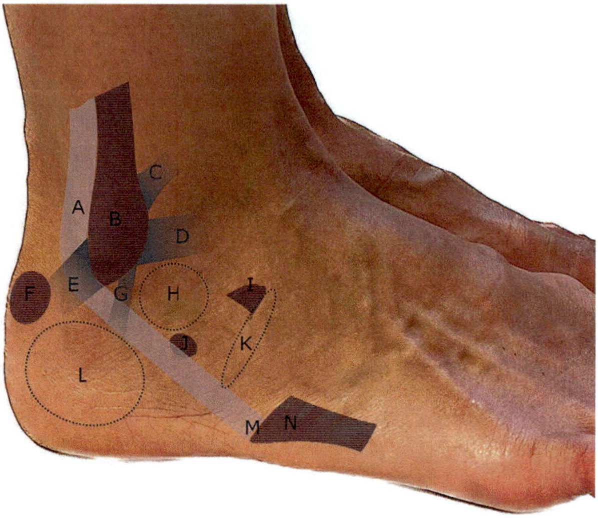

Fig. 9.3 Topographic anatomy of lateral ankle. Bony landmarks (often directly visible, palpable in most cases): (**b**) lateral malleolus, (**f**) Haglund's prominence of calcaneus, (**i**) anterior process of calcaneus, (**j**) peroneal tubercle of calcaneus, (**n**) base of fifth metatarsal. Ligamentous structures: (**c**) distal anterior tibiofibular ligament (ATiFL, anterior syndesmosis ligament), (**d**) anterior talofibular ligament (ATFL), (**e**) superior peroneal retinaculum (SPR), (**g**) calcaneofibular ligament. Joints: (**h**) subtalar joint (also sinus tarsi fat), (**k**) calcaneo-cuboid joint, (**l**) lateral wall of calcaneus, (**a**) peroneus longus and brevis tendon, (**m**) PB tendon insertion

9.2.2 Techniques for Clinical Examination Including Special Tests and Maneuvers

9.2.2.1 Peroneal Compression Test

The peroneal compression test is performed while patient is seated with knees bent at 90° and the foot and ankle relaxed in an inverted and plantar flexed position. The clinician would then place his thumb over the posterior ridge of the fibula and compress the peroneal tendons. The patient is then instructed to forcefully evert and dorsiflex the ankle against resistance. A positive test would reproduce the pain.

In case of tendon instability, the examiner would be sometimes able to feel the tendons jumping over the posterior ridge of the fibula while the patient could be resisting this maneuver, which would be interpreted as a positive "apprehension sign."

9.2.2.2 Peroneal Luxation Test

The examiner would place second, third, and fourth finger against the posterior fibula, pushing the tendons anteriorly. The patient is instructed to evert the ankle, while the examiner resists this movement with his other hand. The tendons might dislocate during this maneuver or the patient might prevent it by exhibiting the positive "apprehension sign."

To conclude, it is very important to exclude the possible diagnosis (Table 9.1). After excluding the possible diagnosis, we can confirm the peroneal tendon disorder.

Table 9.1 Differential diagnosis of lateral ankle pain

Differential diagnosis of lateral ankle pain	Cuboid syndrome
	Subtalar arthritis
	Tarsal coalition
	Subtalar instability
	Sinus tarsi syndrome
	Achilles tendon disorder
	Os trigonum
	Classic ankle sprain
	Peroneal tendon disorder

9.3 Radiological Examination

The initial workup of lateral ankle pain must include a wide range of differential diagnoses. Typical peroneal tendon disorders include tendinitis, tenosynovitis, partial and full thickness tendon tears, peroneal retinacular injuries, and tendon subluxations and dislocations [15–17]. These peroneal tendons pathologies are mostly assessed with magnetic resonance imaging and ultrasound. Although plain film X-ray and computed tomography are less contributing when it comes to the investigation of tendons disorder, they remain a useful tool in the diagnosis of bone abnormalities that can be responsible or associated to tendons disorder [18–20].

9.3.1 Imaging Modalities

9.3.1.1 Plain Radiograph (X-Rays)

Imaging of patients with clinically diagnosed retromalleolar ankle pain should always start with plain films, this should include weight-bearing anteroposterior, lateral, and oblique views of the foot and ankle [15, 21]. In some cases, additional special views may be required to better appreciate some specific structures, for example the Harris view can be used to detect peroneal tubercle deformity; occasionally, on an internal rotation view, one may see a cortical avulsion of the distal tip of the lateral malleolus ("fleck sign"), indicating a rim fracture from an SPR injury (Fig. 9.4b). These images are necessary for a complete assessment of an eventual foot deformity, an accessory bone, or any bone structure abnormality that may be associated or cause peroneal tendon disease (peroneal/retrotrochlear tubercle, soft tissue calcification) (Fig. 9.5c).

9.3.1.2 Computed Tomography (CT)

CT scans have a limited role in the diagnosis of peroneal tendons injuries because of the lack of specificity and the difficulty to assess soft tissues pathologies [20, 22]. The main advantage of CT scan is a better visualization of bone abnormalities in multiple planes and 3D reconstructions. It can also be a useful tool for preoperative planning when there is an associated bone injury [17, 18].

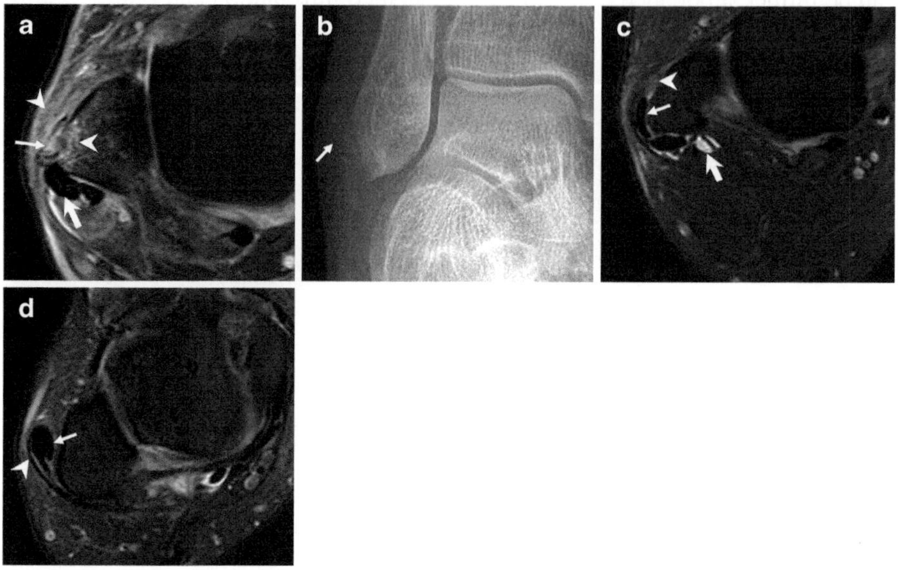

Fig. 9.4 (**a**) Avulsion fracture of the SPR at the distal fibula (arrow), also visible on X-rays (**b**). There is associated on the MRI Proton density image: bone marrow and soft tissue edema(arrowhead) with thickening of the SPR and minimal anterior displacement of the PL. (**c**) Axial Proton density image with fat saturation showing a tear at the fibular attachment of the SPR (arrowhead) and anterior subluxation of the PL (arrow) with minor fluid effusion at the common tendon sheath (thick arrow) in relation with chronic tenosynovitis. (**d**) Axial Proton density weighted image with fat saturation. Chronic tear of the SPR (arrowhead) with complete anterior dislocation of both PL and PB tendons (arrow)

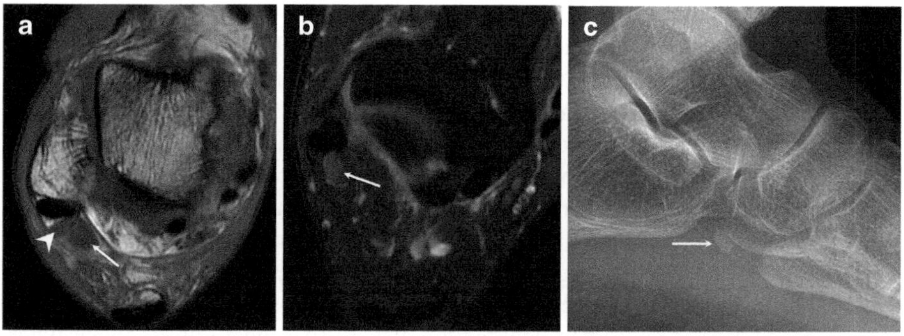

Fig. 9.5 (**a**) Axial T1 weighted image showing a peroneus Quartus accessory muscle(arrow), posteromedial to the PB muscle (arrowhead) at the level of the retromalleolar groove. (**b**) Axial Proton density weighted image with fat saturation showing a low-lying muscle belly of the PB (arrow), still visible distal to the fibular bone tip. (**c**) Ankle lateral view on plan films showing an accessory Os Peroneum (arrow) adjacent to the cuboid

9.3.1.3 Ultrasound (US)

The advantages of the US include high sensitivity and specificity, wide availability, cost-effectiveness, as well as kinematic and Doppler evaluation; however, ultrasound is operator dependent and requires the time of a sonographer or radiologist to perform the examination [22–24]. Tendinosis, tenosynovitis, and tendon tears are frequently seen on ultrasound, but its major role is the dynamic assessment of tendons instabilities. Peroneal tendons are superficial structures that can be easily assessed with ultrasound. It is recommended to use a high-resolution (12 MHz and above) linear "hockey stick" probe with a small footprint [25, 26]. The hockey-stick transducer shows a clear superiority because of multiple curves around the ankle. The patient is positioned in the supine position with the knee semiflexed and the hip internally rotated to expose the lateral side of the ankle and foot [21, 22]. The plantar aspect of the PL tendon is best examined with the patient in the prone position. Imaging of the supramalleolar, retromalleolar, and inframalleolar portions of the PL and PB including the distal portion of their muscle bellies is performed in both the transverse (short-axis) and longitudinal (long-axis) planes [24, 25]. A large amount of US gel is used to accommodate the curved surface of the lateral malleolus and to prevent tendon displacement caused by the application of excessive local pressure with the transducer, and for some regions, the probe must be tilted at the correct angle to follow each of the tendons [18, 26, 27].

9.3.1.4 Magnetic Resonance Image (MRI)

High-resolution ankle MRI examination with dedicated coil is the best tool for assessment of different tendon disorders. Routine protocol in our experience includes multiplanar sequences in T1-weighted images and T2 fat-saturated weighted images, and contrast injection is not necessary in most cases. Axial planes provide the best definition of tendon contours, synovial sheath content, and surrounding structures such as the retinacular and the retromalleolar groove [28, 29]. Sagittal images are useful in assessing the longitudinal extent of tendon disease. A useful sequence for an optimal visualization of the inframalleolar aspect of the tendons is a T2 axial oblique without fat saturation within the tendon's short axis (approximately at 45°). Normal peroneal tendons have low signal intensity in all pulse sequences. Within the fibular retromalleolar groove, the PL have more of a globular configuration, while the PB tendon typically appears flat or mildly crescentic, a finding that should not be misdiagnosed as a tear [26, 28]. Increased signal intensity is associated with tendon disease but may also be due to the magic angle effect. The magic angle effect occurs when a structure is oriented at 55° relative to the static magnetic induction field, which results in increased signal intensity in the structure on short echo time [15, 22, 26]. The peroneal tendons are susceptible to the magic angle effect as they descend the ankle. Thus, imaging with the foot in mild plantar flexion (approximately 20°) can help decrease the magic angle effect [15, 28].

9.3.2 Radiological Findings

Normal appearance of tendons on MRI corresponds to a regular and continuous low signal structure in all sequences (Fig. 9.6b). While on ultrasound we usually see a fibrillary thin structure, hyperechoic to surrounding muscles [21, 30]. Tendon sheaths are not visualized unless pathologic on both imaging modalities.

(a) Tendinosis/Tendinopathy

Peroneal tendinosis is a noninflammatory degenerative process that is most seen in athletes as a result of chronic repetitive trauma, but has also been described in elderly individuals with chronic metabolic diseases [8, 21]. On MRI, tendons are thickened with increased signal intensity on T2-weighted

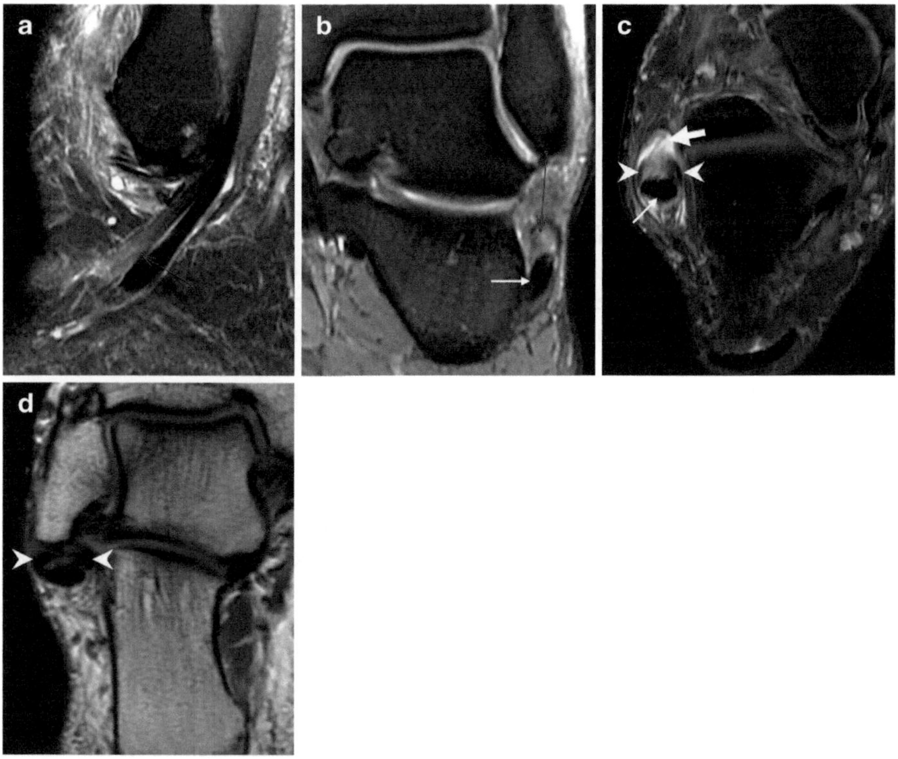

Fig. 9.6 MRI Proton density with fat saturation Images (**a**) (sagittal) and (**b**) (coronal) shows increased signal in the PB tendon(red arrow) related to tendinosis/tendinopathy. The PL (white arrow) posterior and inferior to the PB at that level has normal low signal. Image (**c**) (Proton density with fat saturation in axial plan) and image (**d**) (Proton density in coronal oblique plan without fat saturation) shows PB intra tendinous increased signal with a longitudinal split tear at the retromalleolar level. Note the separated two hemi-tendons (arrowhead) and the PL lying in between (white arrow). There is associated thickening with effusion of the common tendon sheath (thick arrow) consistent with tenosynovitis

images without intratendinous fluid signal (Fig. 9.6a, b). On ultrasound, the affected tendon is thickened with hypoechoic changes and loss of the usual fibrillary aspect [18, 22]. The inframalleolar aspect of the peroneus brevis is the most affected.

(b) Tenosynovitis

Peroneal tenosynovitis may be caused by infection, systemic arthropathies such as rheumatoid arthritis, and by repetitive tendon stress [21]. Effusion around the tendons of more than 3 mm large is considered pathological and related to tenosynovitis (Fig. 9.7c). Thickening with enhancement of the tendon sheath on MRI is highly specific (Fig. 9.6c). This finding corresponds to an increased Doppler signal on ultrasound (Fig. 9.8). Tendon sheath effusion of less than 3 mm large can be physiological as well as a fluid distension of the

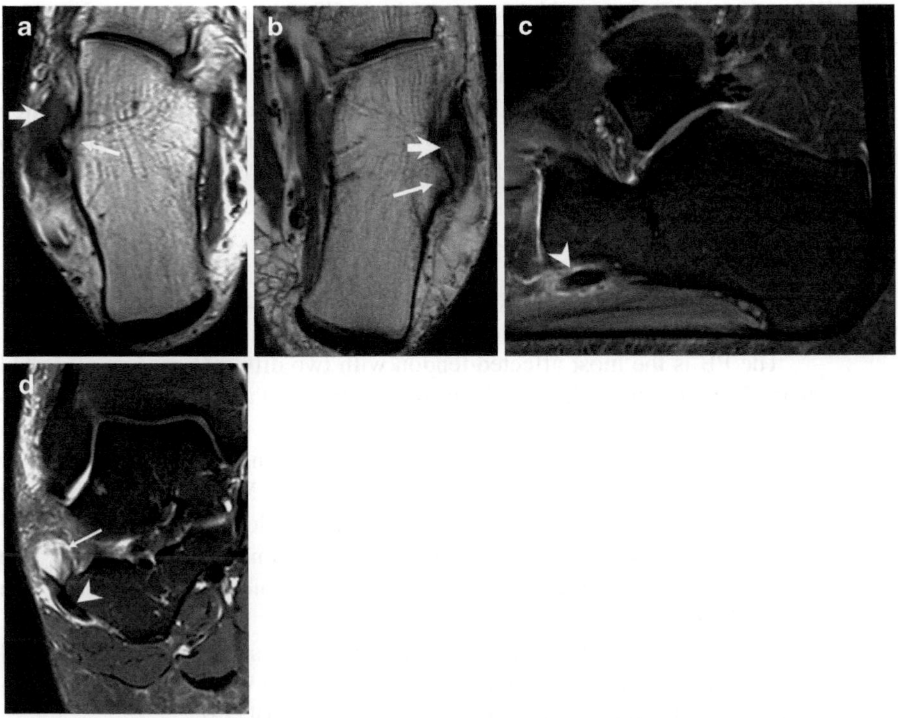

Fig. 9.7 (**a**) Axial T1 weighted image showing hypertrophy of the peroneal tubercle anterior to the peroneal trochlea at the lateral aspect of the calcaneus (arrow) with heterogeneous aspect of the PB tendon (thick arrow). (**b**) Axial T1 weighted image showing this time hypertrophy of the retrotrochlear eminence (arrow) with tenosynovitis of the PL tendon(thick arrow). (**c**) Sagittal Proton density weighted images with fat saturation. Note the thickening and increase signal of the tendon sheath (arrowhead) surrounding the PL tendon at the level of the peroneal trochlea related to tenosynovitis. (**d**) Coronal Proton density weighted images with fat saturation. Same patient in image A, note the non visualization of the PB tendon inside the tendon sheath(arrow), corresponding to a complete tear. The PL tendon, more distal has normal appearance (arrowhead)

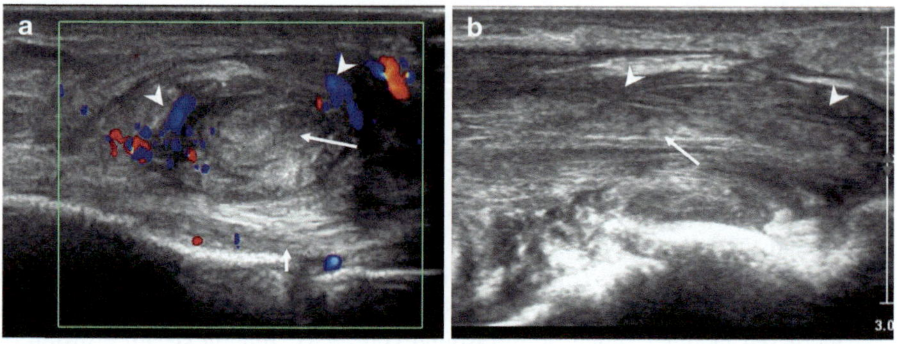

Fig. 9.8 Ultrasound examination. Images A (short axis) and B (long axis) shows a thickened PB tendon at the infra malleolar aspect. Note the heterogeneous tendon with loss of the fibrillary aspect (arrow) associated with thickening and hyperemia of the tendon sheath (arrowhead), consistent with tenosynovitis. Deep to the PB is seen the calcaneofibular ligament(short arrow), intact

tendon sheath after an acute tear of the calcaneofibular ligament owing to the intimate relationship between the calcaneofibular ligament and the common peroneal tendon sheath [22]. The presence of intratendinous changes and a reactive edema at the surrounding soft tissues is not rare in case of active tenosynovitis.

(c) Tendon Tears

- Partial tears:

  Partial tears also evolve from both acute ankle injuries and chronic repetitive trauma with underlying tendinosis.

  The PB is the most affected tendon with two different aspects of partial tears. Longitudinal splits are the most common and arise at the retromalleolar groove due to a shallow or flat posterior aspect of the distal fibula or a chronic compression because of a low-lying peroneus brevis muscle belly or an accessory peroneus quartus muscle [27, 31]. On MRI, a fluid high signal in T2 sequences is then seen separating the PB tendon into two longitudinal components surrounding the PL entrapped in the middle (Fig. 9.6c, d). The same finding is visible on ultrasound with a fluid-like hypoechoic line inside the PB.

  Transversal tears are less frequent and mostly symptomatic with focal thinning and decrease size of the tendon associated with increased intratendinous signal and tenosynovitis but no retraction [19, 22].

  PL partial tears are less common but more distal in location, usually at the level of the peroneal tubercle in the calcaneal bone (Fig. 9.9d). PL tears are chronic mechanic friction with calcaneal tubercle hypertrophy or inside the cuboid tunnel, while more proximal partial tears can occur in the retromalleolar groove similar to the PB lesions with the same MRI and ultrasound findings. The chronic bone disease can be confirmed by the presence of bone marrow edema adjacent to the tear. A pes cavus foot deformity is also a cause of chronic PL tears.

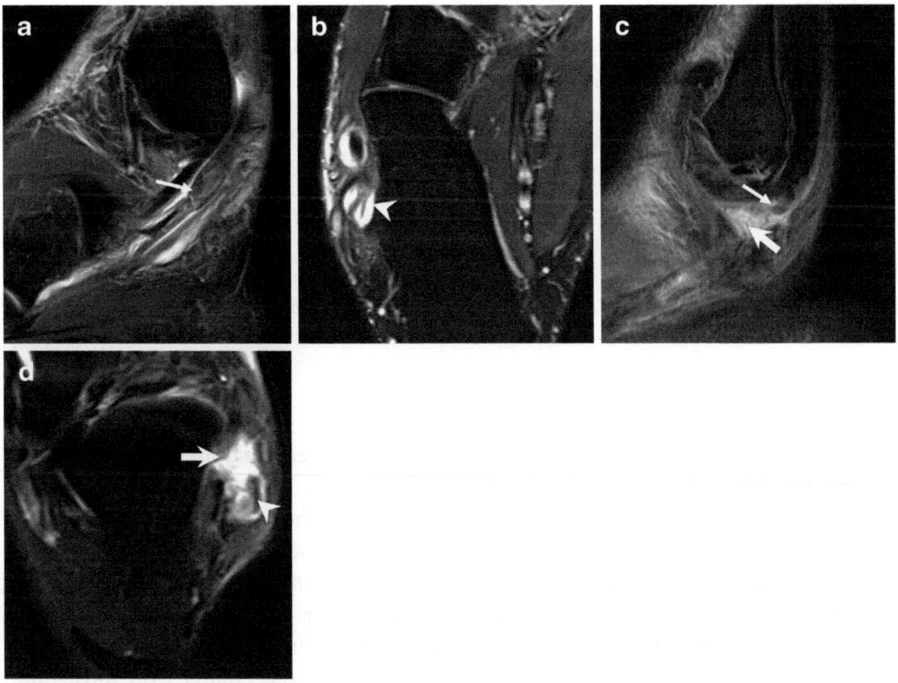

Fig. 9.9 (**a**) (sagittal) and (**b**) (axial) Proton density weighted images with fat saturation: Complete tear of the PL tendon at the peroneal trochlea. Thickening, increased signal and loss of tension at the proximal stamp (**a**, Arrow) with absence of the tendon inside the tendon sheath at the level of the tear (**b**, arrowhead). (**b**) (sagittal) and (**d**) (axial) Proton density weighted images with fat saturation: Complete tear of the PB tendon at the inframalleolar aspect with a clear discontinuity of the tendon on sagittal image (**c**, arrow) and local hematoma (thick arrow) at the level of the tear. The PL tendon, more posterior shows also thinning with increased signal corresponding to a partial tear (**d**, arrowhead)

- Complete tears:

  Complete tears are less common in the PL compared to PB. Complete PB tears are as a result of an acute trauma mainly located distally at the base of M5 or more proximally as the ultimate stage of a chronic degeneration process (Figs. 9.7d and 9.9c) [15, 31].

  In the PL, complete tears occur mainly in the distal aspect of the PL, at the cuboid tunnel. Tendon retraction and lack of visualization of the tendon inside the tendon sheath are the specific signs of a complete tear, frequently associated with chronic situations with diffuse tendinosis and tenosynovitis (Fig. 9.9a,b). The presence of a proximally displaced os peroneum is a sign for a full thickness PL tear. A 6-mm cutoff is considered pathological in case of fracture and partial displacement of the accessory bone [19, 21, 22].

  On ultrasound, the diagnosis of complete tendon tear is confirmed by the non-visualization of the tendon inside the tendon sheath, a retraction of the proximal part and the absence of movement of the distal part during the dynamic dorsal–plantar flexion of the foot.

Table 9.2 Ogden classification of SPR injury

Grade 1	Partial elevation of the SPR off the fibula
Grade 2	Separation of SPR from ridge of lateral malleolus
Grade 3	Cortical avulsion of the SPR off the fibula
Grade 4	SPR tear is off the calcaneus; fibula unaffected
Grade 1	Partial elevation of the SPR off the fibula
Grade 2	Separation of SPR from ridge of lateral malleolus
Grade 3	Cortical avulsion of the SPR off the fibula
Grade 4	SPR tear is off the calcaneus; fibula unaffected

(d) Tendons Instability

At the level of the retromalleolar groove, peroneal tendons are kept in place by the superior peroneal retinaculum (SPR), which is a thin band extending from the anterolateral aspect of the distal fibula to the lateral calcaneus [18, 32].

Any injury of this retinaculum may result in peroneal tendons instability, progressing from intraretinacular subluxation to anterior subluxation or complete anterior dislocation from the retromalleolar groove. Radiological findings confirm the injury of the SPR, showing thickening or discontinuity of the latter with increased signal on MRI and hypoechoic changes on ultrasound (Fig. 9.4a). The presence of a fleck's sign on plain films is indicative of a Grade III SPR injury based on the Ogden classification (Table 9.2, Fig. 9.4b) [16]. Dynamic ultrasound is very useful in showing the anterior subluxation/dislocation on eversion and dorsal flexion of the foot, a finding that can be absent on MRI in neutral position [24, 25, 29]. Local anatomical variants may have a role to facilitate the instability: any low-lying PB muscle, peroneus quartus accessory muscle, and variance of the retromalleolar groove. Frequently, in the chronic phase, tendon abnormalities will be seen, such as tendinosis and tendon tears [29, 32].

- External dislocation/subluxation:

  In case of a partial or complete tear of the SPR, the peroneal tendons are subject to subluxation or complete dislocation outside the retromalleolar groove. The Oden's surgical classification is useful to describe the different injuries of the SPR [18, 22]. SPR tears are best seen on axial MRI images with or without fat saturation (Fig. 9.4c). Edema in the surrounding soft tissues during the acute phase and ATFL tears are frequently associated with SPR tears.

- Internal subluxation:

  Intrasheath subluxation of the peroneal tendons over each other is seen within an intact SPR. Usually not seen on MRI, patients are referred with chronic popping or clicking sound and pain posterior to the lateral malleolus, and ultrasound is therefore the best modality to assess this instability [33]. Dynamic maneuvers with dorsiflexion and eversion can show displace-

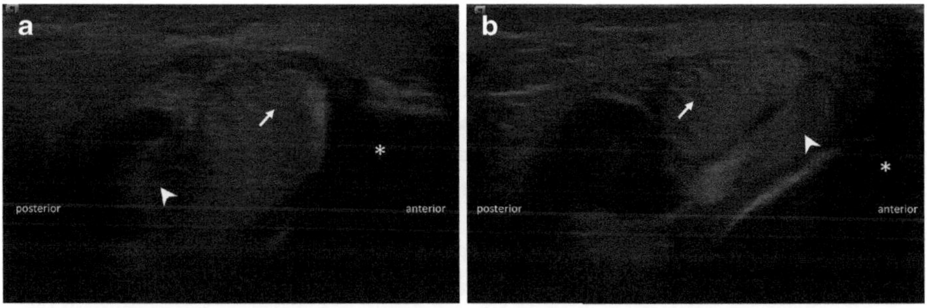

Fig. 9.10 Ultrasound images in short axis of the peroneal tendons showing intra sheath dislocation after dynamic inversion and dorsal flexion. At rest (**a**) the PB (arrowhead) is posterior to the PL (arrow), while on dynamic maneuvers (**b**), the PB becomes anterior to the PL, both tendons remain posterior to the retromalleolar groove (*). Patient was complaining from chronic lateral ankle pain and clicking, MRI was normal

ment of the PB over the PL or the opposite without dislocation outside the retromalleolar groove (Fig. 9.10a, b) [29, 33].

(e) Anatomical Variants, Accessory Bones and Muscles

All anatomical variants related to the peroneus tendons must be described in both MRI and ultrasound reports because of the important role they have in the pathology.

- Retromalleolar groove:

  It is usually concave or flat and regular in 80% of people. A convex or flat and irregular aspect should be described (Fig. 9.5a) as it is a predisposing factor to tendinopathies, PB tendons split tears and dislocation.

- Peroneal tubercule:

  At the lateral aspect of the calcaneal bone, the retrotrochlear eminence lies posterior to the peroneal tendons (PL) and is more often seen, while the peroneal tubercule lies anterior to the tendons and is seen less frequently [22, 26]. Hypertrophy of the peroneal tubercule or in a second order, the retrotrochlear eminence can cause PL tendinopathy or tear and therefore should be described in CT or MRI (Fig. 9.7a, b).

- Accessory bones:

  Os peroneum is an accessory sesamoid bone seen inside the PL tendon, usually at the level of the cuboid tunnel, which can be bipartite or subject to fractures and responsible of PL tendinopathy and tears at that level [22, 28]. A proximal displacement of the os peroneum is a sign of PL tendon complete tear (Fig. 9.5c).

  POPS (painful os peroneum syndrome) is described as pain in the lateral edge of the foot, at the level of the cuboid bone with the presence of an os peroneum attached to the PL tendon. In US or MRI, this correlates with bone edema not only at the accessory bone but also at the cuboid and tenosynovitis of the PL [18, 29]. This condition is usually due to chronic inflammation and friction or can result from an acute injury with contusion/fracture of the os peroneum or partial tear of the PL.

- Accessory muscles:

  Multiples accessory muscles have been described in the lateral aspect of the ankle, and different classifications have been established although most important to describe is the peroneus quartus muscle, which could be present in asymptomatic person but is also responsible of crowding in the retromalleolar groove and therefore cause tendinopathies and eventually tendon tears [22]. On MRI imaging, the PQ is visualized as a normal muscular structure posterior and medial to the PB tendon, with a distinct muscle belly from the PB muscle and a distal tendon inserting variably at the retromalleolar groove, at the PB or PL inframalleolar tendons, or at the lateral side of the calcaneus (Fig. 9.5a).

  A low-lying muscle belly of the PB is considered abnormal variant when the myotendinous junction is reaching the distal tip of the fibula or below [18, 21]. This anatomical variant similar to the peroneus quartus has a crowding effect at the retromalleolar groove (Fig. 9.5b).

Lateral ankle pain from tendon pathology is now a relatively common complaint and is seen in a wide range of patients from young athletes to older sedentary individuals. Imaging of structures in this region should always start with plain radiographs to detect bone abnormalities and deformity such as a fleck sign, cavovarus deformity, or os peroneum migration. Ultrasound imaging is rapidly gaining popularity in the assessment of tendon disorders as it offers multiple advantages including a reported high sensibility and specificity in the assessment of peroneal tendon instability and tears, availability, radiation-free, and affordability, though it is operator dependent. CT may be used to highlight bony anatomy associated to tendon disease such as retromalleolar groove abnormality or enlarged peroneal tubercle. MRI has been the imaging modality of choice for the peroneal tendons as it can elegantly demonstrate features of both normal variants and diseases of the peroneal tendons. MRI is best evaluated with axial views of a slightly plantarflexed ankle to decrease the "magic angle" effect of the curved path of the tendons.

Key Points

- Peroneal muscles are primary ankle evertors and foot pronators. As the tendons are passing posteriorly to ankle axis, they also assist ankle plantarflexion.
- The peroneal tendon region is prone to anatomical variability, which might have a significant role in etiopathogenesis of tendons disorders.
- There are two zones where the peroneal tendons are more frequently affected—the hypovascular or avascular zones where tendons bend around the bones. The first one is behind the retromalleolar groove and the second one affects the PL at the cuboid bone level.
- Pain associated with acute rupture or tear is of sudden onset. Location of pain might differ among both tendons—with PB tear more frequently hurting in retromalleolar region and PL usually hurting at the level of cuboid tunnel.
- Pain associated with chronic tears is usually less specific and resembles the pattern of pain described for tendinopathy.

References

1. Brandes CB, Smith RW. Characterization of patients with primary peroneus longus tendinopathy: a review of twenty-two cases. Foot Ankle Int. 2000;21:462–8.
2. Sammarco GJ, Mangone PG. Review article: diagnosis and treatment of peroneal tendon injuries. Foot Ankle Surg. 2000;6:197–205.
3. Sammarco GJ. Peroneal tendon injuries. Orthop Clin North Am. 1994;25:135–45.
4. Baumhauer JF, Nawoczenski DA, DiGiovanni BF, Flemister AS. Ankle pain and peroneal tendon pathology. Clin Sports Med. 2004;23:21–34.
5. Dombek MF, Lamm BM, Saltrick K, Mendicino RW, Catanzariti AR. Peroneal tendon tears: a retrospective review. J Foot Ankle Surg. 2003;42:250–8.
6. Brodsky JW, Zide JR, Kane JM. Acute peroneal injury. Foot Ankle Clin. 2017;22:833–41.
7. Heckman DS, Gluck GS, Parekh SG. Tendon disorders of the foot and ankle, part 1 peroneal tendon disorders. Am J Sports Med. 2009;37:614–25.
8. Roster B, Michelier P, Giza E. Peroneal tendon disorders. Clin Sports Med. 2015;34:625–41.
9. Slater HK. Acute peroneal tendon tears. Foot Ankle Clin. 2007;12:659–74.
10. Scanlan RL, Gehl RS. Peroneal tendon injuries. Clin Podiatr Med Surg. 2002;19:419–31.
11. Lau BC, Moore LK, Thuillier DU. Evaluation and management of lateral ankle pain following injury. JBJS Rev. 2018;6:1–10.
12. van Dijk PA, Miller D, Calder J, DiGiovanni CW, Kennedy JG, Kerkhoffs GM, Kynsburg Á, Haverkamp D, Guillo S, Oliva XM, Pearce CJ, Pereira H, Spennacchio P, Stephen JM, van Dijk CN. Correction to: the ESSKA-AFAS international consensus statement on peroneal tendon pathologies. Knee Surg Sports Traumatol Arthrosc. 2018;26(10):3108.
13. Selmani E, Gjata V, Gjika E. Current concepts review: peroneal tendon disorders. Foot Ankle Int. 2006;27:221–8.
14. Hamilton WG, Geppert MJ, Thompson FM. Pain in the posterior aspect of the ankle in dancers. Differential diagnosis and operative treatment. J Bone Joint Surg Am. 1996;78:1491–500.
15. Taljanovic MS, Alcala JN, Gimber LH, Rieke JD, Chilvers MM, Latt LD. High-resolution US and MR imaging of peroneal tendon injuries. Radiographics. 2015;35(1):179–99.
16. Sharma A, Parekh SG. Pathologies of the Peroneals: a review. Foot Ankle Spec. 2021;14(2):170–7.
17. Van Dijk PAD, Kerkhoffs GMMJ, Chiodo C, DiGiovanni CW. Chronic disorders of the peroneal tendons: current concepts review of the literature. J Am Acad Orthop Surg. 2019;27(16):590–8.
18. Cotten A. Imagerie musculosquelettique: pathologies locorégionales. Amsterdam: Elsevier Masson; 2008.
19. Khoury NJ, El-Khoury GY, Saltzman CL, Kathol MH. Peroneus longus and brevis tendon tears: MR imaging evaluation. Radiology. 1996;200(3):833–41.
20. Singh K, Hosseini N, Pooyan A, Zadeh FS, Chalian M. Imaging Tendon Disorders in Athletes. Clin Podiatr Med Surg. 2023;40(1):223–38.
21. Davda K, Malhotra K, O'Donnell P, Singh D, Cullen N. Peroneal tendon disorders. EFORT Open Rev. 2017;2(6):281–92.
22. Sobel M, editor. The peroneal tendons: a clinical guide to evaluation and management. Springer Cham; 2020.
23. Pesquer L, Guillo S, Poussange N, Pele E, Meyer P, Dallaudière B. Dynamic ultrasound of peroneal tendon instability. Br J Radiol. 2016;89(1063):20150958.
24. Molini L, Bianchi S. US in peroneal tendon tear. J Ultrasound. 2014;17(2):125–34.
25. Bianchi S, Delmi M, Molini L. Ultrasound of peroneal tendons. Semin Musculoskelet Radiol. 2010;14(3):292–306.
26. Lee SJ, Jacobson JA, Kim S-M, Fessell D, Jiang Y, Dong Q, et al. Ultrasound and MRI of the peroneal tendons and associated pathology. Skeletal Radiol. 2013;42(9):1191–200.
27. Smoker BA, Giakas AM, Mills FB 4th, Williams KA, Bornemann PH, Benjamin Jackson J 3rd. Prevalence of abnormal Ultrasonographic findings in asymptomatic peroneal tendons. Foot Ankle Orthop. 2023;8(1):24730114231164146.

28. Tuite MJ. MR imaging of the tendons of the foot and ankle. Semin Musculoskelet Radiol. 2002;6(2):119–31.
29. Schubert R. MRI of peroneal tendinopathies resulting from trauma or overuse. Br J Radiol. 2013;86(1021):20110750.
30. Wang X-T, Rosenberg ZS, Mechlin MB, Schweitzer ME. Normal variants and diseases of the peroneal tendons and superior peroneal retinaculum: MR imaging features. Radiographics. 2005;25(3):587–602.
31. Danna NR, Brodsky JW. Diagnosis and operative treatment of peroneal tendon tears. Foot Ankle Orthop. 2020;5(2):2473011420910407.
32. Rosenberg ZS, Bencardino J, Astion D, Schweitzer ME, Rokito A, Sheskier S. MRI features of chronic injuries of the superior peroneal retinaculum. AJR Am J Roentgenol. 2003;181(6):1551–7.
33. Guillin R, Marchand AJ, Roux A, Niederberger E, Duvauferrier R. Imaging of snapping phenomena. Br J Radiol. 2012;85(1018):1343–53.

Achilles Tendon Disorders

10

Venkataram Vellaipandi and Ankit Khurana

The Achilles tendon is the strongest, largest, and thickest tendon in the body, which plays an important role in the biomechanics of walking. This tendon is named after the legendary Greek hero Achilles. It is said that an arrow shot through Achilles' heel brought about his end, who was otherwise considered invincible [1]. This chapter will provide readers with an overview of the pertinent anatomy, biomechanics, radiology, and clinical examination of Achilles tendon disorders.

10.1 Anatomy of the Achilles Tendon

The Achilles tendon is formed by the confluence of the gastrocnemius and soleus muscles, with some contribution from the plantaris tendon as well, as shown in Fig. 10.1.

The tendon is approximately 15 cm long, starting at the musculotendinous junction of the gastrocnemius and soleus. It is initially flat at its origin and becomes rounded until it reaches 4 cm proximal to its insertion, where it becomes rounded again [2].

The blood supply of the tendon can be divided into three zones: the musculotendinous junction, the tendon proper, and the insertional portion. There is an area of relatively poor perfusion 2–6 cm proximal to its insertion [3] (Fig. 10.2). The relevance of this zone will be discussed later in the chapter.

V. Vellaipandi (✉)
Department of Orthopedics, Mahatma Gandhi Medical College and Research Institute, Pondicherry, India

A. Khurana
Department of Orthopedics, Dr BSA Medical College and Hospital, New Delhi, India

S. Sharma et al. (eds.), *Clinical and Radiological Examination of the Foot and Ankle*, https://doi.org/10.1007/978-981-97-4202-8_10

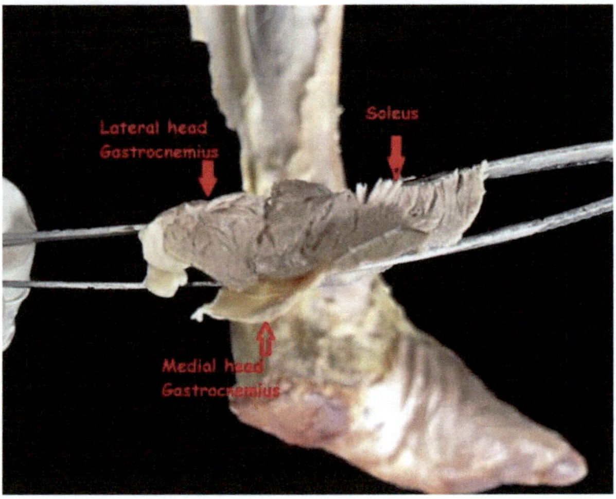

Fig. 10.1 Cross-section of the Achilles tendon at the musculotendinous junction

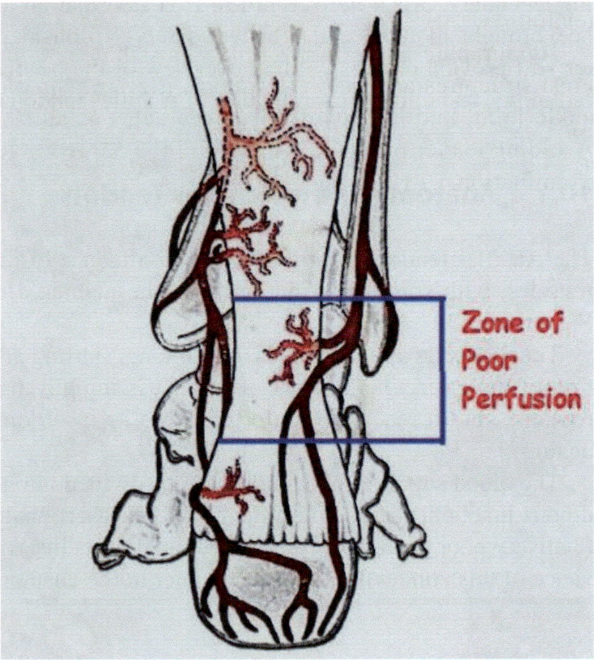

Fig. 10.2 Poor perfusion zone 2–6 cm proximal to insertion

Unlike the other tendons of the foot and ankle, the Achilles tendon has no true synovial sheath. However, it is enclosed by a thin layer of mucopolysaccharides, termed as the "paratenon." It is continuous proximally with the fascial envelope of the muscles and distally with the periosteum of the calcaneus [2]. The paratenon provides mechanical and biological support. The fluid between the paratenon and

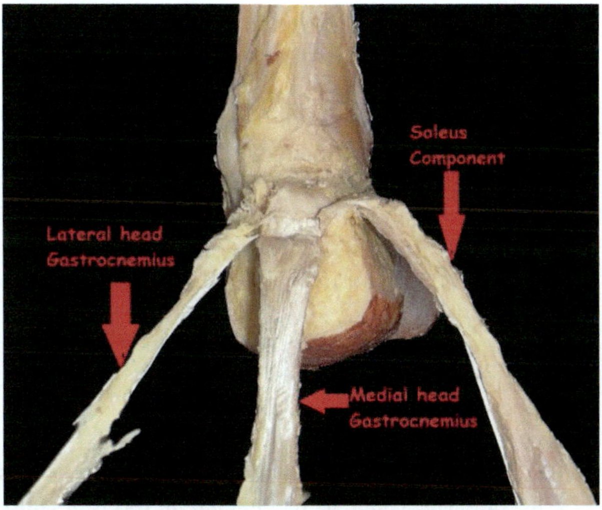

Fig. 10.3 Insertional anatomy of the Achilles tendon showing various points of insertion of tendon subcomponents onto the calcaneal facets

epitenon ensures smooth gliding of the tendon [4]. It is also rich in nerve fibers and provides blood supply to the tendon.

The tuberosity of the calcaneus on which the Achilles tendon inserts can be divided into the superior, middle, and inferior facets [5]. The superior facet marks the location of the retrocalcaneal bursa. The middle facet is further divided into medial and lateral parts. The soleus component inserts medially and the lateral gastrocnemius component inserts laterally. The inferior facet receives the medial gastrocnemius component (Fig. 10.3).

Understanding of the insertional anatomy of the Achilles tendon plays a role in visualizing the various pathologies involving the terminal part of the tendon. This will be discussed in more detail in the imaging section.

10.2 Biomechanics

The Achilles tendon transmits the force generated by the contraction of the gastrosoleus complex across the ankle joint. The soleus acts in conjunction with the tibialis anterior to stabilize the tibia while standing. During walking, the gastrosoleus performs two functions: (1) eccentric contraction to control the forward progression of the tibia during the mid-stance phase and (2) concentric contraction during the terminal stance phase to propel the leg from the ground.

The Achilles tendon exhibits viscoelastic properties and adapts to cyclical loading and eccentric training exercises [6]. However, the strain patterns that develop due to cyclical loading is not uniform along the length of the tendon with compression stresses more concentrated near the insertion [7]. This explains the higher incidence of tendon degeneration at this location.

10.3 Achilles Tendon Pathologies and Their Clinical Evaluation

Most disorders of the Achilles tendon can be diagnosed by appropriate history and a detailed clinical examination. History begins with the onset of presenting complaints, presence of any prodromal symptoms and their duration, the activities of daily living, and the limitations owing to the underlying condition. The patient's medical history and drug history should also be elicited, as there is a well-established association between various medical illnesses, drugs such as fluoroquinolones, and tendon pathologies.

A relevant medical history should include, but is not limited to, the evaluation of:

- Diabetes mellitus [8].
- Hypothyroidism [9].
- Autoimmune conditions and inflammatory disorders [10].
- Forefoot issues such as metatarsalgia, forefoot callosities, and ulcers [11].
- Corticosteroids, especially local injection and prolonged use of fluoroquinolone antibiotics [12].

A thorough clinical examination goes a long way in not only understanding the etiopathology of the underlying tendon disorder but also ruling out other pathologies that present with posterior heel pain.

The skin over the posterior heel needs to be inspected for ecchymosis, swelling, old scars, and bony prominences.

The entire length of the tendon should be palpated for tenderness, fullness, irregularity, or gapping. Both the active and passive range of movements of the ankle joint should be tested. Appropriate muscle strength testing should be done. The patient's gait must be observed, with adequate exposure of both legs.

Special tests such as the Heel raise test, Thompson test, and Silfverskiöld test also need to be performed, depending on the condition under evaluation.

The following flowchart will help to form an algorithm to clinically evaluate a patient with posterior heel pain secondary to Achilles tendon pathology.

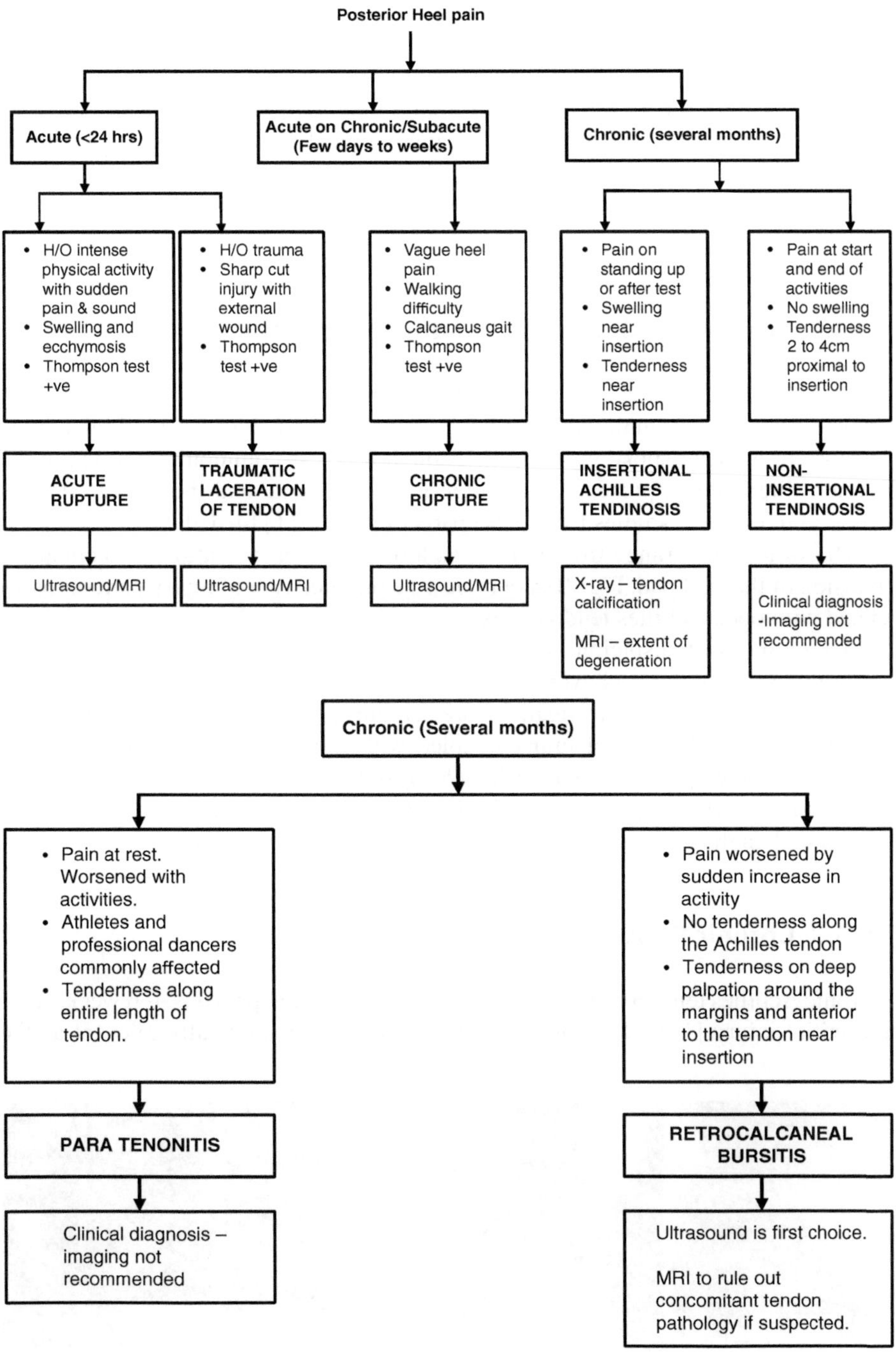
Posterior Heel pain
Acute (<24 hrs)
Acute on Chronic/Subacute (Few days to weeks)
Chronic (several months)
• H/O intense physical activity with sudden pain & sound
• Swelling and ecchymosis
• Thompson test +ve
• H/O trauma
• Sharp cut injury with external wound
• Thompson test +ve
• Vague heel pain
• Walking difficulty
• Calcaneus gait
• Thompson test +ve
• Pain on standing up or after test
• Swelling near insertion
• Tenderness near insertion
• Pain at start and end of activities
• No swelling
• Tenderness 2 to 4cm proximal to insertion
ACUTE RUPTURE
TRAUMATIC LACERATION OF TENDON
CHRONIC RUPTURE
INSERTIONAL ACHILLES TENDINOSIS
NON-INSERTIONAL TENDINOSIS
Ultrasound/MRI
Ultrasound/MRI
Ultrasound/MRI
X-ray – tendon calcification
MRI – extent of degeneration
Clinical diagnosis -Imaging not recommended
Chronic (Several months)
• Pain at rest. Worsened with activities.
• Athletes and professional dancers commonly affected
• Tenderness along entire length of tendon.
• Pain worsened by sudden increase in activity
• No tenderness along the Achilles tendon
• Tenderness on deep palpation around the margins and anterior to the tendon near insertion
PARA TENONITIS
RETROCALCANEAL BURSITIS
Clinical diagnosis – imaging not recommended
Ultrasound is first choice.
MRI to rule out concomitant tendon pathology if suspected.

10.3.1 Acute Tears

Acute rupture of the Achilles tendon typically presents with sudden onset of pain following a high-impact activity. The patient often presents with a history of snapping sound and difficulty in weight-bearing. Clinical signs will include swelling, ecchymosis, and tenderness to palpation. Ultrasound imaging is the investigation of choice and is generally recommended over an MRI to aid in the clinical diagnosis of a tendon rupture [13].

10.3.2 Posttraumatic Achilles Tendon Lacerations

Open tears of the tendon can occur following sharp cuts on the posterior aspect of the heel. The most common mechanism is from the sharp edges of the Indian-style lavatory or a wheel spoke injury [14]. Though it is less commonly reported in the Western literature, it presents quite often in Africa, Asia, and the Middle East as reported in multiple studies from these parts of the world [14–16].

The wounds can range from innocuous looking minor lacerations to a complete avulsion of the heel pad. Fig. 10.4 demonstrate the spectrum of injury patterns associated with open Achilles tendon tears.

A high index of suspicion, whenever a patient presents with a laceration on the posterior aspect of the heel over the tendon, is needed to avoid missing an open tear of the tendon. Imaging investigations are seldom required. A detailed history surrounding the mechanism of injury coupled with a thorough clinical examination will reveal the diagnosis. *The positive Thompson test is a very useful clinical tool whenever there is a suspicion of an Achilles tear.* Ultrasound imaging can be useful in cases when the clinical findings are equivocal and diagnosis is unsure.

10.3.3 Chronic Rupture

Chronic Achilles tendon rupture is defined as a rupture that presents 4–6 weeks after the initial injury [17]. Defining it as acute or chronic is clinically relevant as the

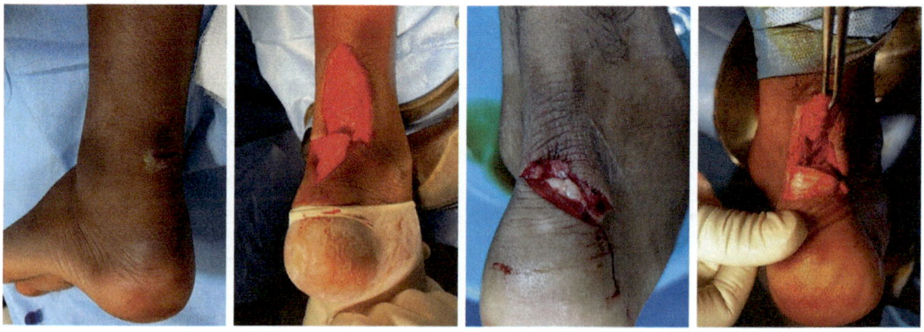

Fig. 10.4 Open wounds with underlying tendon laceration

management is significantly more complex and controversial in these injuries. Chronic injuries often present with a palpable gap and significant defect owing to the retraction of the proximal stump of the tendon. Patients present with difficulty in walking and climbing stairs. Pain is often not the chief complaint though a few may complain of dull aching type of pain.

On examination, the patient walks with a calcaneus gait. In some cases, a gross instability of the hindfoot may also be apparent. Palpation will reveal the presence of a palpable defect. Weakness of plantar flexion can be demonstrated by the inability of the patient to perform a heel raise test. Thompson test in the prone position is often diagnostic. Even before the test is performed, one can appreciate the affected ankle goes into dorsiflexion due to a combination of gravity and the absence of plantar flexion as shown in Fig. 10.5. A positive Thompson test is shown in Fig. 10.6.

Chronic rupture of the Achilles tendon is a clinical diagnosis. Imaging investigations mainly help in determining the extent of the defect, which is crucial in planning for the appropriate surgery. Ultrasound can be used to determine the extent of tendon retraction, and MRI can help to understand the quality of the tendon. Torn

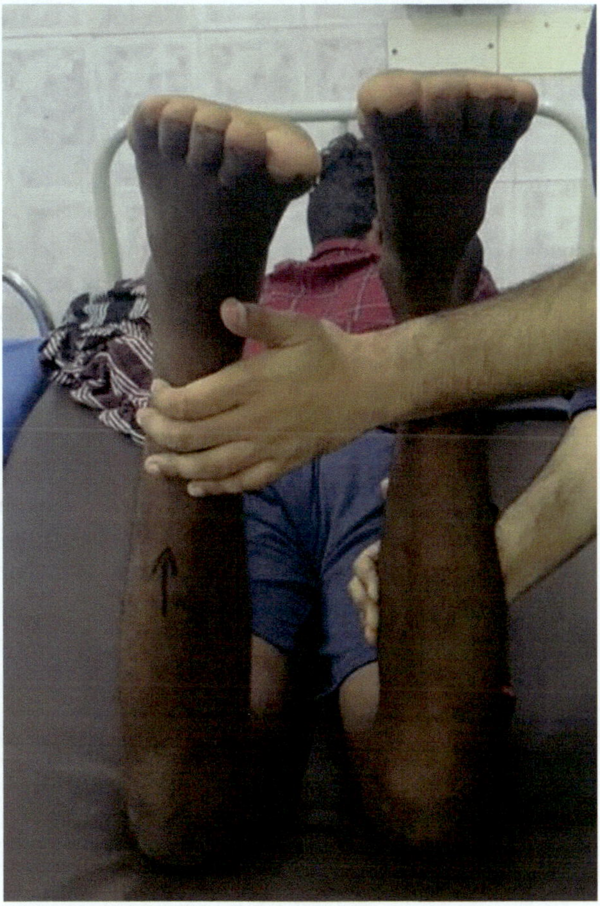

Fig. 10.5 The left ankle in dorsiflexion under the effect of gravity due to chronic rupture of the Achilles tendon

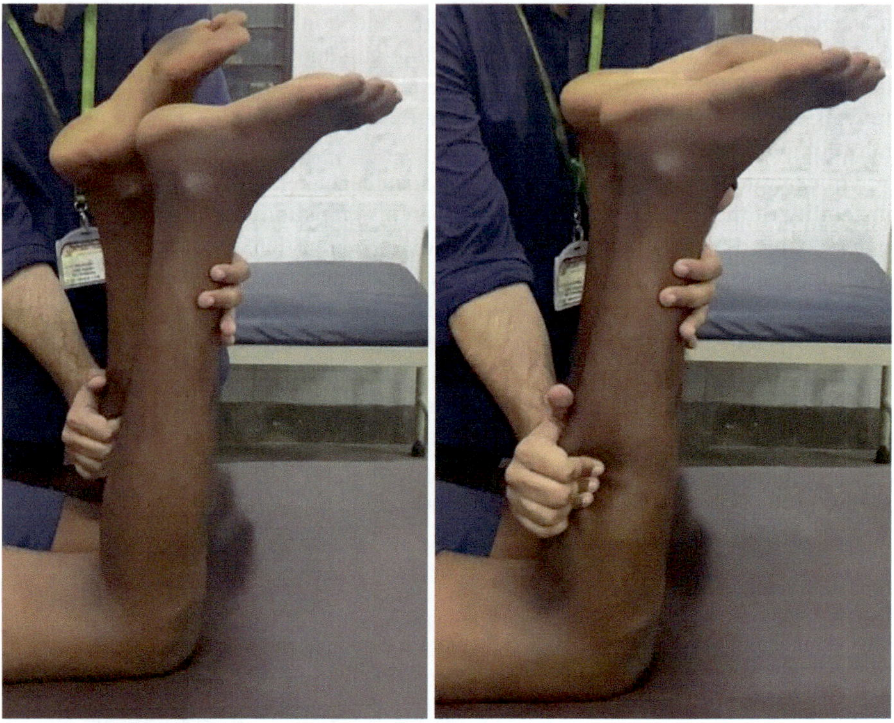

Fig. 10.6 A positive Thompson test of the affected ankle

tendon ends that show signs of degeneration, calcification, or fatty degeneration may need debridement, which will further increase the size of the defect. Fig. 10.7 show the MRI image of the tendon defect and the intraoperative picture of the same patient.

10.3.4 Achilles Tendinosis

Achilles tendinosis was previously termed "tendinitis." However, histological examination of the diseased tendon seldom reveals inflammation. Hence, the term tendinopathy or tendinosis is considered to be more appropriate [18].

Achilles tendinosis can be of three types:

- Isolated tendinosis without paratenonitis.
- Tendinosis with paratenonitis.
- Isolated paratenonitis.

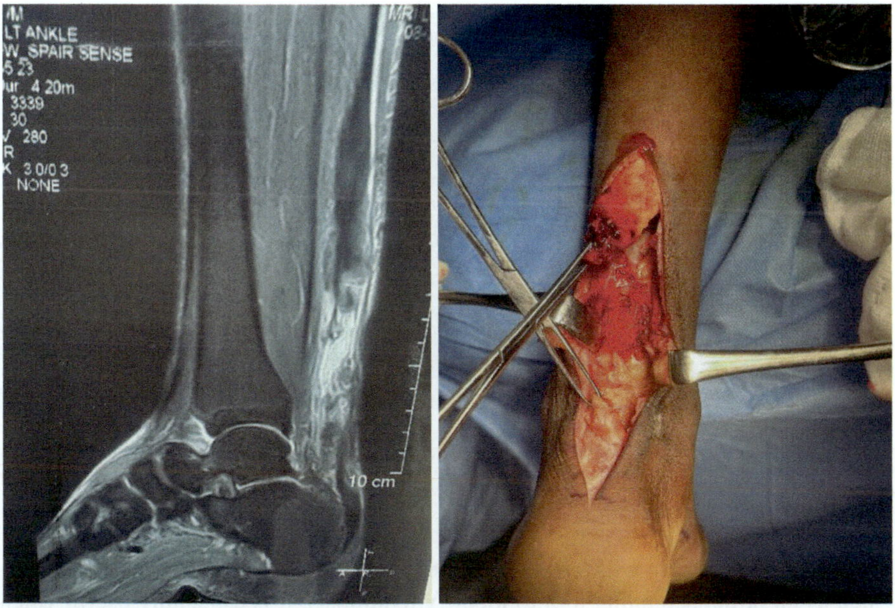

Figs. 10.7 A chronic rupture of the Achilles tendon with a large defect in MRI and intra-op, respectively

10.3.4.1 Isolated Tendinosis

The most common type of Achilles tendinopathy is isolated Achilles tendon degeneration without any inflammation. Based on the location of the tendinosis, this can be further classified into:

- Insertional tendinosis.
- Non-insertional tendinosis.

10.3.4.2 Insertional Tendinosis

This is typically characterized by pain at the back of the heel following a period of rest or while standing from a sitting position. Patients present with swelling on the posterior aspect of the heel. On examination, tenderness is elicited near the insertion of the tendon, as well as in the surrounding area of the calcaneus. Difficulty in performing heel raise is noted. The disease process is often bilateral, with different degrees of severity. Radiographic evaluation of the heel in the lateral view is the preliminary investigation to be done in patients suspected of insertional tendinopathy. X-rays will demonstrate calcification of the tendon near its insertion as shown in Fig. 10.8. The superior facet of the calcaneus is often enlarged and prominent. It is termed the Haglund deformity or the "pump-bump" deformity. This deformity is often seen as a part of the insertional Achilles tendinopathy complex but does not correlate with disease progression or severity [19].

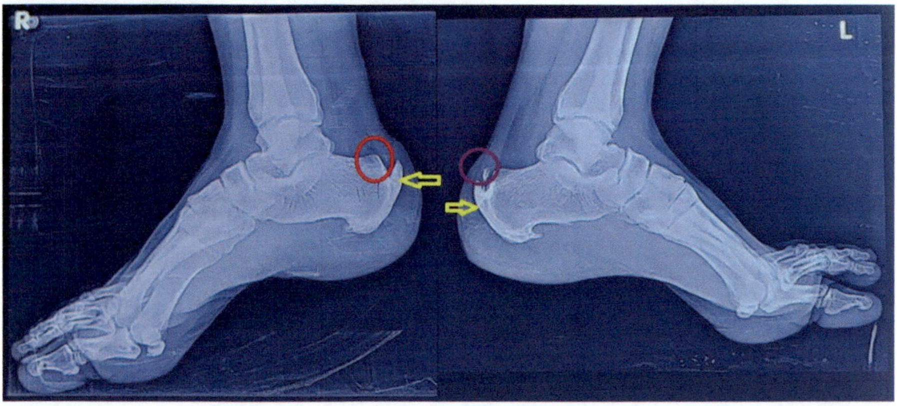

Fig. 10.8 The red circle marks the Haglund deformity. The purple circle shows the intratendinous calcification, which is a hallmark of the disease. The yellow arrows point out the heel spurs that occur secondarily

Ultrasound imaging will reveal the increased thickness of the tendon and is indicated as an initial screening tool to identify tendinosis. It is especially useful when there is a clinical suspicion of degenerative tendon tear.

MRI is indicated when surgical intervention is planned. Increased intratendinous signal change as well as thickening are notable findings. MRI also reveals the extent of tendon involvement, which is crucial to decide the appropriate line of management in these patients. Fig. 10.9 shows the normal MRI appearance of the Achilles tendon, whereas Fig. 10.10 shows the intra-tendinous signal changes that are seen in tendinosis.

10.3.4.3 Non-insertional Tendinosis

Non-insertional tendinosis is less commonly encountered as compared to insertional disorders. Patients complain of pain during the start and at the end of activities. Clinically, there is no swelling near the insertion of the tendon. Difficulty in performing a heel raise test is also noted. Tenderness is typically seen >2 cm from the insertion point. This corresponds to the "watershed zone" of the tendon, which shows poor vascularity in comparison with the musculotendinous junction and the insertional zone. The diagnosis of non-insertional tendinosis is primarily clinical; imaging investigations are indicated in the event of a rupture. Both MRI and ultrasound are equally effective in confirming the clinical diagnosis whenever required. The prognosis is generally better with non-insertional tendinopathy as compared to insertional tendinopathy [20].

10.3.4.4 Achilles Tendon Paratenonitis

Paratenonitis is inflammation of the paratenon that surrounds the Achilles tendon. It is mostly seen in athletes and professional dancers. Pain is felt even during rest and worsened with activities. Diagnosis is clinical with tenderness along the entire length of the tendon. Imaging is generally not recommended as the diagnosis is

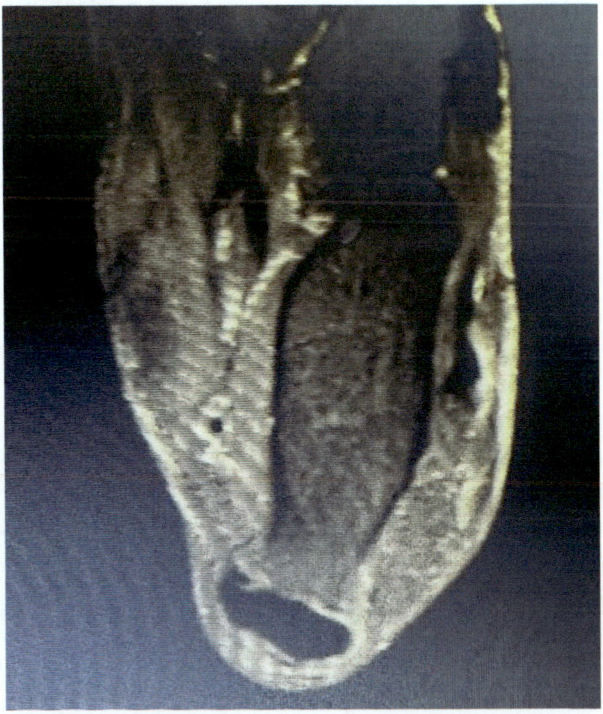

Fig. 10.9 T2W MRI cross-sectional image, showing a normal Achilles tendon

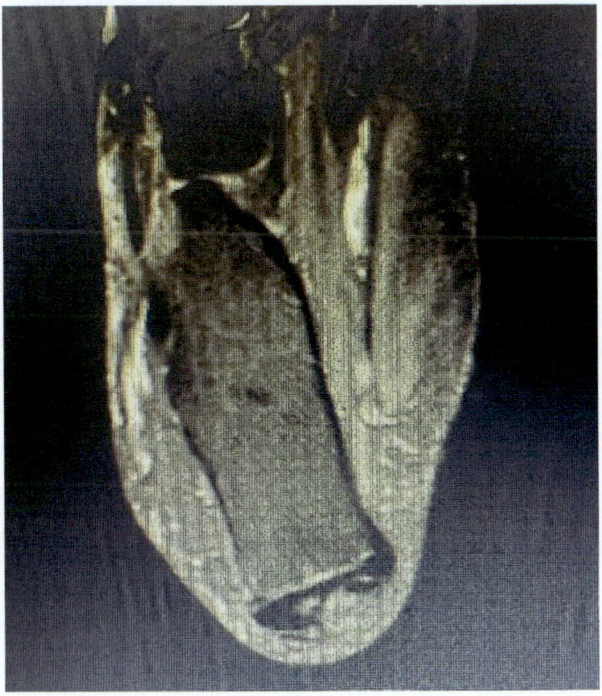

Fig. 10.10 T2W cross-sectional MRI image, showing signal changes within the tendon characteristic of tendinosis

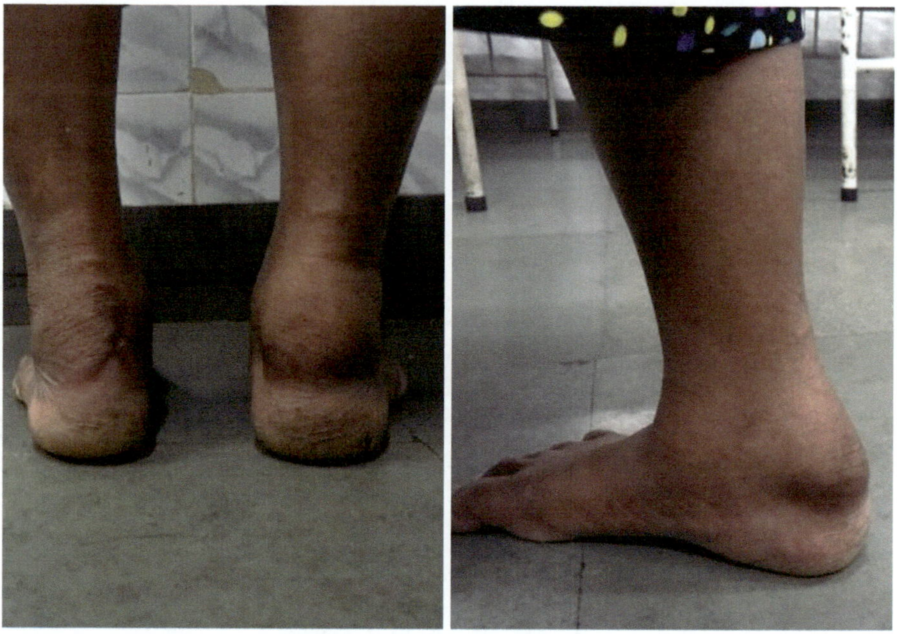

Fig. 10.11 The posterior aspect of the ankle with swelling near the insertion of the Achilles tendon

mostly clinical. However, both ultrasound and MRI are useful to differentiate it from tendinopathy. Tendon is usually uninvolved except very late in the disease process. The prognosis is the best among all Achilles tendon disorders.

10.3.4.5 Retrocalcaneal Bursitis

Retrocalcaneal bursitis is an independent clinical entity characterized by inflammation of the bursa that lies anterior to the Achilles tendon [21]. It can be isolated or can be seen along with Haglund deformity and tendinopathy changes. A patient typically presents with pain in the posterior aspect of the heel and is worsened with activity. On examination, swelling is present near the insertion of the tendon as seen in Fig. 10.11. Tenderness can be felt on either side of the Achilles tendon at the level of its insertion.

Radiographs are recommended to rule out Haglund deformity. Ultrasound or MRI imaging is indicated when there is suspicion of tendinopathy.

10.4 Conclusion

In conclusion, a focused clinical examination is paramount in diagnosing Achilles tendon disorders. Diagnosis of Achilles tendon ruptures may be straightforward most of the time and can be confirmed by ultrasonography or MRI. On the other hand, diagnosis of Achilles tendinosis necessitates meticulous attention to the site of tenderness, and imaging investigations such as MRI are often necessary to achieve a definitive diagnosis.

References

1. Anagnostopoulou S, Mavridis I. Achilles' death: anatomical considerations regarding the most famous trauma of the Trojan war. J Trauma Acute Care Surg. 2013;74(3):946–7.
2. O'Brien M. The anatomy of the Achilles tendon. Foot Ankle Clin. 2005;10(2):225–38. https://doi.org/10.1016/j.fcl.2005.01.011.
3. Ahmed IM, Lagopoulos M, McConnell P, Soames RW, Sefton GK. Blood supply of the Achilles tendon. J Orthop Res. 1998;16(5):591–6.
4. Yinger K, Mandelbaum BR, Almekinders LC. Achilles rupture in the athlete. Current science and treatment. Clin Podiatr Med Surg. 2002;19(2):231–50. v
5. Mahan J, Damodar D, Trapana E, Barnhill S, Nuno AU, Smyth NA, et al. Achilles tendon complex: the anatomy of its insertional footprint on the calcaneus and clinical implications. J Orthop. 2020;17:221–7.
6. Yu C, Deng L, Li L, Zhang X, Fu W. Exercise effects on the biomechanical properties of the Achilles tendon-a narrative review. Biology. 2022;11(2):172.
7. Maganaris CN, Narici MV, Maffulli N. Biomechanics of the Achilles tendon. Disabil Rehabil. 2008;30(20–22):1542–7.
8. Batista F, Nery C, Pinzur M, Monteiro AC, de Souza EF, Felippe FHZ, et al. Achilles tendinopathy in diabetes mellitus. Foot Ankle Int. 2008;29(5):498–501.
9. Knopp WD, Bohm ME, McCoy JC. Hypothyroidism presenting as tendinitis. Phys Sportsmed. 1997;25(1):47–55.
10. Ames PRJ, Longo UG, Denaro V, Maffulli N. Achilles tendon problems: not just an orthopaedic issue. Disabil Rehabil. 2008;30(20–22):1646–50.
11. Cortina RE, Morris BL, Vopat BG. Gastrocnemius recession for Metatarsalgia. Foot Ankle Clin. 2018;23(1):57–68.
12. Knobloch K. Drug-induced tendon disorders. Adv Exp Med Biol. 2016;920:229–38.
13. Dams OC, Reininga IHF, Gielen JL, van den Akker-Scheek I, Zwerver J. Imaging modalities in the diagnosis and monitoring of Achilles tendon ruptures: a systematic review. Injury. 2017;48(11):2383–99.
14. SasankaS C, Arindam S, Ashutosh M. Management of acute open tendo-Achilles injuries in Indian lavatory pans. Indian J Plast Surg. 2006;39:29.
15. Awe AA, Esezobor EE, Aigbonoga QO. Experience with managing open ACHILLES tendon injuries in a tertiary hospital in southern Nigeria. J West Afr Coll Surg. 2015;5(4):30–40.
16. Said MN, Ateeq A, Al Dosari M, Al Subaii N, Kawas A, Al Mas A, Al Ser Y, et al. Open Achilles tendon lacerations. Eur J Orthop Surg Traumatol Orthop Traumatol. 2015;25(3):591–3.
17. Flint JH, Wade AM, Giuliani J, Rue JP. Defining the terms acute and chronic in orthopaedic sports injuries: a systematic review. Am J Sports Med. 2014;42(1):235–41.
18. Ferguson A, Christophersen C, Elattar O, Farber DC. Achilles tendinopathy and associated Disorders. Foot Ankle Orthop. 2019;4(2):2473011419838294.
19. Lee W, Giro ME, Crymes C. Association of Haglund deformity size and insertional Achilles tendinopathy. Foot Ankle Int. 2023;44:719.
20. Pearce CJ, Tan A. Non-insertional Achilles tendinopathy. EFORT Open Rev. 2016;1(11):383–90.
21. Weinfeld SB. Achilles tendon disorders. Med Clin North Am. 2014;98(2):331–8.

Heel Pain

11

Ankit Khurana and Venkataram Vellaipandi

11.1 Introduction

Heel pain refers to pain around the calcaneus, typically felt at the plantar or posterior aspect of the heel. This chapter will focus on the potential sources of heel pain, extending from the calcaneus to both the lateral and medial peri-malleolar regions. It should be noted that anatomically, the heel comprises of the fibrofatty tissue, which forms a cushion beneath and around the calcaneus. This plays an important role in cushioning and protecting the neurovascular structures, bones, and joints during standing and locomotion. However, in clinical practice, patients may perceive a larger area of their foot as their "heel." This underscores the significance of a comprehensive history, clinical examination, and relevant investigations to identify the underlying cause of pain. Similarly, tenderness, when used in isolation, is an inadequate diagnostic test, as many structures can be painful to the touch without necessarily being the source of symptoms. Hence, a comprehensive history and thorough clinical examination remains the cornerstone for accurate diagnosis and management [1].

A. Khurana (✉)
Department of Orthopaedics, Dr BSA Medical College and Hospital, Govt of NCT Delhi, New Delhi, India

V. Vellaipandi
Department of Orthopaedics, Mahatma Gandhi Medical College and Research Institute, Pondicherry, India

S. Sharma et al. (eds.), *Clinical and Radiological Examination of the Foot and Ankle*, https://doi.org/10.1007/978-981-97-4202-8_11

11.2 Potential Sources of Heel Pain

There are several anatomical structures that can contribute to heel pain. Table 11.1 explores the various underlying local causes and mechanisms responsible for plantar heel pain, encompassing conditions such as plantar fasciitis, Achilles tendinitis, heel bursitis, and calcaneal stress fractures (Table 11.1).

Tendons which can contribute to heel pain include the Achilles tendon at its insertion, the flexor hallucis longus (FHL) tendon, the tibialis posterior, and the peroneal tendons (Fig. 11.1). Tibialis posterior tendon pain is more common in older women, whereas FHL tenosynovitis is prevalent in younger individuals, especially dancers who repeatedly flex and extend their ankles and toes to extremes.

Pain in the Achilles insertion, including the retrocalcaneal bursa, should be treated with tendinopathy, as the bursa is rarely affected in isolation. Achilles tendinopathy refers to inflammation and damage to the Achilles tendon, which can result from acute overexertion, blunt trauma, or chronic overuse. It is commonly seen in endurance runners and individuals with tight calf muscles. The mid-substance and insertion of Achilles tendon undergo similar histopathological changes. This is often due to excess compression against the superior calcaneus in dorsiflexion (Haglund deformity). Haglund deformity (heel bump) occurs just lateral to the Achilles tendon and, apart from causing slow degeneration of the Achilles tendon, also interferes with shoe wear and leads to painful bursitis due to rubbing against footwear. Peroneal tendinopathy is a less frequent cause of pain around the heel and can occur either after an acute ankle sprain or due to an insufficient retinaculum at the lateral malleolus, leading to increased compression and friction loads.

The plantar fascia, which is histologically similar to a tendon, can also cause heel pain, especially in older, more obese women. Neural sources of heel pain may mimic soft tissue pain and may also result from entrapment either proximally in the spine and gluteal region beneath the pyriformis or distally at the ankle retinaculum. Plantar fasciitis is a prevalent cause of plantar heel pain, characterized by irritation and inflammation of the plantar fascia ligament that spans the longitudinal arch of the foot. It is particularly common in joggers and individuals with risk factors such as excessive pronation, reduced ankle dorsiflexion, occupations involving obesity, poor footwear cushioning, excessive standing or walking, or unaccustomed running. In some cases, plantar fasciitis may be bilateral and associated with rheumatologic conditions like rheumatoid arthritis, systemic lupus erythematosus, or gout. It is common in the general population and is more frequent in joggers and individuals with risk factors like tight gastrosoleus, over-pronation, or occupations involving excessive walking or standing [3–5].

Conditions like tarsal tunnel syndrome and sural nerve symptoms can also be related to neural heel pain. Similarly, heel neuritis is caused by compression of a branch of the lateral plantar nerve, leading to numbness, tingling, or pain in the heel. This is often related to ankle sprain, fracture, or varicose veins. Heel spurs, abnormal bony growths at the insertion of the plantar fascia on the heel bone, can also cause heel pain, often associated with plantar fasciitis.

Table 11.1 This table outlines the potential sources of heel pain at different sites. The common sources of pain in each location are listed, along with less common sources that may also contribute to heel discomfort

Source of pain	Posterior	Inferior	Medial	Lateral	Deep, vague pain
Common sources of pain	Achilles tendon insertion	Plantar fascia	Tibialis posterior tendon and sheath	Lateral ligaments of the ankle	Subtalar joint
	Superficial calcaneal bursa	Calcaneal fat pad	Tibialis posterior insertion and apophysis in adolescents	Sinus tarsi	FHL, flexor hallucis longus
	Posterior impingement of soft tissues/os trigonum in active people	Medial or lateral calcaneal nerve, especially as they split from the tibial branch	FHL and sheath	Peroneus brevis insertion/ apophysis of base of the fifth metatarsal in adolescents or after an ankle sprain	Bone pain: Calcaneus, talus, navicular
Less common sources of pain	Calcaneal apophysis in adolescents [2]		Abductor hallucis	Cubometatarsal joint	
	Sural nerve		Deltoid and spring ligaments	Peroneal tendinopathy or tenosynovitis associated with subluxation	
			Posterior tibial nerve in tarsal tunnel (associated with neural symptoms such as tingling)		
			Bone: Medial malleolus		

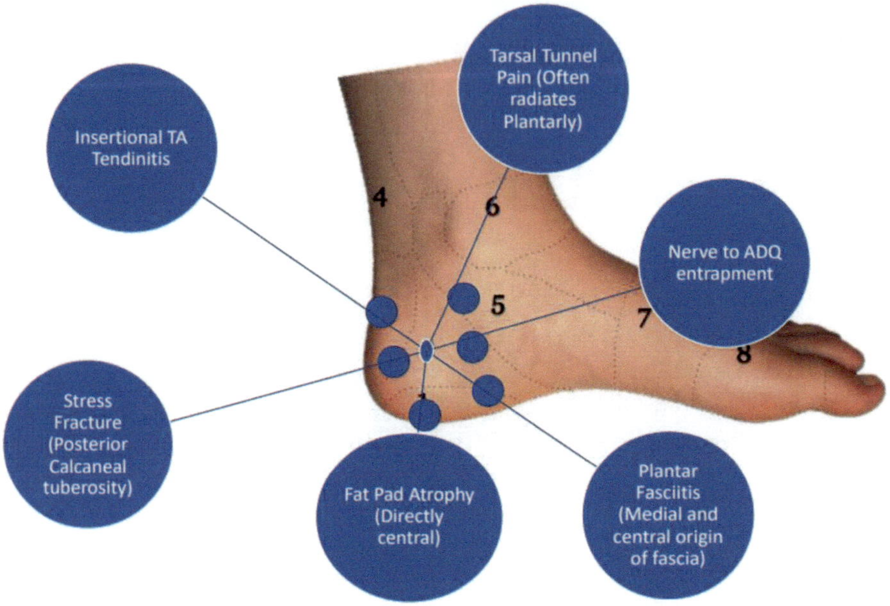

Fig. 11.1 Potential sources of heel pain at different sites, categorized by different anatomical sites of tenderness

Heel bursitis is an inflammation of the heel bursa, causing pain deep inside or behind the heel, worsened by pressure from footwear or landing awkwardly on the heels. Sever's disease (calcaneal apophysitis) is a common cause of heel pain in children and adolescents, occurring between 5 and 11 years old [2]. This condition is characterized by inflammation of the growth plate in the heel due to mechanical stress and overuse in active children.

11.2.1 Diagnosing the Source of Pain

11.3 History

A detailed history and thorough clinical examination is the primary method for diagnosing heel pain, with the location of pain and the absence of associated systemic symptoms providing strong indications for diagnosis (Table 11.2). Patients may present their symptoms, undergo physical examination, and, if necessary, undergo imaging studies such as X-rays, ultrasound, and MRI scans. In some cases, blood analysis tests may be required for further diagnostic evaluation [3–5].

While taking a history, gathering detailed information about the pain's characteristics, including its onset, location, radiation, diurnal variations, modifying factors,

Table 11.2 History, examination, and investigations for heel pain diagnosis. Note: Clinical evaluation remains crucial for accurate diagnosis, while imaging serves to identify potential pathologies and structural abnormalities. Relying solely on imaging may lead to misdiagnosis and ineffective treatment plans

Aspect	Description
History	Detailed history focusing on pain characteristics (onset, location, radiation, modifying factors, timing, activities)
	Most common cause: Proximal plantar fasciitis, gradual onset, no prior injury, or fever
	Other possible causes: Traumatic rupture, fat pad atrophy, neurologic origin, inflammatory disorders, neoplasms
Examination	General examination to rule out systemic causes
	Spine examination if pain radiates or indicates radiculopathy
	Local examination to inspection of the foot and heel for abnormalities and tenderness
	Palpation to identify the site of maximum tenderness and bony deformities
	Percussion over tibial nerve and scars to detect hypersensitivity or neuroma
	Range of motion examination at the ankle joint and Silfverskiöld test for gastrocnemius/triceps surae complex
	Tenderness in plantar and medial aspects of the heel
	Bilateral heel pain with general symptoms indicates inflammatory association
	Elderly/atypical presentations may necessitate investigation for deficiency fractures or tumors
Investigation	Blood analysis tests for further evaluation (e.g., serum uric acid, vitamin D3, TSH, ESR, CRP, rheumatological workup)
	Radiological imaging (X-rays, ultrasound, MRI) to assess heel spurs, calcification, Achilles tendon, plantar fascia
	MRI scans for Achilles tendinosis and possible tendon ruptures
	Bone scan for increased activity or discrete uptake in stress fractures
	Advanced imaging for specific pathologies (e.g., os trigonum, peroneal tendon)
	Importance of clinical assessment as the primary diagnostic tool
	Caution in relying solely on imaging findings, as they may not always correlate with pain or pathology

and connection to activities, is vital in distinguishing between various potential causes. The most prevalent reason for plantar heel pain is proximal plantar fasciitis. This condition is commonly found in individuals whose occupations require prolonged standing. The pain is typically experienced on one side but occasionally can affect both feet, with one side being more painful than the other. In most cases, the pain arises gradually and without any prior injury or fever. Some patients may recall stepping on small objects or starting a new exercise routine involving walking or running. Additionally, weight gain could be linked to the pain [4].

The discomfort is localized to the plantar medial aspects of the heel, and it is particularly worse during the first few steps in the morning. This often causes a patient to limp for about 30 min after waking. It also worsens after periods of rest. Walking and stretching tend to alleviate the pain, but prolonged walking and standing can aggravate it. Consequently, patients may develop a limp and adjust their walking pattern, putting more weight on the forefoot and lateral aspect of the foot. This can further worsen the problem.

Sudden onset of pain, particularly after intense physical activity, may suggest a traumatic rupture of the plantar fascia. Elderly patients and individuals who have received multiple cortisone injections may experience fat pad atrophy, which results in more diffuse pain under the heel, affecting most of the plantar weight-bearing surface. This pain intensifies on hard surfaces and with hard-soled footwear. Unlike plantar fasciitis, patients with fat pad atrophy do not show initial improvement in walking [1, 3].

Heel pain which radiates either proximally or distally and is accompanied by burning, tingling sensation or numbness following either rest or may indicate a neurologic origin. This could be caused by a compressive neuropathy which occurs locally, such as tarsal tunnel syndrome, or in the spine possibly associated with low back pain. Bilateral heel pain and pain at tendon insertions, along with other symptoms like recurrent fever, malaise, bowel dysfunction, or polyarthralgia, may suggest an association with inflammatory conditions like RA (rheumatoid arthritis), spondyloarthropathies, Behcets' syndrome, or Reiter syndrome. Moreover, significant weight loss, loss of appetite, pain at night, or atypical presentations in elderly patients may necessitate investigation for deficiency fractures or tumors.

11.4 Clinical Examination

During the physical examination, the foot is assessed both while bearing weight and while at rest. Visual inspection may reveal bony deformities, swelling, contusion, or skin conditions. A comprehensive examination of the whole body is necessary to rule out any systemic etiology. If the pain radiates or there are indications of radiculopathy, a spine examination is also recommended. During the local examination, careful inspection of the foot and heel is conducted to identify any abnormalities such as scars, swelling, or foot deformities like pes cavus or flatfeet. Palpation is carried out to pinpoint the site of maximal tenderness while considering the "windlass mechanism" of the plantar fascia. The condition of the fat pad is checked, and any defects or growths in the plantar fascia as well as deformities due to old fractures are identified [3].

Percussion is employed to assess heightened sensitivity or tingling sensations along the tibial nerve within the tarsal tunnel (Fig. 11.2) [4, 6–8].

Neuroma in scars from previous surgeries can also be detected through percussion. The assessment further involves checking the range of motion in the ankle joint and conducting a Silfverskiöld test to evaluate any stiffness in the gastrocnemius and/or triceps surae complex (as shown in Fig. 11.3a, b) [9, 10].

In individuals with proximal plantar fasciitis, tenderness is typically concentrated around the medial calcaneal tuberosity where the plantar fascia originates (as illustrated in Fig. 11.4).

Additional characteristics may encompass a contracture in the triceps surae, reduced subtalar mobility, high arches (pes cavus), or flat feet (pes planus), all of

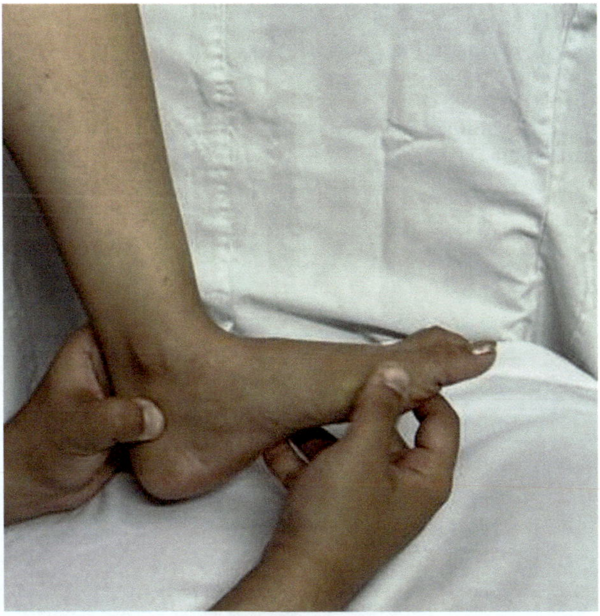

Fig. 11.2 Clinical examination demonstrating percussion over the tibial nerve within the tarsal tunnel to detect hypersensitivity or neuroma

which can increase tension on the plantar fascia. Importantly, pain does not intensify during a clinical test that stretches the plantar fascia by dorsiflexing the toes in these individuals. Conversely, those with an acute plantar fascia rupture may experience pain aggravation during this maneuver, along with localized bruising or a palpable defect.

Tenderness when pressure is applied to both the inner and outer sides of the back of the heel (posterior calcaneal tuberosity) strongly indicates a stress fracture of the calcaneus (as depicted in Fig. 11.5). This condition is often accompanied by localized swelling.

In instances of compressive neuropathy, which can affect either the tibial nerve in the tarsal tunnel or the first branch of the lateral plantar nerve, the most tender point in the heel tends to be situated more toward the inner side of the front of the heel (as shown in Fig. 11.6).

Percussion (tapping) over the branches of the tibial nerve can elicit sensations of tingling, burning, or numbness. Excessive stretching of the tibial nerve can occur in cases of a valgus heel associated with flat arch (pes planus) and may lead to tarsal tunnel syndrome [4, 6–8].

In elderly patients or those who have received multiple cortisone (steroid) injections in the heel, the pain and tenderness are most pronounced over the central part (weight-bearing area) of the heel (Fig. 11.7).

Dorsiflexion of the toes does not worsen the pain. The heel feels soft and lacks the usual firmness, with the underlying calcaneus being more palpable. The examination involves looking for any tenderness or abnormalities around the prominent bony points and tendon attachments near the heel and midfoot. Passive movement of the foot, subtalar, and ankle joint is assessed to check for any limitations in

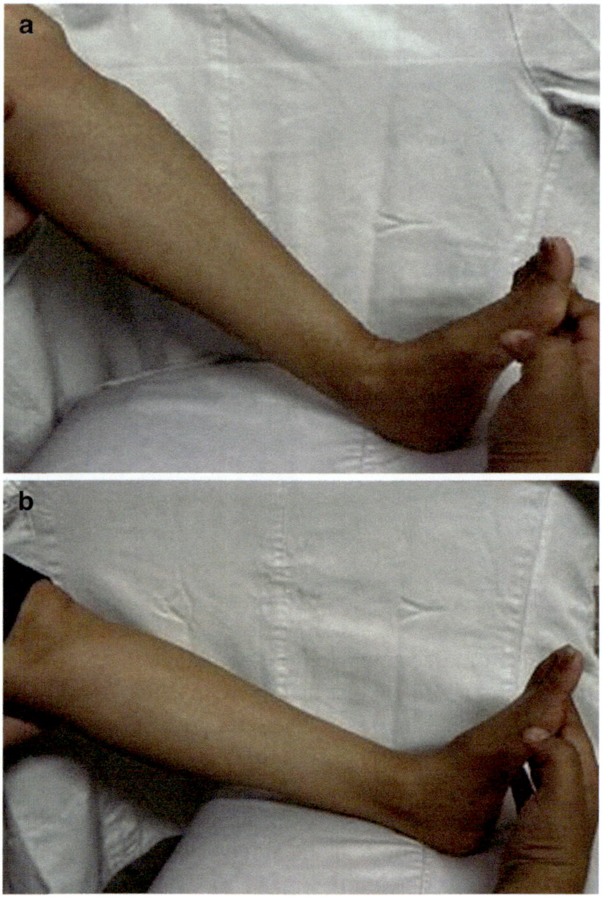

Fig. 11.3 (**a**) Silfverskiöld test: evaluating stiffness in the gastrocnemius muscle. (**b**) Silfverskiöld test: evaluating stiffness in the triceps surae complex

mobility. The patient's foot posture and arch structure are visually inspected while they stand, with special attention given to detecting unusual pronation or other biomechanical irregularities. Additionally, observing the patient's gait can offer additional diagnostic clues. Following a thorough review of the patient's medical history, potential sources of heel pain are pinpointed, and additional objective evaluations are conducted to validate the diagnosis.

Blood investigations including serum uric acid, serum vitamin D3 levels, serum TSH, ESR, CRP, and rheumatological workup should be considered especially in bilateral heel pain. Although some diagnoses are less common, such as Achilles tendon rupture, progressive tibialis posterior lengthening, seronegative arthropathies, stress fractures, osteoid osteoma, and tarsal coalition, they are still be recognizable on a thorough clinical evaluation (e.g., night pain in osteoid osteoma, distal

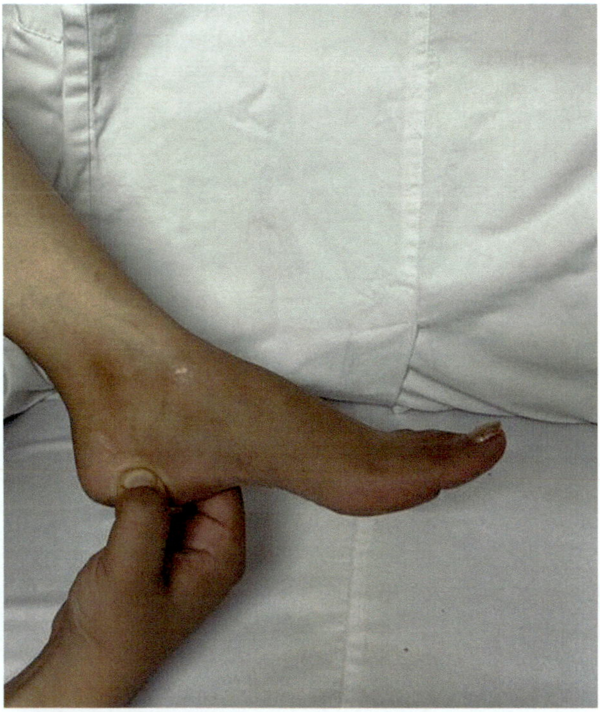

Fig. 11.4 Localization of tenderness at the medial calcaneal tuberosity, a common site for proximal plantar fasciitis

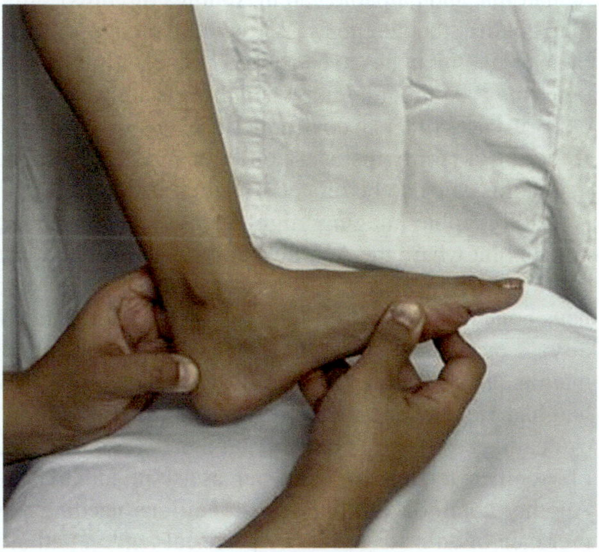

Fig. 11.5 Tenderness when pressure is applied to both the inner and outer sides of the back of the heel, indicating a potential stress fracture of the calcaneus

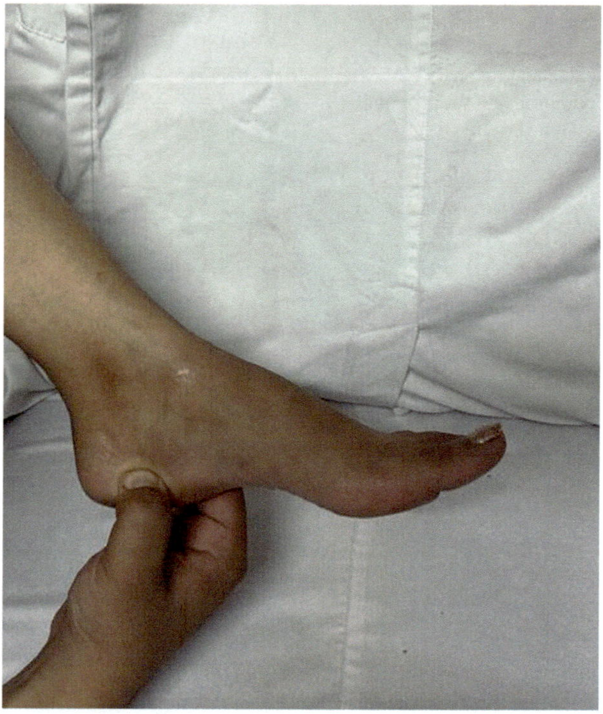

Fig. 11.6 Localization of tenderness on the inner side of the front of the heel, suggesting possible compressive neuropathy

hematoma, and palpable gap in Achilles ruptures), although additional imaging will help confirm the diagnosis. While these diagnoses are rare, it is crucial to recognize them as treatment has to be carried out accordingly. Moreover, primary bony pathologies may require imaging when there is a high suspicion of these conditions, especially in cases of tarsal coalition, which may present with flatfoot and pain especially if there is a family history. Conditions like tarsal tunnel syndrome and sural nerve symptoms can also be related to neural heel pain.

Radiographic imaging might uncover the presence of heel spurs or calcification at the point where the Achilles tendon attaches or at the origin of the plantar fascia. Additionally, an excessively prominent area at the back and upper part of the calcaneus bone may extend into the Achilles tendon, a condition referred to as Haglund disease or Haglund deformity. This can result in symptoms stemming from inflammation in the retrocalcaneal bursa. A bone scan may reveal a localized area of heightened uptake, as seen in cases of stress fractures. MRI scans can be useful in evaluating the extent of degeneration in the Achilles tendon, particularly in instances of Achilles tendinosis, and can assist in identifying potential tendon ruptures. Another common examples of conditions needing advanced imaging include the presence of an os trigonum or peroneal tendon pathology, which may be observed as increased signal on MRI following an ankle inversion injury. While these findings are evident on imaging, they may not necessarily be the actual cause of the patient's pain. Instead, other factors like posterior ankle impingement syndrome or subtalar joint synovitis, which could result from an ankle sprain, are more likely to be the true sources of pain [11].

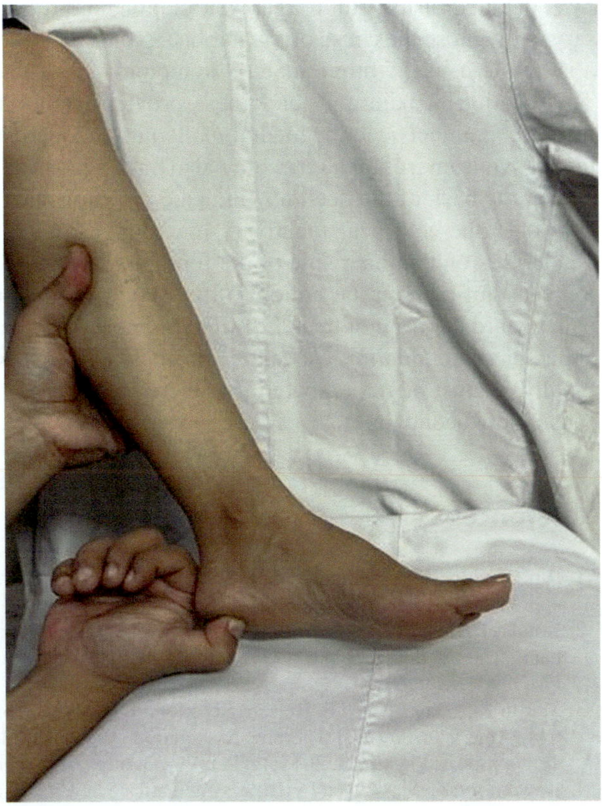

Fig. 11.7 Tenderness over the central weight-bearing area of the heel, indicating potential fat pad atrophy, common in elderly patients or those who have received multiple cortisone injections

The significance of diagnostic imaging in the assessment of heel pain lies in its ability to identify various pathologies and structural abnormalities. However, it is crucial to note that clinical assessment remains the primary diagnostic tool, as imaging can sometimes show pathology or abnormalities without any associated pain. This can lead clinicians to mistakenly attribute the identified imaging findings as the definitive source of pain. Similarly, imaging may reveal pathology in the retrocalcaneal bursa and Achilles tendon, along with a Haglund deformity, in individuals who are entirely free of pain. The absence of pain indicates that the presence of such findings on imaging does not necessarily correlate with the presence of pain or pathology. Relying solely on imaging findings to determine the source of heel pain can lead to misdiagnosis and potentially ineffective treatment plans.

11.5 Conclusion

It's crucial for the treating surgeon to answer three important questions to determine the cause of plantar heel pain. Firstly, they need to figure out if the pain is localized to the foot or if it's referred from elsewhere. Secondly, they should identify if the pain is related to a systemic health issue or if it is specific to the foot. Lastly, the surgeon must pinpoint the cause of the local heel pain.

A detailed history, thorough examination, and relevant investigations are necessary to diagnose the cause of plantar heel pain. In most cases, bilateral pain suggests a systemic cause, while unilateral symptoms are usually due to local foot issues. The location of pain and tenderness provides important clues for diagnosis. For instance, pain between the medial malleolus and tendoachilles could indicate tarsal tunnel syndrome, and patients may also have a positive Tinel sign. Pain and tenderness at the back of the heel may indicate insertional TA tendinopathy. Deep tenderness over the calcaneal tuberosity may suggest a stress fracture due to overuse. Tenderness on the inner side of the foot may indicate plantar fasciitis, while a central cause of pain on the bottom of the foot could be due to fat pad atrophy. More pain toward the front of the heel might be caused by compression of the nerve to the abductor digiti minimi, especially in obese patients with adult-acquired flatfoot. Finally, it must be remembered that there can be a combination of different issues causing the heel pain.

References

1. Rio E, Kidgell D, Purdam C, Gaida J, Moseley GL, Pearce AL, et al. Heel pain: a practical approach. Aust Fam Physician. 2015b;44(3):96–101.
2. Kothari EA, Padgett AM, Young SM, Ray J, Shah A, Conklin MJ. A review of pediatric heel pain. Cureus. 2023;15(1):e34228. https://doi.org/10.7759/cureus.34228. PMID: 36852370; PMCID: PMC9960861.
3. Martin RL, Davenport TE, Reischl SF, McPoil TG, Matheson JW, Wukich DK, McDonough CM, American Physical Therapy Association. Heel pain-plantar fasciitis: revision 2014. J Orthop Sports Phys Ther. 2014;44(11):A1–33. https://doi.org/10.2519/jospt.2014.0303.
4. Agyekum EK, Ma K. Heel pain: a systematic review. Chin J Traumatol. 2015;18(3):164–9. https://doi.org/10.1016/j.cjtee.2015.03.002.
5. Morrissey D, Cotchett M, Said J'Bari A, Prior T, Griffiths IB, Rathleff MS, Gulle H, Vicenzino B, Barton CJ. Management of plantar heel pain: a best practice guide informed by a systematic review, expert clinical reasoning and patient values. Br J Sports Med. 2021;55(19):1106–18. https://doi.org/10.1136/bjsports-2019-101970. Epub 2021 Mar 30. PMID: 33785535; PMCID: PMC8458083.
6. Scher DL, Belmont PJ Jr, Bear R, Mountcastle SB, Orr JD, Owens BD. The incidence of plantar fasciitis in the United States military. J Bone Joint Surg Am. 2009;91(12):2867–72.
7. Bolgla LA, Malone TR. Plantar fasciitis and the windlass mechanism: a biomechanical link to clinical practice. J Athl Train. 2004;39(1):77–82.
8. Cheng HY, Lin CL, Wang HW, Chou SW. Finite element analysis of plantar fascia under stretch-the relative contribution of windlass mechanism and Achilles tendon force. J Biomech. 2008;41(9):1937–44.
9. Singh D. Nils Silfverskiöld (1888-1957) and gastrocnemius contracture. Foot Ankle Surg. 2013;19(2):135–8.
10. Patel A, DiGiovanni B. Association between plantar fasciitis and isolated contracture of the gastrocnemius. Foot Ankle Int. 2011;32(1):5–8.
11. Prokuski LJ, Saltzman CL. Challenging fractures of the foot and ankle. Radiol Clin North Am. 1997;35(3):655–70.

Progressive Collapsing Foot Deformity 12

Rohan Bhimani

12.1 Introduction

Acquired adult flatfoot deformity (AAFD), more recently termed as progressive collapsing foot deformity (PCFD) elucidates a syndrome encompassing initial physical symptoms in the medial retromalleolar area of the posterior tibial tendon leading to the gradual collapse of the medial longitudinal arch in multiple planes, resulting in hindfoot valgus, midfoot abduction, forefoot supination, and potentially causing ankle joint valgus due to weakening of the ankle ligament complex [1–3]. In 1989, Johnson and Strom first proposed the classification system, which relied mainly on the dysfunction of the posterior tibial tendon [4]. Additional anatomical structures such as the spring ligament complex, the deltoid ligament, and the intraosseous ligaments within the subtalar joint have been identified over the past decade with an improved understanding of the pathology. However, these structures undergo secondary changes secondary to stress as opposed to the primary cause of the deformity. This has led to the emergence of various modified classification systems, including a recent one aimed at assisting clinicians in effectively characterizing deformities and aiding appropriate treatment recommendations [2].

The deformity is not always symptomatic, with less than 15% of adults with flatfoot experiencing symptoms [5]. The majority of symptomatic patients are middle-aged females with a history of obesity, diabetes, hypertension, equinus, or those involved in high-impact sports [6–9]. Pain along the medial longitudinal arch, as well as the posteromedial ankle, subfibular area, and over the sinus tarsi, is common in symptomatic patients, and can be exacerbated by prolonged standing and weight-bearing activity. Patients may also observe a collapse of their medial

R. Bhimani (✉)
Department of Orthopaedic Surgery, Massachusetts General Hospital, Harvard Medical School, Boston, MA, USA

S. Sharma et al. (eds.), *Clinical and Radiological Examination of the Foot and Ankle*, https://doi.org/10.1007/978-981-97-4202-8_12

longitudinal arch during weight-bearing activities. The presence of edema above the medial ankle may be observed in deformities that have progressed to a more advanced stage, and in such cases, the deformity may become irreducible. Tenderness along the course of the posterior tibial (PT) tendon and occasionally in the lateral heel, distal to the lateral malleolus, can be elicited on physical examination. Additionally, the collapse of the medial longitudinal arch and the presence of the "too many toes sign" can be observed during weightbearing. The hindfoot may exhibit valgus alignment, while the forefoot compensates by adopting a varus alignment [6, 7]. The patient may experience difficulty in performing single heel rise. As the deformity advances, a lack of varus rotation of the heel with the heel rise may be observed, ultimately resulting in an inability to perform a single heel rise [2, 6, 7, 9, 10].

The use of imaging modalities plays a crucial role in the evaluation of PCFD as it allows for the identification and exclusion of underlying conditions that contribute to the development of the deformity. Additionally, imaging aids in the assessment of the soft tissue structures responsible for the disorder, the identification of deformity resulting from their dysfunction, and the determination of the stage of the disease. With the advent of new imaging modalities, the imaging findings can now serve as a guide in selecting the most appropriate treatment option for PCFD. This chapter will focus on clinical evaluation and parameters that can be assessed using traditional and newer imaging techniques to aid the clinician in the diagnosis and management of patients with PCFD.

12.2 Clinical Evaluation

Patients with PCFD may exhibit a range of symptoms, varying from noticeable pain to an absence of apparent discomfort. Typically, patients initially complain of pain located along the medial side of the foot and ankle, particularly along the course of the posterior tibial tendon, which runs behind and below the ankle. This pain is usually exacerbated during physical activities and abnormal shoe wear and is often accompanied by an unsteady gait. As the pathology progresses, the medial-sided pain along the posterior tibial tendon diminishes, usually due to tendon rupture, and the deformity advances. Consequently, pain is observed more laterally in the calcaneofibular region due to impingement.

Clinical examination reveals specific indicators of PCFD, such as valgus hindfoot alignment when the patient stands (Fig. 12.1a). This can be associated with forefoot abduction, a "too many toes" sign which is observed from the rear and refers to the visibility of three or more toes beyond the lateral border of the foot (Fig. 12.1b), and a collapsing of the medial longitudinal arch. Evaluating the patient's ability to perform the double and single heel rise tests, especially the latter, is crucial for staging the condition (Fig. 12.2). In the advanced stages, the valgus

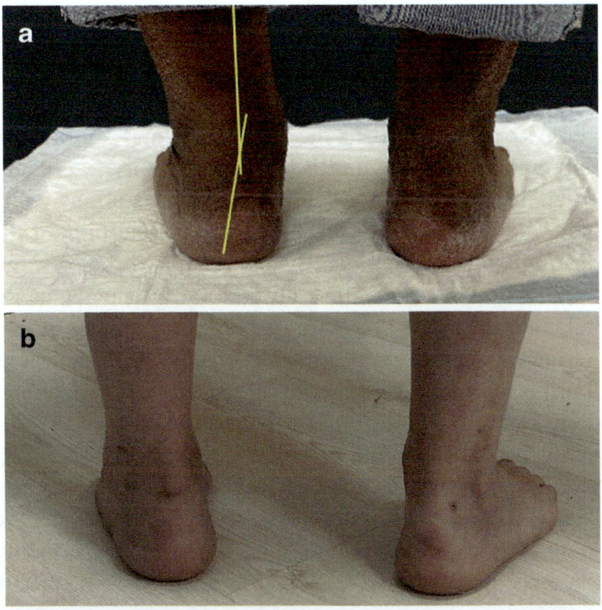

Fig. 12.1 (**a**) Posterior weight-bearing clinical image demonstrating hindfoot valgus bilaterally with left greater than right (yellow angle). (**b**) "Too many toes sign" is visible from posterior on the right side

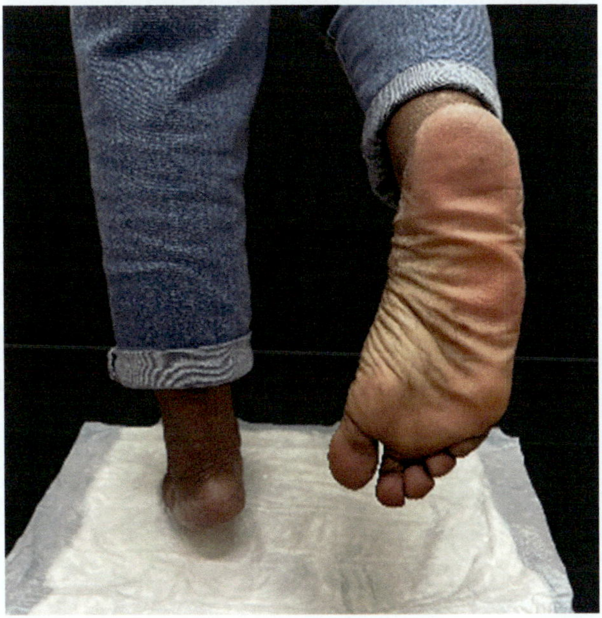

Fig. 12.2 Posterior clinical image demonstrating single heel rise test with hindfoot valgus correction on the left and ability to reach full rise

deformity of the hindfoot becomes more pronounced, often indicating deltoid ligament insufficiency. In the seated position, tenderness to palpation and swelling are assessed along the posterior tibial tendon, typically from 3 to 5 cm proximal to the ankle joint up to its insertion over the navicular bone. In more advanced cases, tenderness may extend to the sinus tarsi and subfibular area, suggesting subfibular impingement. Assessing hindfoot flexibility is crucial and is done by applying a medially directed force until the calcaneus aligns with the tibia with the ankle in a neutral dorsiflexed position. This helps determine if the subtalar motion is flexible or rigid and can also identify any residual forefoot varus.

Pain may also manifest around the talonavicular joint due to dorsolateral peritalar subluxation [11]. Chronic eccentric loading of the ankle joint in valgus results in lateral compartment arthritic changes and pain within the lateral ankle gutter. Furthermore, it is essential to rule out acute tarsal tunnel symptoms, considering the long-standing valgus tilt of the hindfoot and ankle [12, 13]. Finally, the assessment includes checking for gastrocnemius tightness and Achilles contracture. This is done with the knee both extended and flexed. Improved ankle dorsiflexion with knee flexion, compared to dorsiflexion with the knee extended, suggests gastrocnemius tightness. On the other hand, when ankle dorsiflexion is equally limited with the knee flexed and extended, it indicates Achilles contracture.

12.3 Staging of Progressive Collapsing Foot Deformity

Johnson and Strom first described three stages of posterior tibial tendon dysfunction in 1989 which was later elaborated by Myerson [4, 14] (Table 12.1).

Lately, a novel PCFD classification has been formulated through a collaborative effort, with panel members comprising experts from various parts of the world [2] (Table 12.2). The PCFD classification system relies on both anatomical and functional considerations to categorize deformities. It divides these deformities into two main stages: flexible (stage 1) and rigid (stage 2). The system specifically documents deformities that exist, while those that are absent are not included. The PCFD system emphasizes five specific deformity patterns, which include hindfoot valgus, midfoot/forefoot abduction, forefoot varus/medial column instability, peritalar subluxation/dislocation, and ankle instability (Table 12.2).

Table 12.1 Johnson and Strom modified by Myerson classification system for posterior tibial tendon dysfunction

Stage	Deformity	Clinical findings	Radiographic findings
1	– PTT tenosynovitis – May have relative deformity	– Able to perform single heel rise test – Medial ankle/midfoot pain	– None – May have tendinosis on MRI
2A	– Moderate flexible hindfoot valgus – Normal forefoot – Partial tears of the PTT – Decreased height of the medial longitudinal arch	– Unable to perform single heel rise test – May have mild sinus tarsi pain	– Talonavicular uncoverage <30% – Break in Meary's line (talo–first metatarsal line)
2B	– Severe flexible hindfoot valgus – Forefoot abduction (too many toes sign) – High grade tears of the PTT – Failure of spring ligament	– Unable to perform single heel rise test – Mild sinus tarsi pain	– Talonavicular uncoverage >30% – Break in Meary's line (talo–first metatarsal line)
3	– Rigid hindfoot valgus – Rigid forefoot abduction – Lateral hindfoot impingement – Complete loss of the medial longitudinal arch	– Unable to perform single heel rise test – Severe sinus tarsi pain	– Subtalar, talonavicular, and calcaneocuboid arthritis
4	– Rigid hindfoot valgus – Rigid forefoot abduction – Tibiotalar valgus – Deltoid ligament insufficiency	– Unable to perform single heel rise test – Severe sinus tarsi pain – Ankle pain	– Lateral talar tilt – Ankle arthritis – Hindfoot arthritis

Table 12.2 Progressive collapsing foot deformity classification 2020

Stage	Deformity
1	Flexible
2	Fixed

Types of deformity (classes—isolated or combined)	Deformity type/location	Consistent clinical/radiographic findings
A	Hindfoot valgus deformity	– Hindfoot valgus alignment – Increased hindfoot moment arm, hindfoot alignment angle, foot and ankle offset
B	Midfoot/forefoot abduction deformity	– Decreased talar head coverage – Increased talonavicular coverage angle. Presence of sinus tarsi impingement
C	Forefoot varus deformity/ medial column instability	– Increased talus-first metatarsal angle—plantar gapping first TMT joint/ NC joints – Clinical forefoot varus
D	Peritalar subluxation/ dislocation	– Significant subtalar joint subluxation/ subfibular impingement
E	Ankle instability	– Valgus tilting of the ankle joint

12.4 Imaging Modalities

12.4.1 Radiographs

Weight-bearing radiographs are the initial imaging tool performed on all patients presenting with PCFD. Anteroposterior (AP), lateral, oblique and Saltzman views of both feet are obtained to assess the alignment, even in patients with unilateral symptoms. Multiple radiographic parameters have been described to evaluate PCFD. The measurements assessed on AP views include the following [9, 15, 16]:

1. *Talar-first metatarsal angle*: The angle formed by the lines drawn along the long axis of the talus and the first metatarsal (Fig. 12.3). Normal value = 0°; mild deformity ≥4°; moderate deformity ≥8°; and severe deformity ≥30°.
2. *Talonavicular coverage angle*: This angle evaluates the presence of forefoot abduction and is formed between the line that joins the medial and lateral articular margins of the talus and the line that joins the medial and lateral articular margins of the navicular (Fig. 12.4). Normal value ≤7° and PCFD with forefoot abduction ≥7°.
3. *Talonavicular uncoverage percentage*: This parameter also assesses forefoot abduction and is quantified as the percentage of the talus that is not in contact with the navicular on the medial side (Fig. 12.5). Normal value = 10–30% and PFCD with presence of forefoot abduction ≥30%

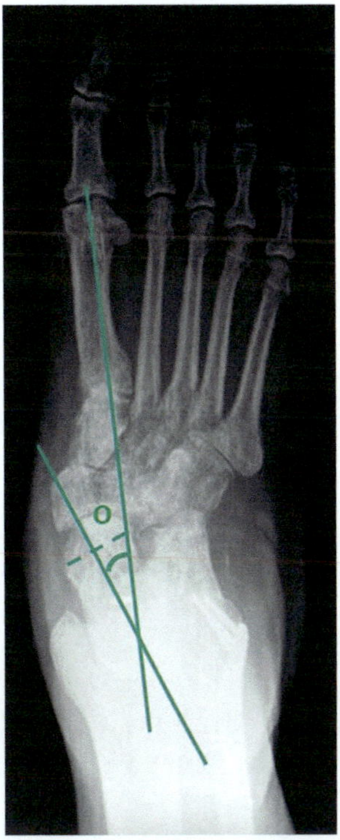

Fig. 12.3 Weight-bearing anteroposterior (AP) radiograph of the right foot demonstrating assessment of talar-first metatarsal angle in a patient with progressive collapsing foot deformity

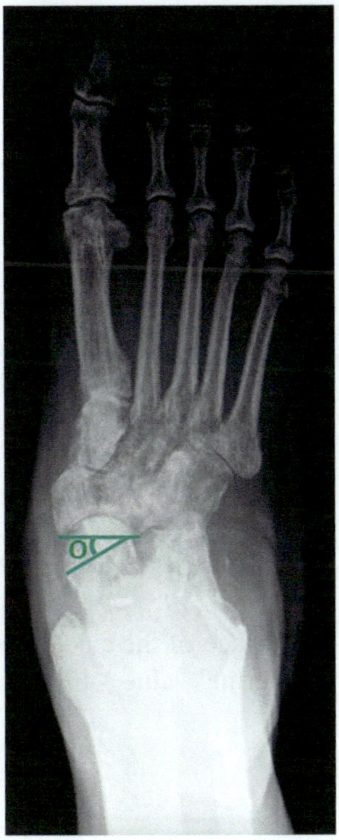

Fig. 12.4 Weight-bearing anteroposterior (AP) radiograph of the right foot demonstrating assessment of talonavicular coverage angle in a patient with progressive collapsing foot deformity

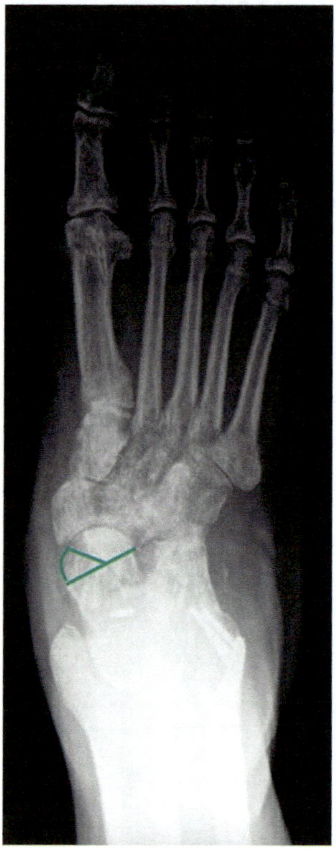

Fig. 12.5 Weight-bearing anteroposterior (AP) radiograph of the right foot demonstrating assessment of talonavicular uncoverage percentage in a patient with progressive collapsing foot deformity

4. *Talocalcaneal angle (kite angle):* The angle is measured between the line bisecting the head and neck of the talus and the line parallel to the lateral surface of the calcaneus (Fig. 12.6). Normal: 25°–40°; heel varus ≤25° and heel valgus ≥40°.

On the lateral radiographic image, the following measurements are performed [9, 15, 16]:

1. *Meary's line (Talar-first metatarsal line):* This is defined as the angle between the long axis of the talus and the long axis of the first metatarsal (Fig. 12.7). Normal value = 0; mild deformity ≥4°; moderate deformity ≥15°; and severe deformity ≥30°.
2. *Calcaneal pitch:* This angle is measured between the line parallel to the ground and the line along the plantar surface of the calcaneus (Fig. 12.8). Normal value = 20°–30°; PCFD ≤20°.
3. *Calcaneal-fifth metatarsal angle:* This angle is measured between the plantar surface of the calcaneus and along the inferior fifth metatarsal shaft (Fig. 12.9). Normal value ≤170° and PCFD ≥170°.

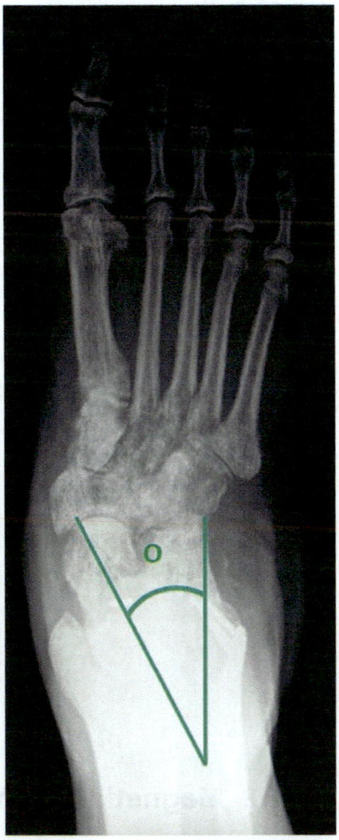

Fig. 12.6 Weight-bearing anteroposterior (AP) radiograph of right foot demonstrating assessment of talocalcaneal angle (kite angle) in a patient with progressive collapsing foot deformity

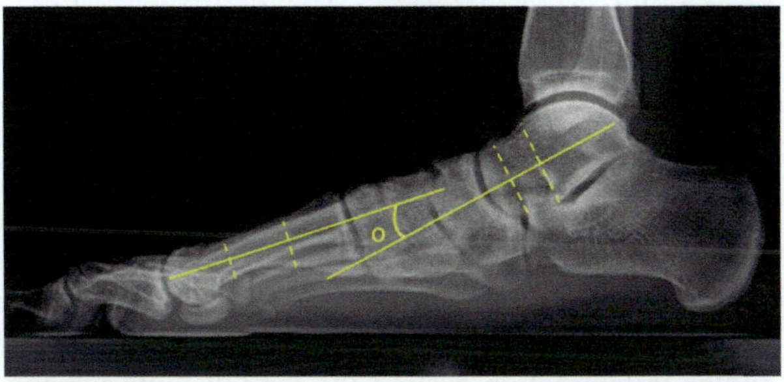

Fig. 12.7 Weight-bearing lateral radiograph of the right foot demonstrating assessment of Meary's line in a patient with progressive collapsing foot deformity

Prior studies have tried to quantify the reliability of angular measurements using radiographs as they are commonly obtained in practice. The study carried out by Sensiba et al. examined the reliability of intra- and inter-observer measurements using three digital radiographs—AP, lateral, and hindfoot alignment views [17]. They also evaluated the impact of different levels of observer experience on the reliability of the measurements. The authors reported a positive correlation between the level of experience possessed by the observer and intra-observer reliability.

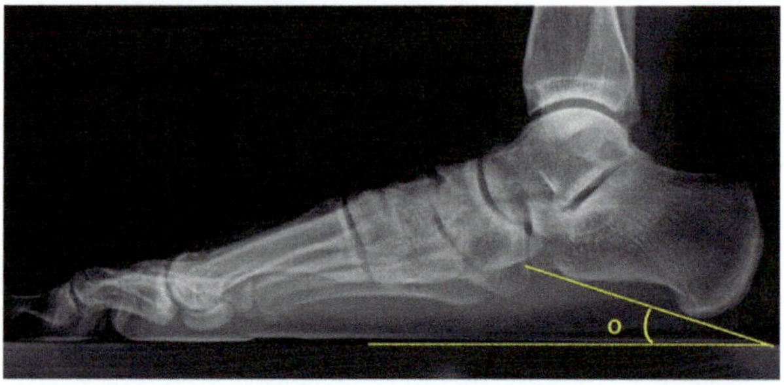

Fig. 12.8 Weight-bearing lateral radiograph of the right foot demonstrating assessment of calcaneal pitch in a patient with progressive collapsing foot deformity

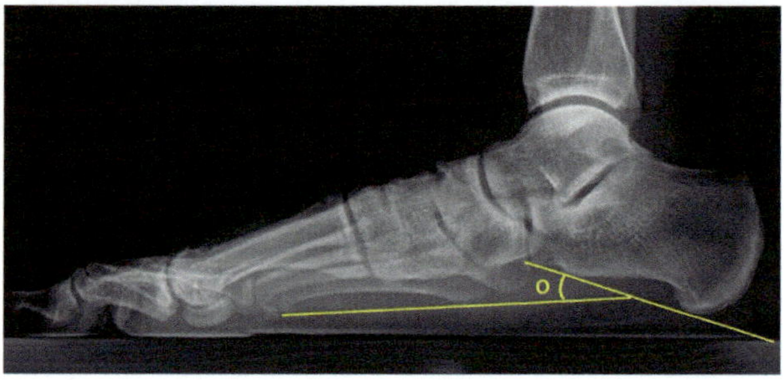

Fig. 12.9 Weight-bearing lateral radiograph of the right foot demonstrating assessment of calcaneal-fifth metatarsal angle in patient with progressive collapsing foot deformity

12.4.2 Magnetic Resonance Imaging (MRI)

MRI allows for the evaluation of soft tissues, tendons, and osseous structures. MRI has a high level of accuracy in detecting abnormalities in the posterior tibial tendon (PTT) [18–20]. The normal PTT is an oval-shaped structure and is the largest and most medial of the three flexor tendons [20–22]. The tendon expands in size and has an uneven appearance near its insertion over the navicular due to the presence of fibrocartilage or connective tissue within its divisions [23, 24]. The degree of PTT involvement can range from degenerative tendinosis to tenosynovitis to partial or complete tears. Tendinosis is the most commonly observed pathology, which presents as thickening, surface irregularities, and a heterogeneous intratendinous signal intensity [9]. Tenosynovitis, on the other hand, manifests as the presence of fluid and/or inflammation around the tendon with surrounding edema, and the signal intensity is greater than in tendinosis and surrounds 50% or more of the PTT circumference (Fig. 12.8) [9]. The presence of tenosynovitis without concurrent tendon abnormalities does not exhibit a significant correlation with quantitative evaluations of foot misalignment [25, 26]. In order to identify a tear in the PTT using MRI, increased T2 and T1 signal intensity, as well as an enlargement of the tendon, is seen. Other findings within the tendon include linear fluid signal within the tendon, which is referred to as a longitudinal split tear or complete disruption of tendon fibers indicating a total rupture (Fig. 12.9) [26].

The spring ligament, also known as the plantar calcaneonavicular ligament, is widely acknowledged as the primary stabilizer of the medial arch, with its significance secondary only to that of the PTT. Together with the PTT and deltoid ligament, it forms a soft tissue confluence at the medial navicular bone to prevent excessive descent of the talar head [9]. The talocalcaneonavicular joint, along with the anterior and middle facets of the calcaneus and the navicular bone, creates a deep cavity that articulates with the talus. The spring ligament consists of three segments arranged in a medial-to-lateral orientation: the superomedial, medioplantar oblique, and inferoplantar longitudinal bundles [27, 28]. Typically, injury to the spring ligament is primarily attributed to PTT insufficiency, resulting in repetitive and excessive downward movement of the talar head. This excessive movement places undue stress on the superomedial bundle, leading to overload and subsequent damage [29]. Spring ligament alterations are frequently observed in patients with PTT tears, with prevalence rates ranging from 74% to 92%, making them the most commonly associated secondary abnormalities in this patient population [26, 29–31] (Fig. 12.10).

Evaluating the spring ligament clinically poses challenges, underscoring the importance of employing imaging techniques. MRI allows visualization and differentiation of all the components of the spring ligament. The superomedial bundle is the most commonly affected component of the spring ligament and is best observed in coronal and axial oblique planes [28, 32]. The superomedial bundle becomes thicker than 5 mm and may demonstrate increased signal intensity, waviness, edema around the ligament, and discontinuity of the fibers, which may result in direct contact between PTT and the talar head [30, 33, 34].

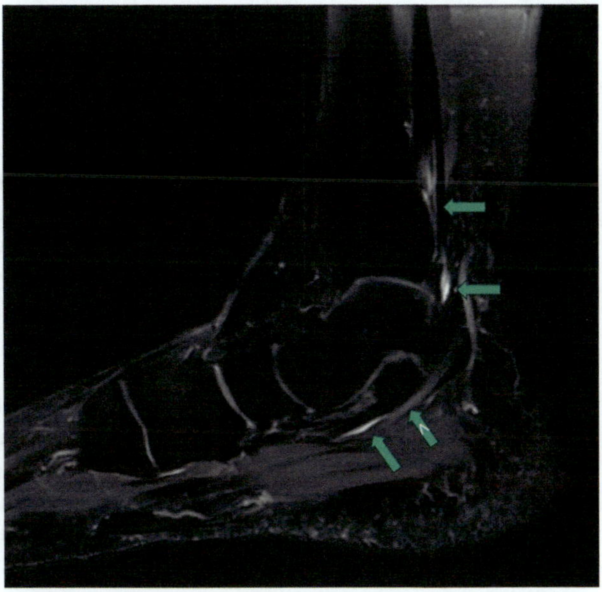

Fig. 12.10 MRI sagittal image of the right ankle obtained at the medial malleolus demonstrates the posterior tibial tendon outlined by excessive fluid in its tendon sheath (arrowheads) and surrounding soft-tissue inflammation

12.4.3 Ultrasound

Ultrasound (US) is a cost-effective, radiation-free diagnostic tool, which is available at the point of care and has an accuracy comparable to MRI to evaluate PTT and spring ligament [16]. Ultrasound imaging reveals the manifestation of tendinosis as an increase in the thickness of the affected tendon, accompanied by the existence of irregular hypoechoic areas that displace the usual fibrillar arrangement. Furthermore, color Doppler ultrasound imaging has the capability to detect augmented vascularity within the region. In contrast, tears in a tendon affected by tendinosis manifest as discontinuities or disruptions, often accompanied by changes in its diameter [35]. Dynamic US is a valuable diagnostic tool in cases where patients are suspected of presenting with friction syndrome, which is characterized by a thickened retinaculum and tendon instability [36]. The occurrence of this instability is commonly associated with the disturbance of the flexor retinaculum, resulting in the anterior subluxation of the tendon [19, 35, 37]. However, US is operator-dependent and is associated with a learning curve.

12.4.4 Computed Tomography and Weight-Bearing Computed Tomography

Computed tomography (CT) offers excellent three-dimensional bony alignment including assessment of arthritis and tarsal coalitions (Fig. 12.11) [38]. However, conventional CT does such evaluation without load and does not allow dynamic evaluation of such deformity. To bridge this concern, the utilization of weight-bearing CT (WBCT) enables the assessment of bony alignment in a dynamic manner under load (Fig. 12.12). Additionally, WBCT facilitates the quantification of measurements in one-dimensional, two-dimensional, and three-dimensional planes [39–41]. The use of WBCT allows identification of inherent anatomical variations in the subtalar joint that result in peritalar subluxation and PCFD [3] (Fig. 12.13). The application of imaging techniques can aid in identifying specific deformity locations, such as the collapse of the medial arch at the talonavicular or naviculocuneiform joint [42]. This, in turn, can help identify areas of impingement, arthritis, or significant calcaneal subluxation, which may necessitate hindfoot fusion, particularly subtalar fusion [43]. WBCT has proven its utility in evaluating rigid flatfoot conditions, allowing for the examination of potential fusion in joints like talocalcaneal or calcaneonavicular joints in younger individuals and detecting osteoarthritis in subtalar, tibiotalar, talonavicular, or Lisfranc joints in older patients. Additionally, WBCT facilitates the measurement of foot and ankle offset (FAO), which assesses the relationship between the center of the ankle joint and the weight-bearing tripod of the foot, consisting of the first and fifth metatarsal heads and the calcaneus [44, 45]. WBCT has emerged as a significant diagnostic imaging modality for assessing the severity of PCFD, as well as the widening and instability of distal tibiofibular

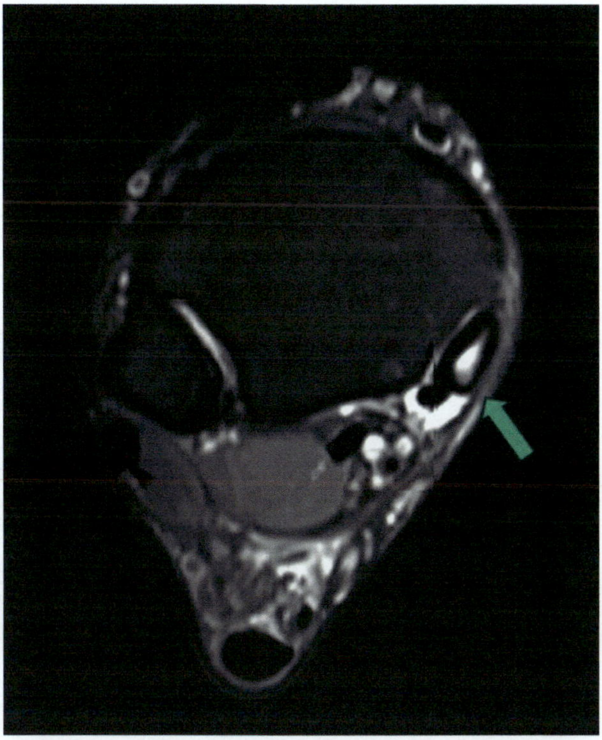

Fig. 12.11 MRI axial image of the right ankle demonstrating thickening of the posterior tibial tendon (arrow), with fluid-filled clefts in the tendon, indicating intrasubstance degenerative tearing and tenosynovitis

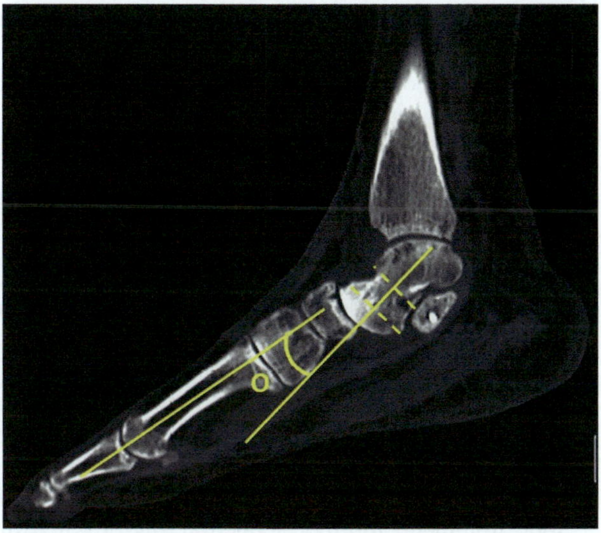

Fig. 12.12 Computed tomography of the right foot demonstrating assessment of Meary's line in patient with progressive collapsing foot deformity

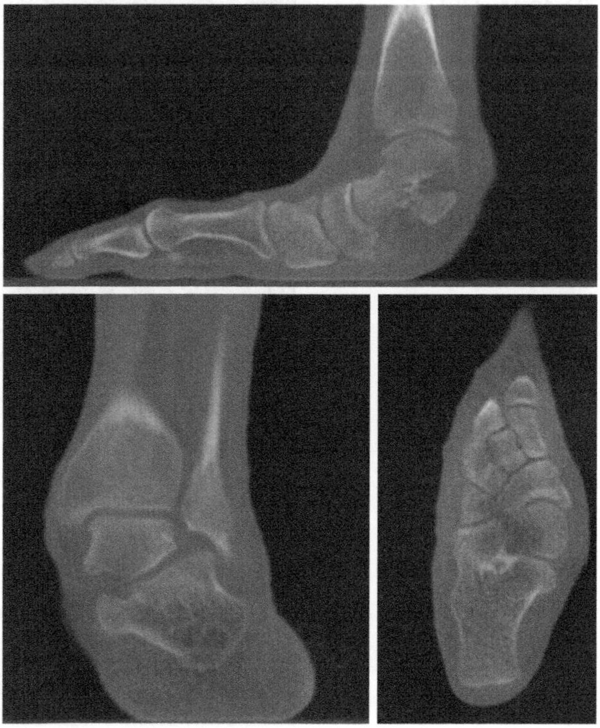

Fig. 12.13 Weight-bearing computed tomography (WBCT), showing hindfoot valgus, decreased talonavicular coverage, and collapsed arch

syndesmotic (DTFS) injuries [1]. In individuals with chronic hindfoot valgus deformity and progressive peritalar subluxation, particularly those diagnosed with PCFD, impingement between the talus and/or calcaneus with the distal fibula can lead to increased stress on the DTFS ligaments. This, in turn, may result in syndesmotic space widening and syndesmotic instability.

Key Points

- Patients with PCFD may exhibit a range of symptoms, varying from noticeable pain to an absence of apparent discomfort.
- Typically, patients initially complain of pain located along the medial side of the foot and ankle, particularly along the course of the posterior tibial tendon, which runs behind and below the ankle.
- As the pathology progresses, the medial-sided pain along the posterior tibial tendon diminishes, usually due to tendon rupture, and the deformity advances. Consequently, pain is observed more laterally in the calcaneofibular region due to impingement.
- Clinical examination reveals specific indicators of PCFD, such as valgus hindfoot alignment, forefoot abduction, and a collapsing of the medial longitudinal arch.
- Evaluating the patient's ability to perform the double and single heel rise tests, especially the latter, is crucial for staging the condition.

- Multiple imaging tools are available for PCFD evaluation. In clinical practice, radiographs are commonly employed to quantify deformities; however, they exhibit limitations in terms of reproducibility and accuracy.
- Advanced cross-sectional imaging techniques like MRI, CT, and WBCT offer valuable insights into understanding and assessing deformities across multiple dimensions.

References

1. Auch E, Barbachan Mansur NS, Alexandre Alves T, Cychosz C, Lintz F, Godoy-Santos AL, et al. Distal tibiofibular syndesmotic widening in progressive collapsing foot deformity. Foot Ankle Int. 2021;42(6):768–75.
2. Myerson MS, Thordarson DB, Johnson JE, Hintermann B, Sangeorzan BJ, Deland JT, et al. Classification and nomenclature: progressive collapsing foot deformity. Foot Ankle Int. 2020;41(10):1271–6.
3. de Cesar Netto C, Myerson MS, Day J, Ellis SJ, Hintermann B, Johnson JE, et al. Consensus for the use of weightbearing CT in the assessment of progressive collapsing foot deformity. Foot Ankle Int. 2020;41(10):1277–82.
4. Johnson KA, Strom DE. Tibialis posterior tendon dysfunction. Clin Orthop Relat Res. 1989;239:196–206.
5. Capuzzi MJ, Miller JR, Siwy T. Approach to the ankle in adult acquired flatfoot deformity. Clin Podiatr Med Surg. 2023;40(2):341–9.
6. Jackson JB, Pacana MJ, Gonzalez TA. Adult acquired flatfoot deformity. J Am Acad Orthop Surg. 2022;30(1):e6–e16.
7. Deland JT. Adult-acquired flatfoot deformity. J Am Acad Orthop Surg. 2008;16(7):399–406.
8. Vulcano E, Deland JT, Ellis SJ. Approach and treatment of the adult acquired flatfoot deformity. Curr Rev Musculoskelet Med. 2013;6(4):294–303.
9. Flores DV, Mejía Gómez C, Fernández Hernando M, Davis MA, Pathria MN. Adult acquired flatfoot deformity: anatomy, biomechanics, staging, and imaging findings. Radiographics. 2019;39(5):1437–60.
10. Ling SK, Lui TH. Posterior tibial tendon dysfunction: an overview. Open Orthop J. 2017;11:714–23.
11. Bluman EM, Title CI, Myerson MS. Posterior tibial tendon rupture: a refined classification system. Foot Ankle Clin. 2007;12(2):233–49.
12. Daniels TR, Lau JT, Hearn TC. The effects of foot position and load on tibial nerve tension. Foot Ankle Int. 1998;19(2):73–8.
13. Francis H, March L, Terenty T, Webb J. Benign joint hypermobility with neuropathy: documentation and mechanism of tarsal tunnel syndrome. J Rheumatol. 1987;14(3):577–81.
14. Myerson MS. Adult acquired flatfoot deformity: treatment of dysfunction of the posterior tibial tendon. Instr Course Lect. 1997;46:393–405.
15. Henry JK, Shakked R, Ellis SJ. Adult-acquired flatfoot deformity. Foot Ankle Orthop. 2019;4(1):2473011418820847.
16. Abousayed MM, Alley MC, Shakked R, Rosenbaum AJ. Adult-acquired flatfoot deformity: etiology, diagnosis, and management. JBJS Rev. 2017;5(8):e7.
17. Sensiba PR, Coffey MJ, Williams NE, Mariscalco M, Laughlin RT. Inter- and intraobserver reliability in the radiographic evaluation of adult flatfoot deformity. Foot Ankle Int. 2010;31(2):141–5.
18. Conti S, Michelson J, Jahss M. Clinical significance of magnetic resonance imaging in preoperative planning for reconstruction of posterior tibial tendon ruptures. Foot Ankle. 1992;13(4):208–14.

19. Arnoldner MA, Gruber M, Syré S, Kristen KH, Trnka HJ, Kainberger F, et al. Imaging of posterior tibial tendon dysfunction–comparison of high-resolution ultrasound and 3T MRI. Eur J Radiol. 2015;84(9):1777–81.
20. Rosenberg ZS, Cheung Y, Jahss MH, Noto AM, Norman A, Leeds NE. Rupture of posterior tibial tendon: CT and MR imaging with surgical correlation. Radiology. 1988;169(1):229–35.
21. de Cesar NC, Saito GH, Roney A, Day J, Greditzer H, Sofka C, et al. Combined weightbearing CT and MRI assessment of flexible progressive collapsing foot deformity. Foot Ankle Surg. 2021;27(8):884–91.
22. Kunas GC, Probasco W, Haleem AM, Burket JC, Williamson ERC, Ellis SJ. Evaluation of peritalar subluxation in adult acquired flatfoot deformity using computed tomography and weightbearing multiplanar imaging. Foot Ankle Surg. 2018;24(6):495–500.
23. Delfaut EM, Demondion X, Bieganski A, Cotten H, Mestdagh H, Cotten A. The fibrocartilaginous sesamoid: a cause of size and signal variation in the normal distal posterior tibial tendon. Eur Radiol. 2003;13(12):2642–9.
24. Chhabra A, Soldatos T, Chalian M, Faridian-Aragh N, Fritz J, Fayad LM, et al. 3-Tesla magnetic resonance imaging evaluation of posterior tibial tendon dysfunction with relevance to clinical staging. J Foot Ankle Surg. 2011;50(3):320–8.
25. Lim PS, Schweitzer ME, Deely DM, Wapner KL, Hecht PJ, Treadwell JR, et al. Posterior tibial tendon dysfunction: secondary MR signs. Foot Ankle Int. 1997;18(10):658–63.
26. Lin YC, Mhuircheartaigh JN, Lamb J, Kung JW, Yablon CM, Wu JS. Imaging of adult flatfoot: correlation of radiographic measurements with MRI. AJR Am J Roentgenol. 2015;204(2):354–9.
27. Taniguchi A, Tanaka Y, Takakura Y, Kadono K, Maeda M, Yamamoto H. Anatomy of the spring ligament. J Bone Joint Surg Am. 2003;85(11):2174–8.
28. Mengiardi B, Zanetti M, Schöttle PB, Vienne P, Bode B, Hodler J, et al. Spring ligament complex: MR imaging-anatomic correlation and findings in asymptomatic subjects. Radiology. 2005;237(1):242–9.
29. Deland JT, de Asla RJ, Sung IH, Ernberg LA, Potter HG. Posterior tibial tendon insufficiency: which ligaments are involved? Foot Ankle Int. 2005;26(6):427–35.
30. Toye LR, Helms CA, Hoffman BD, Easley M, Nunley JA. MRI of spring ligament tears. AJR Am J Roentgenol. 2005;184(5):1475–80.
31. Balen PF, Helms CA. Association of posterior tibial tendon injury with spring ligament injury, sinus tarsi abnormality, and plantar fasciitis on MR imaging. AJR Am J Roentgenol. 2001;176(5):1137–43.
32. Yao L, Gentili A, Cracchiolo A. MR imaging findings in spring ligament insufficiency. Skeletal Radiol. 1999;28(5):245–50.
33. Omar H, Saini V, Wadhwa V, Liu G, Chhabra A. Spring ligament complex: illustrated normal anatomy and spectrum of pathologies on 3T MR imaging. Eur J Radiol. 2016;85(11):2133–43.
34. Mansour R, Teh J, Sharp RJ, Ostlere S. Ultrasound assessment of the spring ligament complex. Eur Radiol. 2008;18(11):2670–5.
35. Kong A, Van Der Vliet A. Imaging of tibialis posterior dysfunction. Br J Radiol. 2008;81(970):826–36.
36. Gwani AS, Asari MA, Mohd Ismail ZI. How the three arches of the foot intercorrelate. Folia Morphol (Warsz). 2017;76(4):682–8.
37. Jain NB, Omar I, Kelikian AS, van Holsbeeck L, Grant TH. Prevalence of and factors associated with posterior tibial tendon pathology on sonographic assessment. PM R. 2011;3(11):998–1004.
38. Meehan RE, Brage M. Adult acquired flat foot deformity: clinical and radiographic examination. Foot Ankle Clin. 2003;8(3):431–52.
39. Bhimani R, Ashkani-Esfahani S, Lubberts B, Guss D, Hagemeijer NC, Waryasz G, et al. Utility of volumetric measurement via weight-bearing computed tomography scan to diagnose syndesmotic instability. Foot Ankle Int. 2020;41(7):859–65.

40. Bhimani R, Sornsakrin P, Ashkani-Esfahani S, Lubberts B, Guss D, De Cesar NC, et al. Using area and volume measurement via weightbearing CT to detect Lisfranc instability. J Orthop Res. 2021;39(11):2497–505.
41. Ashkani Esfahani S, Bhimani R, Lubberts B, Kerkhoffs GM, Waryasz G, DiGiovanni CW, et al. Volume measurements on weightbearing computed tomography can detect subtle syndesmotic instability. J Orthop Res. 2022;40(2):460–7.
42. Polichetti C, Borruto MI, Lauriero F, Caravelli S, Mosca M, Maccauro G, et al. Adult acquired flatfoot deformity: a narrative review about imaging findings. Diagnostics. 2023;13(2):225.
43. Conti MS, Garfinkel JH, Ellis SJ. Outcomes of reconstruction of the flexible adult-acquired flatfoot deformity. Orthop Clin N Am. 2020;51(1):109–20.
44. Toullec E. Adult flatfoot. Orthop Traumatol Surg Res. 2015;101(1):S11–7.
45. Day J, de Cesar Netto C, Nishikawa DR, Garfinkel J, Roney A, O'Malley J, et al. Three-dimensional biometric weightbearing CT evaluation of the operative treatment of adult-acquired flatfoot deformity. Foot Ankle Int. 2020;41(8):930–6.

13 Cavovarus Foot

Tahir Ogut and Necip Selcuk Yontar

13.1 Introduction

13.1.1 Definition of Cavovarus and Etiological Factors

Cavovarus deformity is defined with a high plantar arch, deep first ray, varus position of the heel, and forefoot equinus [1]. It is mostly related to neurological disorders, but it can be seen as a sequelae of congenital disorders or post-traumatic conditions as well [2]. Among the neurological etiologies, hereditary motor sensory neuropathies are the most common causes of pes cavovarus, and the most common form of hereditary motor sensory neuropathies is Charcot-Marie-Tooth disease (CMT); thus CMT disease is usually used as a template for the understanding of this deformity [3] (Fig. 13.1).

CMT results from the mutation of the genes that affect Schwann cells and neurons [3]. CMT is mostly inherited in an autosomal dominant form, but autosomal recessive or X-linked forms also exist. There are more than 50 genetic subtypes of CMT, but the most commonly encountered ones are Type I and Type II [4].

The disease has demyelinating (Type 1 and Type 4), axonal (Type 2 and Type 4), and intermediate (CMTX, CMT2E) forms [3]. The classical one and also the most common one is CMT Type 1A, and it is related with the abnormality in the peripheral protein 22 on chromosome 17 which affects the myelin that surrounds the nerves and manifests itself as slowing in the electromyogram (EMG). On the other hand, gene defect for CMT Type 1B is on chromosome 1, and gene defect for CMT Type 2 is on chromosomes 1 and 3 [2]. CMT Type 2 has abnormal axonal function, and EMG shows normal impulse transfer with diminished magnitude [5].

T. Ogut · N. S. Yontar (✉)
Nisantasi Orthopaedics Center, Foot and Ankle Surgery Clinic, Istanbul, Turkey

S. Sharma et al. (eds.), *Clinical and Radiological Examination of the Foot and Ankle*, https://doi.org/10.1007/978-981-97-4202-8_13

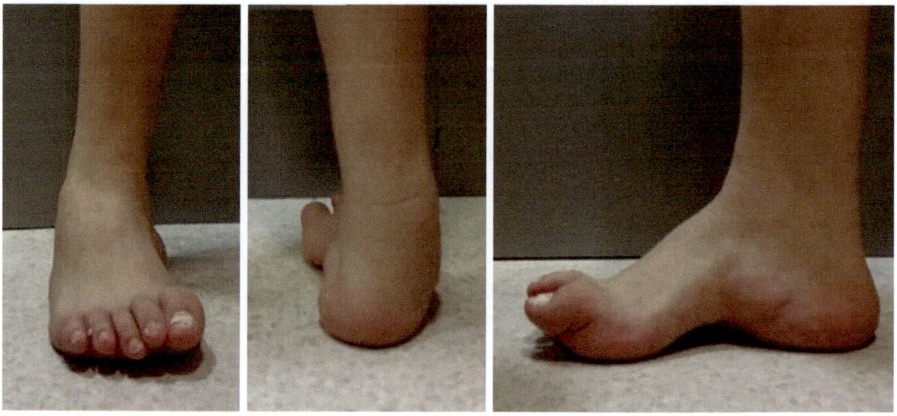

Fig. 13.1 Classical cavovarus foot. Note deep first ray, cock-up deformity along with the clawing of the toes and hindfoot varus

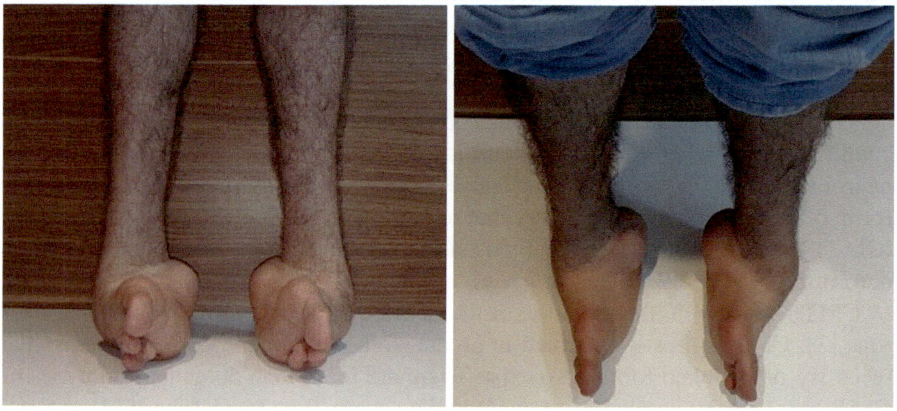

Fig. 13.2 A 22-year-old male patient with neglected, bilateral pes equinovarus

Other neurological conditions that may lead to cavovarus are amyotrophic lateral sclerosis, Huntington disease, cerebral palsy, poliomyelitis, and spinal cord lesions. Pes equinovarus and arthrogryposis are listed as the congenital causes of cavovarus, and they can be diagnosed from birth (Fig. 13.2).

Trauma is another etiological factor. Compartment syndrome, superficial peroneal nerve injury-related drop foot, malunited talus or calcaneus fractures, or burn injuries that cause muscle contractions can lead to cavovarus deformity.

There are also idiopathic cavovarus cases. This subtle form may present without an underlying cause [6] (Table 13.1).

Patients with CMT often present with abnormalities before puberty. But depending on the etiology, affected muscles, and age at presentation, the appearance of the deformity varies in between the patients (Fig. 13.3).

Table 13.1 Etiological factors that are related to cavovarus deformity development

Neurological
Hereditary motor sensory neuropathies
Cerebral palsy
Amyotrophic lateral sclerosis
Poliomyelitis
Huntington disease
Congenital
Pes equinovarus
Arthrogryposis
Tarsal coalition
Traumatic
Malunited talar neck or calcaneal fractures
Compartment syndrome
Drop foot after superficial peroneal nerve injury
Burns
Idiopathic

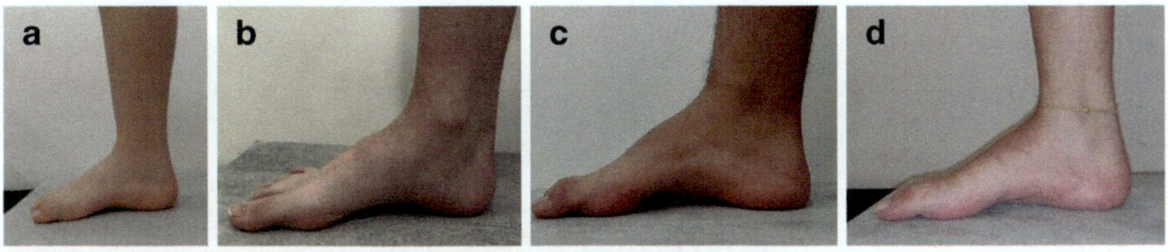

Fig. 13.3 Cavovarus deformity that started to give symptoms at different ages. (**a**) A 4-year-old patient, (**b**) a 13-years-old patient, (**c**) a 17-year-old patient, (**d**) a 23-year-old patient

13.2 Pathoanatomy

In the normal foot, proper function and alignment are maintained by the opposing action of agonist and antagonist muscles, and like in a rope-pulling contest, if the power of these muscles match, the foot is properly aligned and functional. But if this equation is disturbed, because of the emerging muscular imbalance, cavovarus deformity develops.

In CMT patients, pathology affects the tibial nerve first [7]. As a result, the intrinsic muscles of the foot are affected, and this leads to plantar wasting, contractures, and fibrosis [2]. In this process, long flexors and extensors overpower the affected intrinsic muscles, and the longitudinal arch elevates and the foot shortens. This is followed by the involvement of the peroneus brevis, tibialis anterior, and finger extensors. Weakness of these muscles and relatively strong tibialis posterior and peroneus longus cause plantarflexion of the first metatarsal, and when the plantar flexed first ray touches the ground, hindfoot varus occurs to restore the tripod that is formed by the first and fifth metatarsals and calcaneal tuberosity [4, 8]. With the hindfoot in varus alignment, the Achilles tendon migrates medially and acts as a secondary invertor of the hindfoot. In the disease process, flexible hindfoot varus gets rigid.

As the weakness of the tibialis anterior increases, patients start to use extensor hallucis longus and extensor digitorum longus as ankle dorsiflexors. Recruitment of these muscles results in cock-up deformity of the toe and clawing of the little toes. Clawing of the toes is further increased by the flexor muscle contraction and deficiency of the interosseous muscles. This in turn shortens the plantar fascia, and contracture of the fascia further depresses the metatarsal heads (Fig. 13.4).

Long-standing hindfoot varus and supination of the foot increase the stress on the lateral structures and along with the peroneal weakness and altered proprioception cause chronic instability and may lead to varus ankle arthritis [8, 9].

In poliomyelitis patients pathoanatomy differs. In these patients, weakness of gastrocnemius and sparing of tibialis anterior lead to elevated calcaneal pitch, and because of the weak push-off strength, patients recruit the long toe flexors which lead to forefoot plantarflexion and development of the cavus deformity [3]. This cavus is called as "calcaneocavus," and it is a separate anatomical and treatment entity [4] (Fig. 13.5).

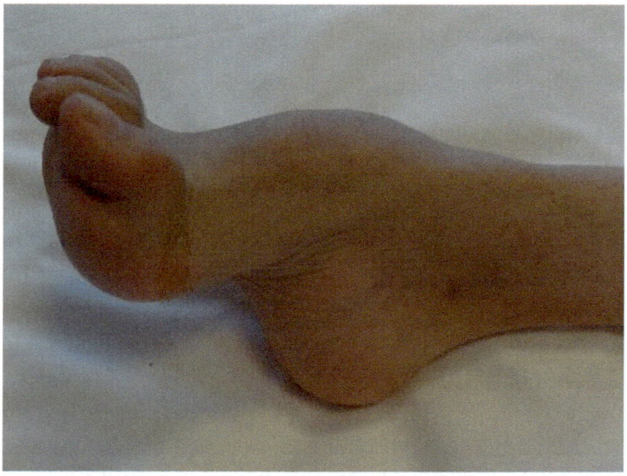

Fig. 13.4 Cock-up deformity of the hallux

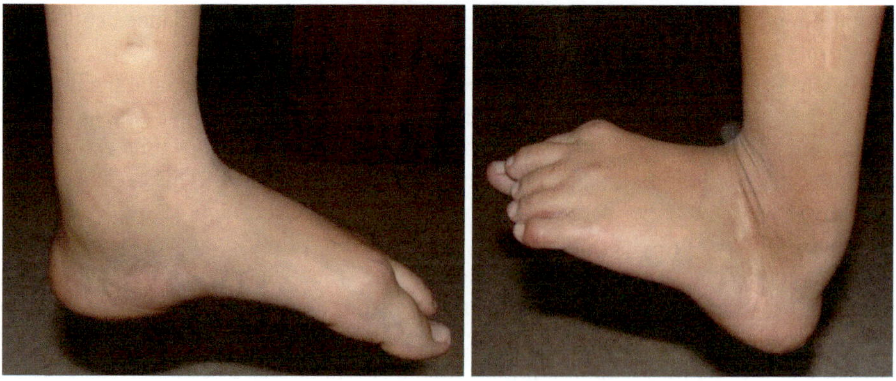

Fig. 13.5 Poliomyelitis-related calcaneocavus

Trauma and related malunions are another cause of hindfoot-driven cavovarus deformity [10]. To obtain a plantigrade foot, subtalar joint tends to compensate the trauma-induced deformity, but this compensation may lead to progressive cavovarus deformity with time.

13.3 Clinical Examination

13.3.1 Common Clinical Findings and Their Significance

The evaluation of a patient with pes cavovarus begins with a careful history of the condition and a detailed family history. Lateral foot pain is usually the most common symptom, but lateral instability and frequent falls may be reported as well.

Physical examination of a patient with cavovarus deformity should include inspection, sensory and neurological examination, motor function evaluation, active and passive range of motion assessment, and finally evaluation of hindfoot alignment, rigidity of the deformity, and ankle stability [9].

13.3.1.1 Inspection

- It should be done while the patient is sitting, barefoot walking, and standing.
- While walking, the ability to dorsiflex the ankle, alignment of the foot and ankle, toe deformities, and presence of cock-up deformity of the first MTP joint are evaluated. Aside from these, CMT patients may have equinus during swing phase and a step-page gate [4].
- Cock-up deformity of the first MTP is a finding which shows extensor hallucis longus is used as an accessory dorsiflexor of the ankle because of foot drop [11] (Fig. 13.6).

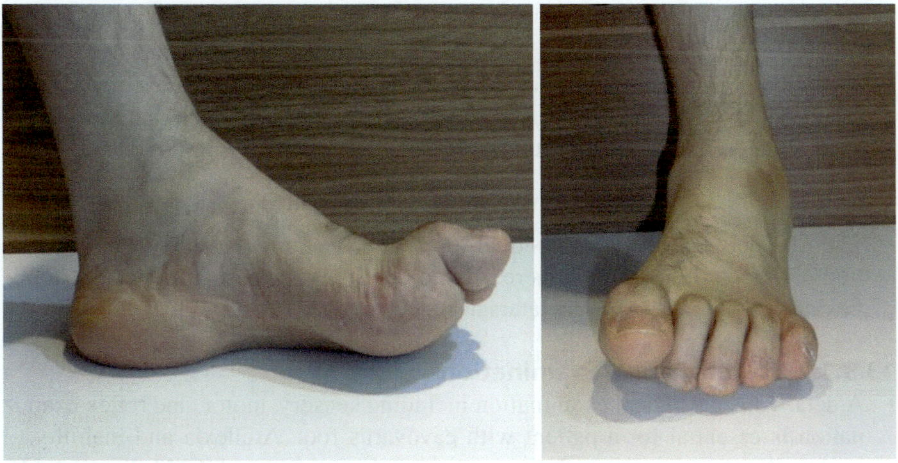

Fig. 13.6 Cock-up deformity that is apparent while standing

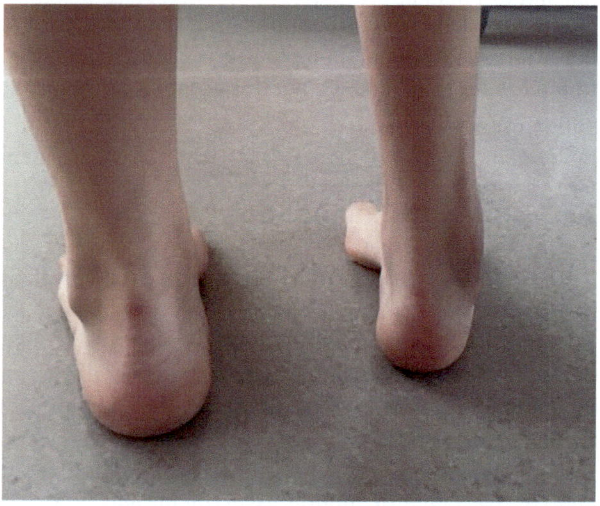

Fig. 13.7 Unilateral cavovarus deformity. The patient had a history of spinal tumor surgery

- On the standing position, lower extremity alignment from the pelvis to the foot, posture of both feet (to check whether the deformity is unilateral or bilateral), and weight-bearing position of the foot (cavus, metatarsus adductus, toe deformities, varus heel) are inspected from the front and back (Fig. 13.7).
- "Peek-a-boo" heel sign is another finding that can be seen on standing inspection [5]. If the heel is visible on the medial side from the front of a neutrally rotated feet, it confirms hindfoot varus and called as "peek-a-boo" heel (Fig. 13.8).
- A unilateral deformity can be related with neuroaxial abnormalities such as diastematomyelia, tethered cord, or myelomeningocele, and also post-traumatic incidents like pilon fracture, talus fracture, or compartment syndrome may be the underlying cause [1, 4]. Thus, in case of unilateral deformities, findings of an underlying cause like scarring around the leg, spinal skin dimples, or a hair patch should be looked for.
- While sitting, resting position of the foot is checked first. Plantar aspect of the foot is inspected to evaluate the presence and location of the callosities (Fig. 13.9).
- Reducibility of the present deformities and position of the foot after manual correction of the hindfoot or forefoot should also be checked. Kelikian test can be used to determine the flexibility of the toe deformities [12]. In flexible deformities, the passive dorsiflexion of the metatarsal head can correct the finger deformity.
- The forefoot is checked for residual forefoot pronation after the manual correction of the hindfoot varus. If there is fixed deformity, it shows the need for dorsiflexion osteotomies of the metatarsal(s) [11].

13.3.1.2 Neurological Examination

- A detailed neurological examination including sensory, motor, and reflex examination is essential for a patient with cavovarus foot. Areflexia and inability to heel walk are common presenting symptoms in patients with CMT Type 1A [4].

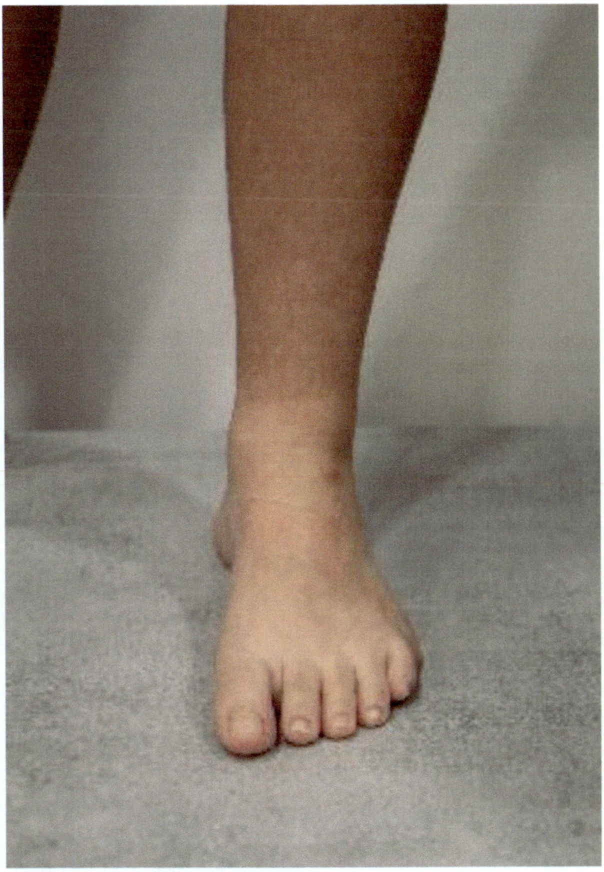

Fig. 13.8 Peek-a-boo heel

- Up to 25% of adult patients may have related sensory loss; thus sensation of the foot should be checked [4].
- Evaluation of the toe proprioception can help differentiate the contribution of sensory dysfunction to gait impairment [9].
- It should be kept in mind that patients who present with sensory loss are more likely to develop Charcot neuroarthropathy in the foot or ankle [2].

13.3.1.3 Motor Examination

- While the patient is sitting, strength testing of the muscles in the lower extremity should be done. Medical Research Council (MRC) manual motor testing scale should be used to grade muscle strength [9]. In this scale, every muscle is graded from 0 to 5; however it is highly subjective, and for a more accurate assessment, the patient should be evaluated several times at multiple examinations.
- Motor examination is the most important step to understand the muscular imbalance. Accurate evaluation in this step affects the choice of tendon transfers and thus directly affects the success of the surgical treatment. Every muscle along

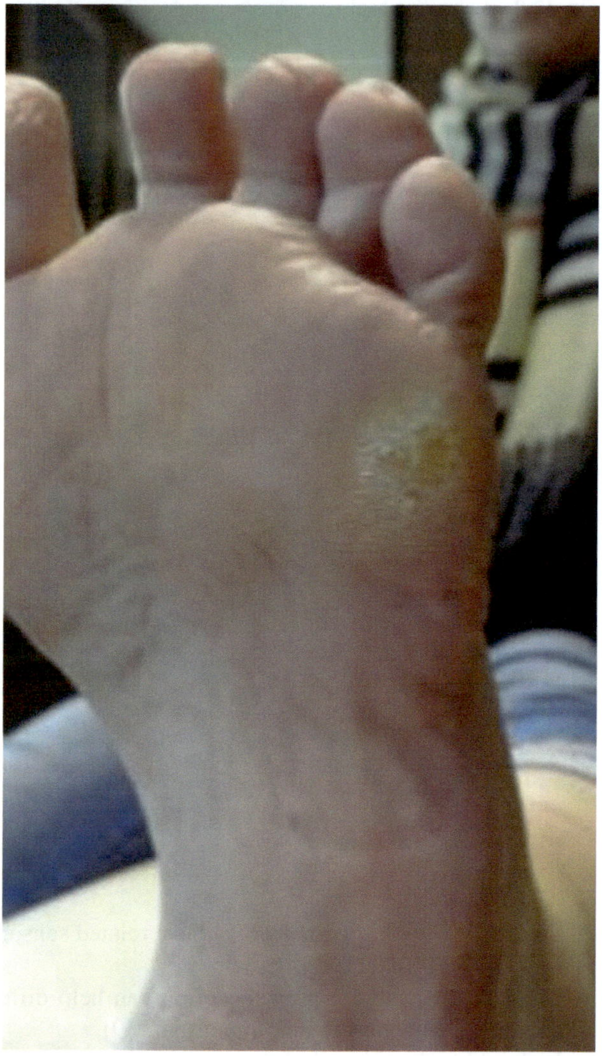

Fig. 13.9 Plantar callosity under the fifth metatarsal head

with its antagonist should be carefully evaluated, and findings of the examination should be clearly noted.

- The tibialis anterior (TA) is the primary dorsiflexor of the ankle, and the peroneus longus (PL) acts as an antagonist to TA. To evaluate the TA muscle strength, the patient is asked to dorsiflex the ankle, while the examiner passively holds the metatarsophalangeal (MTP) joints in slight plantarflexion [9].
- However, it is difficult the evaluate PL separately. To examine the PL, both thumbs should be placed under the heads of the first and fifth metatarsal, and plantarflexion power under the first metatarsal should be checked [9].
- Tibialis posterior (TP) is the antagonist of the peroneus brevis (PB) and acts as the primary invertor of the foot. As TA works as the secondary invertor, it is

important to separate TP function from TA. To do this, the foot should be in the plantarflexion position, and the patient is asked to invert the foot against resistance; while doing so, TP tendon should also be palpated. This can also be done with the patient's legs crossed.

- PB mainly everts the foot, and to evaluate the peroneal power, the patient is asked to evert the foot against resistance.
- Muscle strength of the extensor hallucis longus, flexor hallucis longus, and extensor and flexor digitorum longus should also be checked.

13.3.1.4 Evaluation of Range of Motion

- The next step should be the evaluation of the passive and active range of motion (ROM) of the ankle, subtalar, transvers tarsal, MTP, and IP joints.
- As inability to dorsiflex the ankle is a common finding, it should be differentiated whether it is related to the tightness of Achilles or gastrocnemius, mechanical impingement from the talus, or TA weakness [9].
- Achilles or gastrocnemius tightness should be differentiated with Silfverskiöld test which compares the amount of ankle dorsiflexion with the knee in 90° of flexion and extension. With the flexion of the knee, gastrocnemius relaxes, but soleus stays under tension. If the dorsiflexion is not corrected with knee flexion, the cause may be tight Achilles. If dorsiflexion improves with knee flexion, the gastrocnemius is contracted. It should also be kept in mind that bony impingement related to talar or tibial spurs can also block the dorsiflexion and can imitate Achilles tightness.

13.3.1.5 Assessment of Hindfoot Varus

- An important part of hindfoot varus assessment is the Coleman block test. It is used to determine whether the hindfoot varus is driven by the forefoot and whether the deformity is flexible or not [13].
- In this test, the lateral aspect of the forefoot and the heel are supported with a block, and the first metatarsal is allowed to stay free and touch the floor. On the other hand, some authors prefer to place the heel on the floor instead of a block and place the block under the distal part of first metatarsal [4].
- If the heel is corrected to neutral or valgus alignment, the hindfoot varus is considered flexible, and with the correction of forefoot deformity, hindfoot deformity should be corrected as well. If the hindfoot position does not change, the deformity is considered as rigid, and to correct the deformities, both forefoot and hindfoot procedures will be necessary. But it does not provide information about the expected amount of varus correction after a forefoot procedure (Figs. 13.10 and 13.11).
- To assess the flexibility of the subtalar joint, the patient is also asked to do double- and single-legged heel rises. If the heel moves to an inverted position with the test, this shows that the subtalar joint is flexible and decreases the likelihood of subtalar arthritis [14].

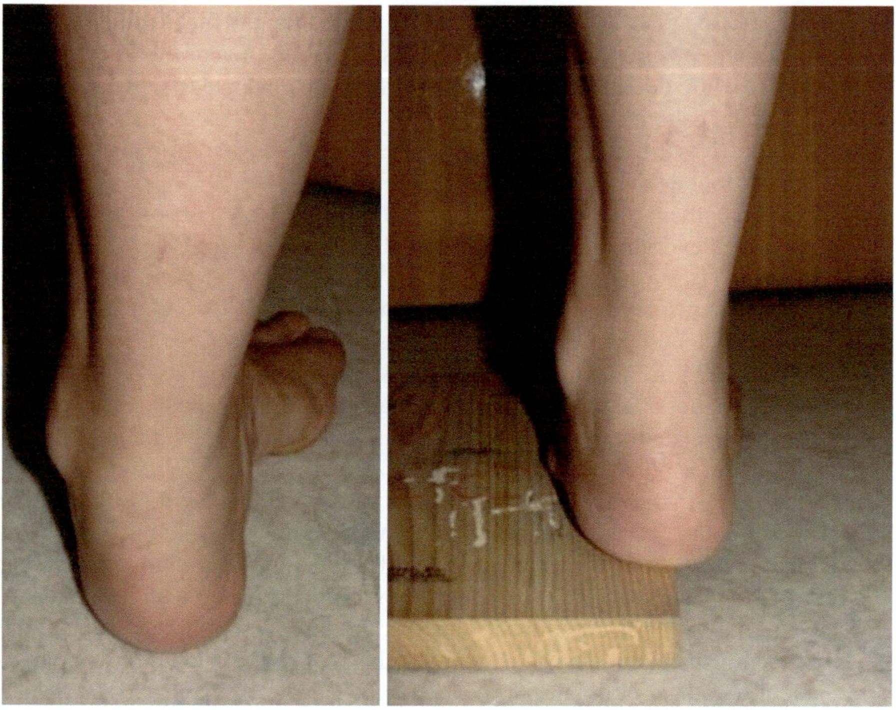

Fig. 13.10 Classical Coleman block test

13.3.1.6 Evaluation of Ankle Stability

- Ankle stability should be regularly checked both in the clinic and in the operating room [9].
- Ankle instability may be related to peroneal weakness as well, but it is usually associated with varus alignment of the heel that overloads the lateral structures [8]. Long-standing instability may also lead to varus ankle arthritis (Fig. 13.12).
- Anterior drawer and talar tilt tests should be used to evaluate the ankle stability [14].

13.3.2 Techniques for Clinical Examination Including Special Tests and Maneuvers

The aforementioned Coleman block test, Silfverskiöld test, and assessment of the residual deformities after manual correction of forefoot or hindfoot are special tests that are used to evaluate the cavovarus foot.

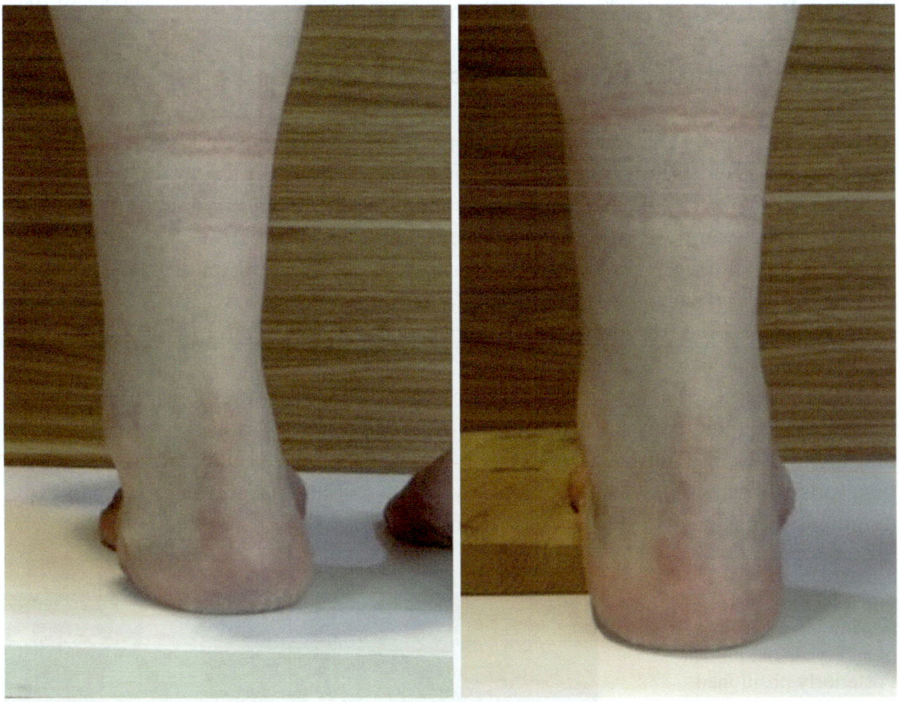

Fig. 13.11 Coleman block test with the block under the fifth ray

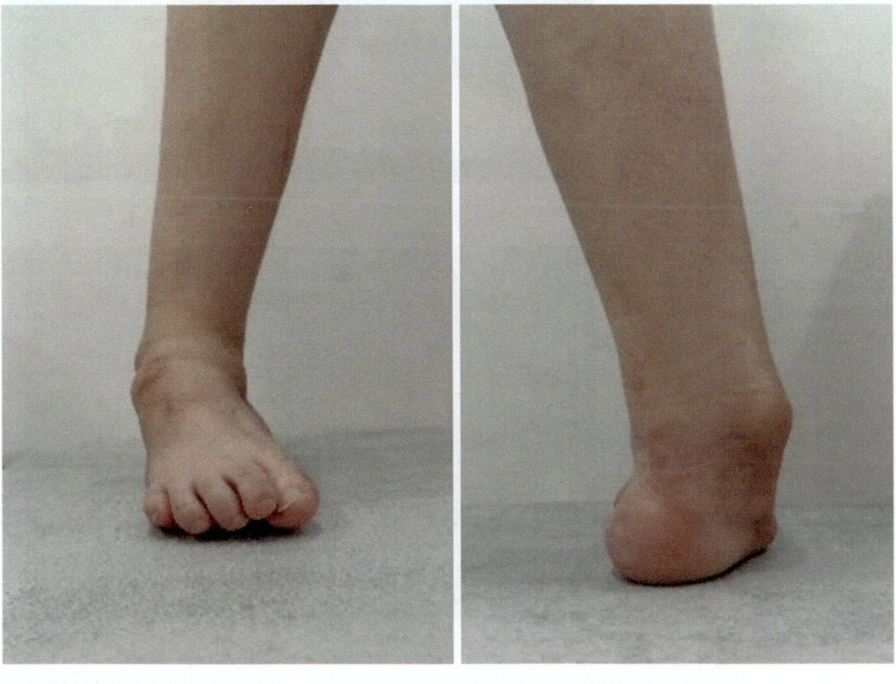

Fig. 13.12 Clinical appearance of a patient with instability-related ankle varus

13.4 Radiological Examination

13.4.1 Common Radiological Findings and Their Significance

The weight-bearing AP and lateral radiographies of the foot and ankle along with ankle mortise and hindfoot alignment view are essential in the evaluation of cavovarus deformity [9]. These are used to identify the location of the deformity, to guide surgical planning, and to determine the required correction [8].

To understand and evaluate the deformity, the evaluating physician should be familiar with normal radiographic landmarks and angular measurements. The most commonly used measurements are Meary angle, calcaneal pitch, Hibbs angle, and lateral talocalcaneal angle.

On the lateral view, aside from the aforementioned measurements, plantarflexed first metatarsal, an open sinus tarsus (see-through canal), double talar dome (shows that the talus is in external rotation), bell- or wedge-shaped cuboid, and posteriorly positioned fibula (indicates an externally rotated tibia) may be seen and lead to a diagnosis of pes cavus [3] (Figs. 13.13 and 13.14).

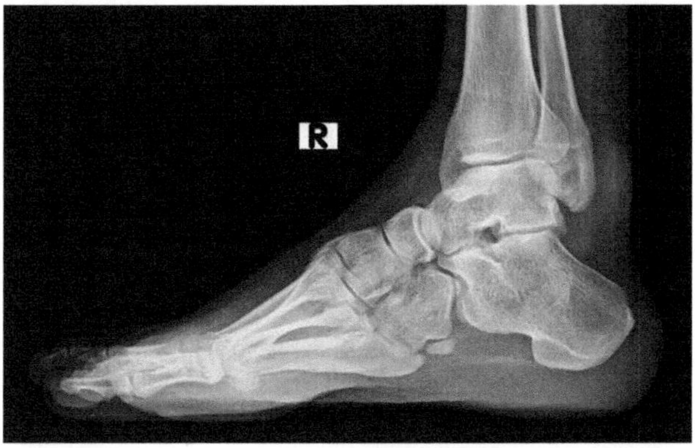

Fig. 13.13 Weight-bearing lateral radiography of a 28-year-old patient. Posteriorly positioned fibula, plantar flexed first ray, and open sinus tarsi can be seen

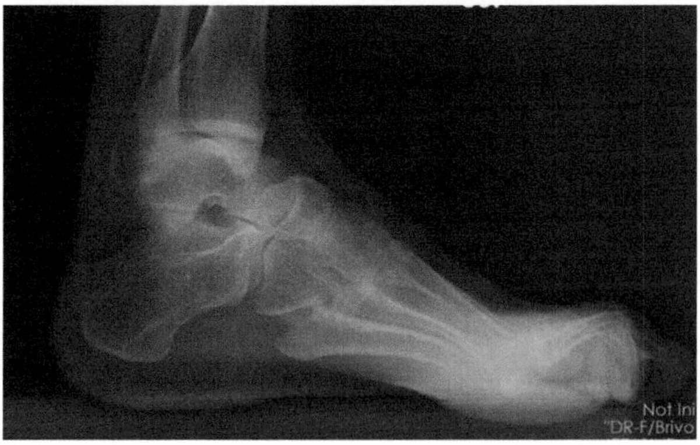

Fig. 13.14 Increased calcaneal inclination in a 43-year-old patient with poliomyelitis history

It may not be possible to evaluate the ankle on lateral foot radiographies because of the rotational deformity [14]. To evaluate the ankle joint for arthritic changes, internal rotation of the leg can help get a proper lateral ankle view.

Foot AP radiographies are used to evaluate the coronal alignment of the forefoot (forefoot adduction or displacement in clubfoot), presence of metatarsus adductus, degree of midfoot and subtalar joint supination, stress fractures of the lesser metatarsals (especially at the fifth metatarsal base), and talonavicular coverage which is indicative of medial peritalar subluxation [15] (Fig. 13.15).

Ankle AP is also evaluated for signs of arthritis and varus alignment with medially localized narrowing of the joint space (Fig. 13.16).

Hindfoot alignment view is used to demonstrate the calcaneal inversion. For these purposes, Saltzman modification of Cobey view or long axial radiographic view described by Lamm and colleagues can be used, but it should be kept in mind that these techniques' accuracy is questionable [15]. In time, weight-bearing CT (WB CT) and 3D reconstructions will probably provide more accurate information about the hindfoot alignment.

As the reliability of WB CT in pes cavovarus is shown by Bernasconi et al., it is obvious that WB CT is going to be used more often to make a detailed analysis of pes cavovarus and 3D reconstructions obtained from these images can be used for a more detailed and precise preoperative treatment plan [15, 16].

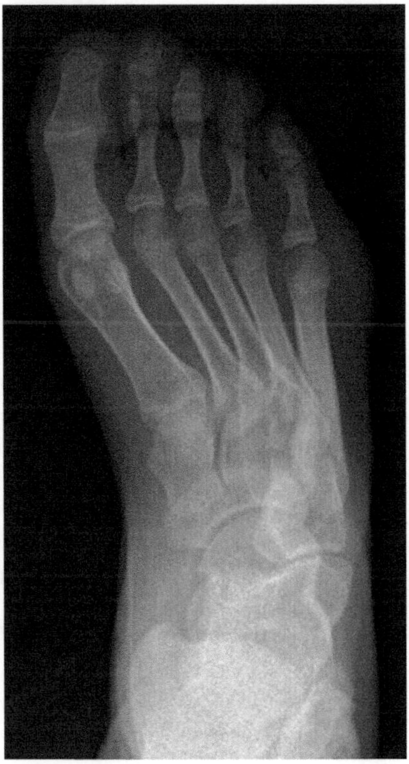

Fig. 13.15 Weight-bearing foot AP view

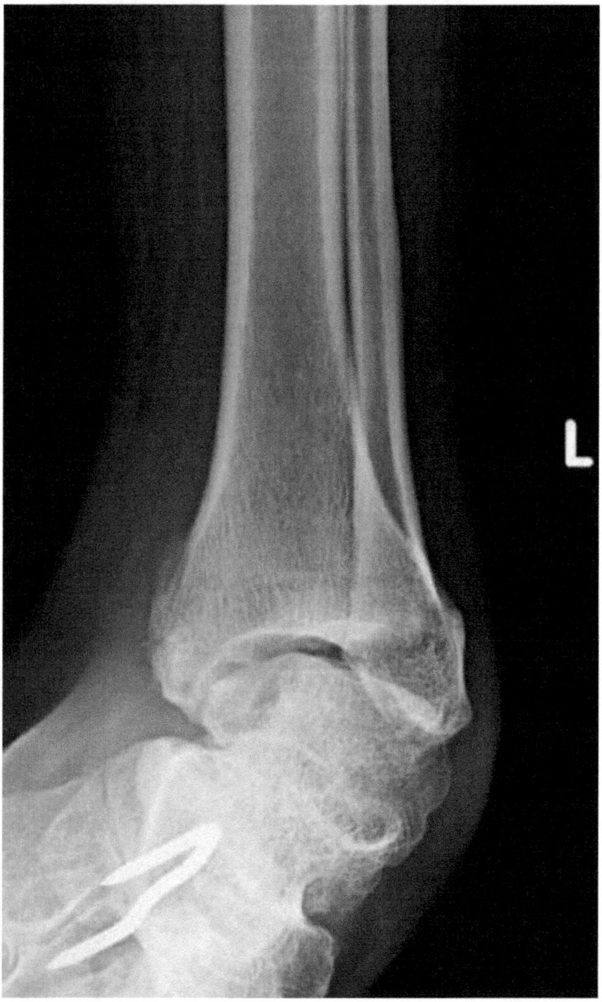

Fig. 13.16 Weight-bearing ankle AP view which demonstrates varus alignment

Even if WB CT is not available, standard CT with 3D reconstruction should be used to evaluate the deformity preoperatively and also to assess the arthritic changes in the joints. It can also be used to determine tarsal coalition-related pes cavus [15].

On the other hand, the use of magnetic resonance in the evaluation of cavovarus foot is not clear. It can be used to evaluate the integrity of lateral ligaments and peroneal tendons and the extent of cartilage damage and arthritis and to assess and monitor the magnitude and progression of neuropathic disease (can be used to see fatty infiltration in the muscles) [15, 17]. But in cases of unilateral cavovarus or spasticity, MR is always needed [4].

13.4.2 Specific Measurements for Each Radiological Modalities, Cut-Off Points with References from Literature, Tables Are Encouraged to Summarize the Normal Values

- Talonavicular coverage: It is the angle between the line that joins the medial and lateral articular margins of the talus and the line that joins the medial and lateral articular margins of the navicular. It is used to describe the alignment of the midfoot. In normal subjects, it is less than 7°, and for flatfoot, it is higher than 7°. In the case of pes cavus, medial peritalar subluxation is the usual finding when examining this angle (Fig. 13.17).

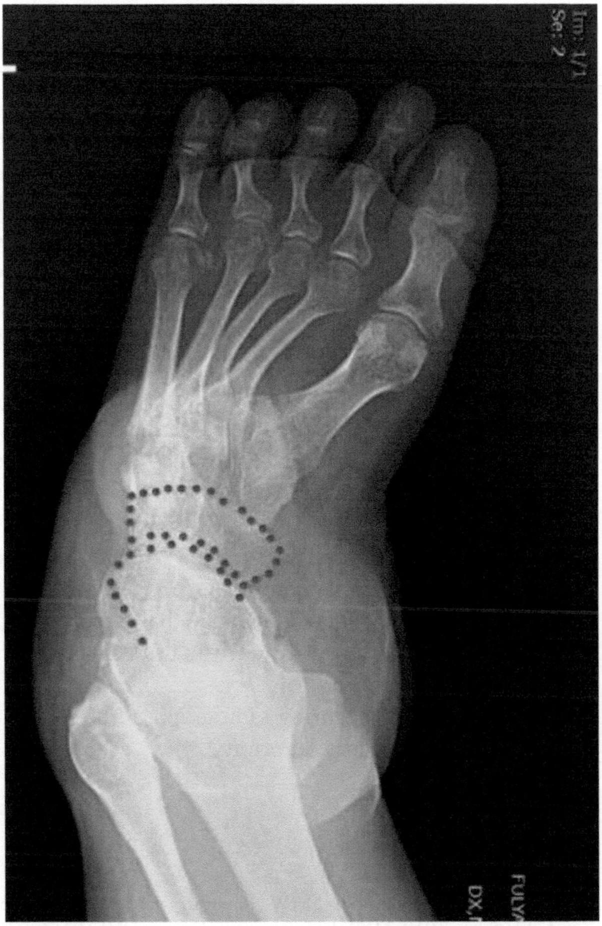

Fig. 13.17 Weight-bearing AP view which demonstrates peritalar subluxation

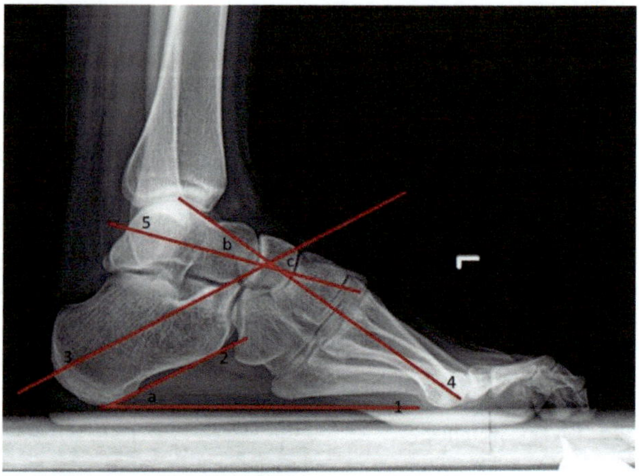

Fig. 13.18 Most commonly measured angles on lateral X-ray. (1) Plane of support. (2) Calcaneal inclination axis. (3) Axis of the calcaneus. (4) First metatarsal axis. (5) Talar axis (a) Calcaneal inclination angle. (b) Lateral Meary angle. (c) Hibbs angle

- Lateral Meary angle: It is the angle formed by the bisection of the talar neck and the anatomic axis of the first metatarsal [18]. Normal value of lateral Meary angle is between 0° and 5°, and it increases as the first metatarsal plantarflexes.
- Calcaneal pitch (inclination): It is the angle formed between the line outlining the plantar aspect of the calcaneus and the weight-bearing surface [18]. Normal values are 25° ± 5°. Values greater than 30° is related with cavus.
- Hibbs angle: It can also be called as calcaneal—first metatarsal angle [15]. It is the angle formed by the intersection of the longitudinal axes of the calcaneus and first metatarsal. If it is less than 150°, it is indicative of pes cavus.
- Lateral talocalcaneal angle: It is the angle formed by the intersection of the calcaneal and talar longitudinal axes on the lateral view. Normal values are 35–50° [15]. It can be less than 25° in residual clubfoot patients [14] (Fig. 13.18).

Key Points

- Cavovarus foot is usually associated with neurological disorders, and CMT is the common reason of this deformity.
- It results from muscular imbalance that affects the foot.
- To clearly define the imbalance is critical in understanding the deformity and treatment planning.
- The lateral weight-bearing radiograph of the foot is the most important imaging modality of cavovarus deformity.

References

1. Krähenbühl N, Weinberg MW. Anatomy and biomechanics of cavovarus deformity. Foot Ankle Clin. 2019;24(2):173–81. https://doi.org/10.1016/j.fcl.2019.02.001.
2. Neumann JA, Nickisch F. Neurologic disorders and cavovarus deformity. Foot Ankle Clin. 2019;24(2):195–203.

3. Johnson AH. Cavovarus deformity. In: Chou LB, editor. Orthopaedic knowledge update: foot and ankle. 6th ed. Rosemont: American Academy of Orthopaedic Surgeons; 2020. p. 51–64.
4. Georgiadis AG, Spiegel DA, Baldwin KD. The cavovarus foot in hereditary motor and sensory neuropathies. JBJS Rev. 2015;3(12):e5.
5. Beals TC, Nickisch F. Charcot-Marie-Tooth disease and the cavovarus foot. Foot Ankle Clin. 2008;13(2):259–74.
6. Deben SE, Pomeroy GC. Subtle cavus foot: diagnosis and management. J Am Acad Orthop Surg. 2014;22(8):512–20.
7. Sabir M, Lyttle D. Pathogenesis of Charcot-Marie-Tooth disease. Gait analysis and electrophysiologic, genetic, histopathologic, and enzyme studies in a kinship. Clin Orthop Relat Res. 1984;184:223–35.
8. Maynou C, Szymanski C, Thiounn A. The adult cavus foot. EFORT Open Rev. 2017;2(5):221–9.
9. Pfeffer GB, Gonzalez T, Brodsky J, Campbell J, Coetzee C, Conti S, Guyton G, Herrmann DN, Hunt K, Johnson J, McGarvey W, Pinzur M, Raikin S, Sangeorzan B, Younger A, Michalski M, An T, Noori N. A consensus statement on the surgical treatment of charcot-marie-tooth disease. Foot Ankle Int. 2020;41(7):870–80.
10. Apostle KL, Sangeorzan BJ. Anatomy of the varus foot and ankle. Foot Ankle Clin. 2012;17(1):1–11.
11. Krause FG, Guyton GP. Pes cavus. In: Coughlin MJ, Saltzman CL, Anderson RB, editors. Mann's surgery of the foot and ankle. 9th ed. Philadelphia: Elsevier; 2014.
12. Nogueira MP, Farcetta F, Zuccon A. Cavus foot. Foot Ankle Clin. 2015;20(4):645–56. https://doi.org/10.1016/j.fcl.2015.07.007.
13. Coleman SS, Chesnut WJ. A simple test for hindfoot flexibility in the cavovarus foot. Clin Orthop Relat Res. 1977;123:60–2.
14. Akoh CC, Phisitkul P. Clinical examination and radiographic assessment of the cavus foot. Foot Ankle Clin. 2019;24(2):183–93.
15. Osher L, Shook JE. Imaging of the pes cavus deformity. Clin Podiatr Med Surg. 2021;38(3):303–21.
16. Bernasconi A, Cooper L, Lyle S, et al. Intraobserver and interobserver reliability of cone beam weightbearing semi-automatic three-dimensional measurements in symptomatic pes cavovarus. Foot Ankle Surg. 2019;26(5):564–72.
17. Ziebarth K, Krause F. Updates in pediatric cavovarus deformity. Foot Ankle Clin. 2019;24(2):205–17.
18. Lamm BM, Stasko PA, Gesheff MG, Bhave A. Normal foot and ankle radiographic angles, measurements, and reference points. J Foot Ankle Surg. 2016;55(5):991–8.

Hallux Disorders

14

Adarsh Aratikatla, Nour Nassour,
and Lorena Bejarano Pineda

14.1 Introduction

The hallux is the most medial digit of the foot, whose bony structure, proximally to distally, consists of the first metatarsal, two sesamoid bones (located on the plantar surface of the first metatarsal head), proximal phalanx, and distal phalanx. While the second to the fifth digits have a proximal and distal interphalangeal joint between the proximal, middle, and distal phalanges, the hallux has a singular interphalangeal joint between the proximal and distal phalanx. The primary movements of the great toe are extension, mediated by the extensor hallucis longus and brevis [1], and flexion, mediated by the flexor hallucis longus, flexor hallucis brevis, abductor hallucis, and adductor hallucis [2, 3].

14.2 Hallux Valgus

14.2.1 Introduction

Hallux valgus (i.e., metatarsus primus varus, bunion) is one of the most common deformities of the forefoot and the most common pathology of the great toe [4]. It is caused by lateral deviation of the proximal phalanx along with medial deviation and pronation of the first metatarsal (MT) head [4, 5]. Epidemiological studies have shown that the prevalence of HV increases with age, with 23% of adults under the age of 65 and 35.7% of those older than 65 years of age presenting with the deformity [6].

A. Aratikatla · N. Nassour · L. B. Pineda (✉)
Foot and Ankle Research and Innovation Lab, Harvard Medical School, Massachusetts General Hospital, Boston, MA, USA
e-mail: lpineda@mgh.harvard.edu

S. Sharma et al. (eds.), *Clinical and Radiological Examination of the Foot and Ankle*, https://doi.org/10.1007/978-981-97-4202-8_14

Risk factors for the development of hallux valgus include wearing high-heeled or narrow-toe box shoes, additional foot and ankle pathologies (i.e., pes planus, rheumatoid arthritis), and generalized ligamentous laxity [7–12]. Furthermore, considering women are more prone to the development of bunions, researchers have sought to determine whether hormonal factors, possibly during pregnancy or menopause, contribute to ligamentous laxity resulting in the formation of HV [13, 14]. Certain studies have suggested genes *TTN*, *COL6A3*, and *LARS* as potential genetic risk factors, as they play a significant role in the anatomical development of the great toe; however, confirmatory studies have not been conducted to definitively prove the correlation between gene mutations and this deformity [15].

14.2.2 History and Physical Examination

The diagnosis of hallux valgus relies mainly on history and physical examination (see Fig. 14.1). Patients commonly report pain over the medial eminence of the MTP joint that can be associated with irritation, blistering, or ulceration, limiting their activities of daily living [5, 16]. Patients may also experience neuropathic symptoms related to compression of the medial dorsal cutaneous nerve, such as a burning and tingling sensation in the dorsal region of the affected foot [5]. Furthermore, the lateral deviation of the hallux can result in an overlapping or cross-over deformity with the lesser toes (see Fig. 14.2), which can lead to concomitant lesser toe deformities (i.e., hammer or claw toes) [17, 18]. The dermatological implications of HV may include calluses, corns, or keratosis formation (see Fig. 14.3) [19]. These anatomic changes may lead to difficulty finding accommodating footwear, as narrow, tight-fitting toe box shoes may exacerbate the patient's pain [20].

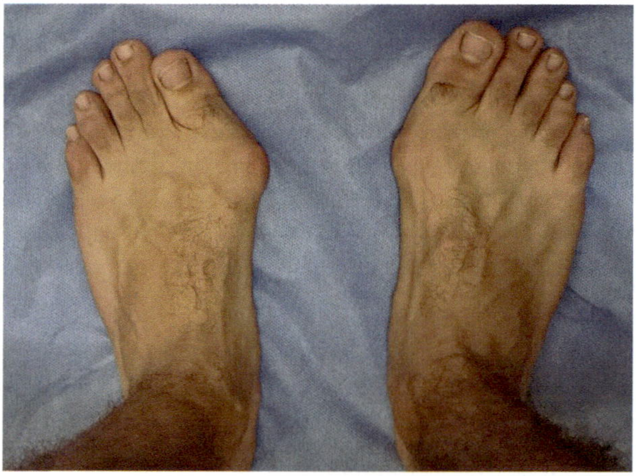

Fig. 14.1 Clinical photograph of hallux valgus

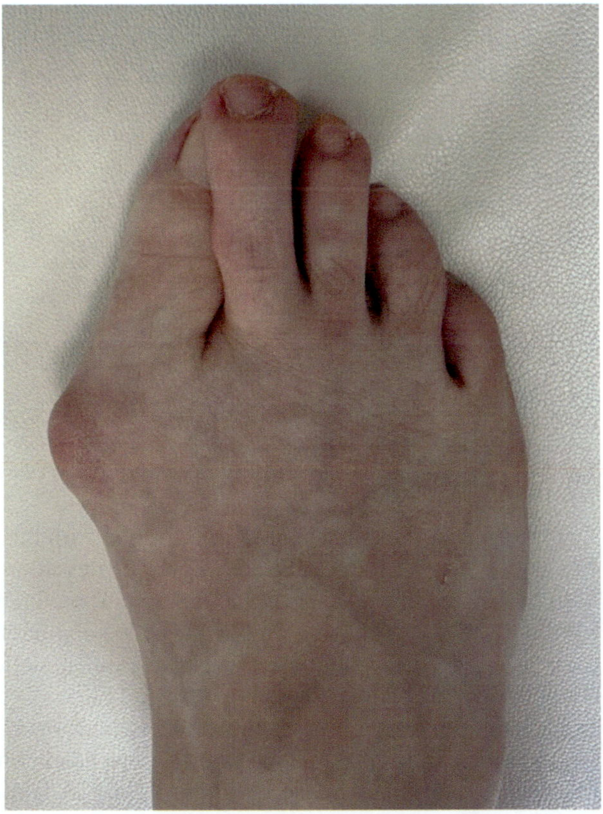

Fig. 14.2 Clinical photograph of crossover deformity seen in association with hallux valgus

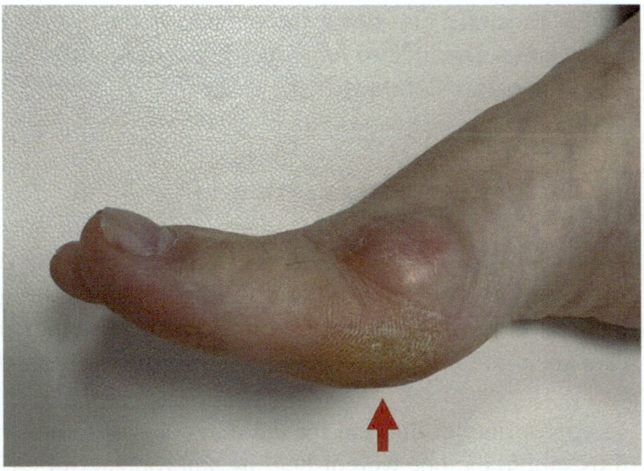

Fig. 14.3 Clinical photograph of a callus due to hallux valgus deformity

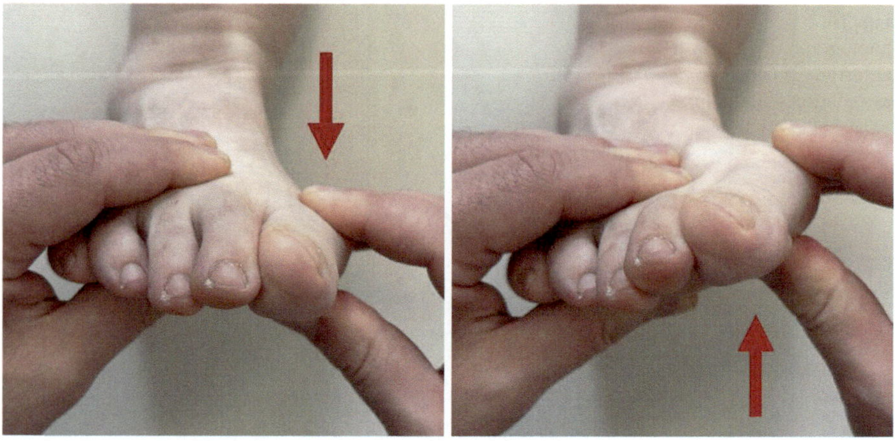

Fig. 14.4 Clinical photographs demonstrating the first TMT instability

Physical examination should be performed with the patient seated and standing. Evaluation starts by assessing the degree of deviation in the coronal plane along with the degree of pronation in the first ray. The clinician should also evaluate the rigidity of the deformity, as well as the active and passive motion of the first MTP joint, as these factors can impact surgical treatment options [5]. Likewise, the stability of the first TMT joint should be assessed clinically and subsequently correlated to radiologic findings. The examination is performed by delivering a dorsal and plantar directed force to the first ray with one hand, while the other hand holds the lesser metatarsals stationary (see Fig. 14.4). Instability was defined as sagittal mobility of 1 cm on clinical examination.

Lastly, the rest of the foot should be evaluated for any concomitant pathologies such as lesser toe deformity, pes planus, metatarsalgia, equinus of the Achilles complex, plantar fasciitis, or any other midfoot/hindfoot conditions that might contribute to the hallux deformity [21, 22].

14.2.3 Radiological Examination

Radiological evaluation is necessary to confirm the diagnosis and determine the severity of the deformity. Plain radiographs are the gold-standard imaging technique used to diagnose this disorder [23], but other modalities such as magnetic resonance imaging (MRI) and computer tomography (CT) have been described as well [24].

Appropriate evaluation for the diagnosis of hallux valgus deformity using plain films should include weight-bearing anteroposterior (AP), lateral, oblique, and sesamoid axial views of the foot (Table 14.1) (Fig. 14.5). Typical radiographic findings in the AP view include lateral subluxation and secondary osteoarthritis of the first MTP joint [21, 25]. This view can also see thickening of the skin or osteophyte

Table 14.1 Described radiographic angles to diagnose hallux valgus and its deformities

Name of angle	Definition	Constitutional range and range indicating pathology
Hallux valgus angle (HVA)/first MTP angle	The angle created between the lines that longitudinally bisect the proximal phalanx and the first metatarsal (see Fig. 14.8)	Less than 15° is normal >20° = abnormal 45–50° = severe
First to second intermetatarsal angle (IMA)	The angle drawn between the first and second metatarsal shafts on an axial view of the foot (see Fig. 14.9)	<9° = normal
Distal metatarsal articular angle (DMAA)—used mostly for operative planning	The angle between the perpendicular line to the longitudinal axis of the first MT and the articular surface of the first metatarsophalangeal joint (see Fig. 14.10)	<10° is normal
Hallux interphalangeus angle (HIA)/proximal phalangeal articular angle (PPAA)	The angle formed by the proximal and distal phalanx (see Fig. 14.11)	<10° is normal
First metatarsal protrusion distance (FMTP)	Difference between the protrusion of the first and second metatarsals from the point where their mid-axial lines intersect (see Fig. 14.12)	Normal range: 0 ± 2 mm

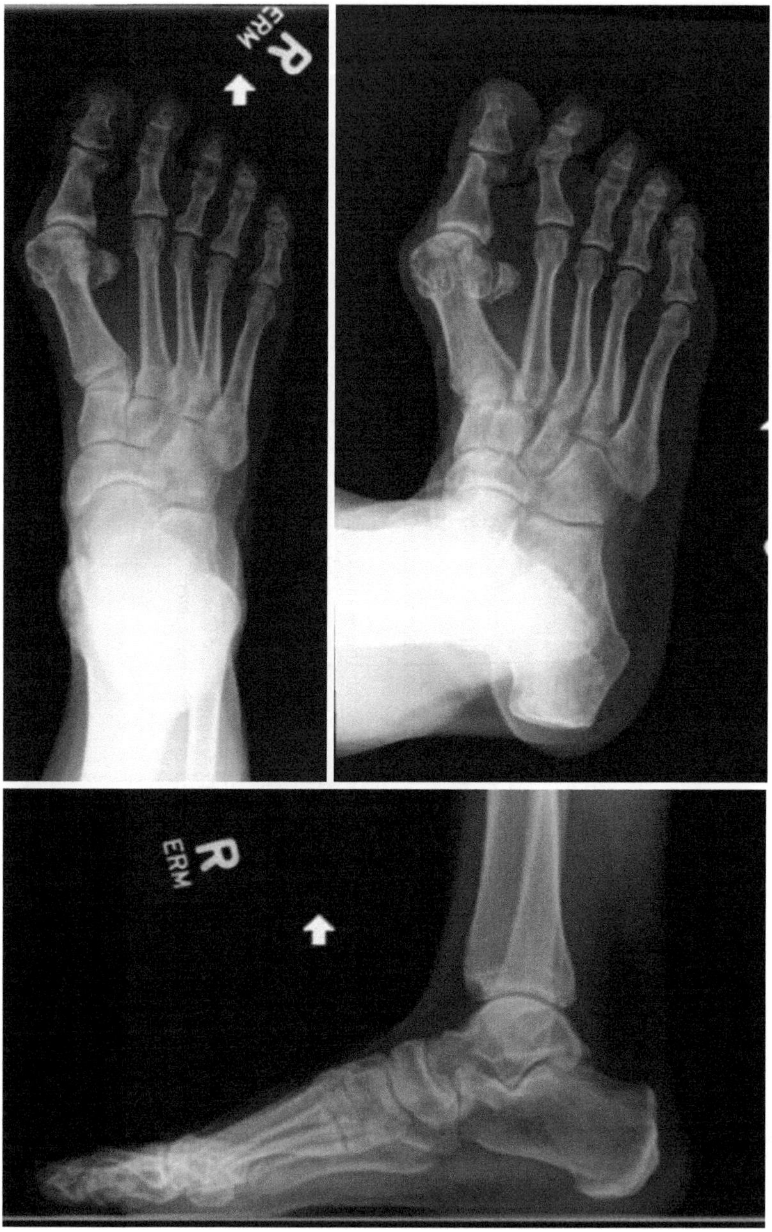

Fig. 14.5 Weight-bearing AP, lateral, and non-weight-bearing oblique views for hallux valgus on plain radiography

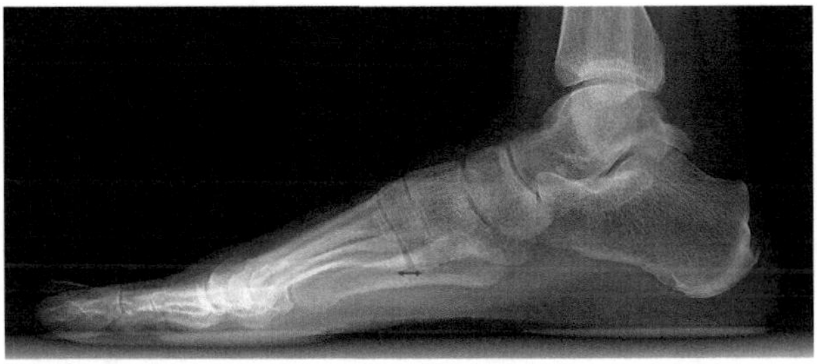

Fig. 14.6 Lateral weight-bearing radiograph showing the first TMT plantar gapping in hallux valgus

formation at the medial eminence of the metatarsal head. On the lateral view, dorsal angulation of the first MTP is usually present, along with dorsal osteophytes of the metatarsal head. The presence of plantar gaping at the first TMT joint should also be evaluated on lateral weight-bearing radiographs and correlated with the clinical evaluation confirming hypermobility of the first ray (see Fig. 14.6) [21, 25].

Several classification systems, represented in Table 14.2, are described to determine the severity of hallux valgus deformity, Coughlin's system being the most prevalent. Recently, the pronation of the hallux over the sesamoids has been described as a pivoting factor to take into consideration for treatment options, and therefore, the four-grade classification system has been developed [26].

Weight-bearing computer tomography (WBCT) is becoming an important imaging tool in the diagnosis of hallux valgus, as it offers a tridimensional approach allowing for better visualization of the foot while the patient is standing (Fig. 14.7). It allows for better observation and assessment of the deformity relative to the position of the sesamoids when compared to regular anteroposterior weight-bearing radiographs; this imaging modality also provides details on the first MTP joint congruency and joint degeneration [12, 27–29].

Several vital measurements observed in patients with hallux valgus deviate from the norm (Figs. 14.8, 14.9, 14.10, 14.11, and 14.12), particularly when assessing the dorsoplantar (DP) thickness, mediolateral (ML) width, and cross-sectional area (CSA). Notably, the abductor hallucis measurements demonstrate atypical findings in HV. Ultrasound (US) guidance has been used to visualize these irregular findings and provide easily attainable radiographic confirmation of this deformity [30].

Table 14.2 Described classification systems for the diagnosis and severity of hallux valgus deformity

Name of classification system	Grades	Associated features
Coughlin (HVA, IMA, respectively)	Normal	<15°, <9°
	Mild	<20°, 9–11°
	Moderate	20–40°, 11–16°
	Severe	>40°, >16°
Triplane hallux valgus classification—based on anatomic findings	Type 1	Increased HVA and IMA, no first MT pronation on radiographs, sesamoids may be subluxed
	Type 2a	Increased HVA and IMA, first MT pronation on AP and sesamoid radiographs, no sesamoid subluxation
	Type 2b	Increased HVA and IMA, first MT pronation on AP and sesamoid radiographs, with sesamoid subluxation
	Type 3	Increased HVA and IMA, >20 DMAA
	Type 4	Increased HVA and IMA ± first MT pronation (clinical ± radiographic evidence of degenerative joint disease)
The Nijmegen classification	0	No clinical changes in the MTP joints, none-mild radiographic changes
	1	– Decreased mobility (particularly plantarflexion) of ≥1 of the joints – The ability to reduce the plantar soft tissues under the MT heads – Adequate quality of the plantar soft tissues ± radiographic erosive changes or evident intra-articular changes
	2a–2b (a) With a hallux valgus of >20° (b) Without a hallux valgus of >20°	– Loss of plantar flexion in ≥1 of the MTP joints (up to 0°) – Loss of the ability to reduce the plantar soft tissues under the MT heads ± inadequate quality of the plantar soft tissues
	3a–3b (a) With a hallux valgus of >20° (b) Without a hallux valgus of >20°	Deep contracture in >1 MTP joint ± radiographic subluxation or dislocation
Four-grade classification	Grade 0	Tibial sesamoid is entirely medial to the first MT long-axis reference line (RL)
	Grade 1	<50% overlap of sesamoid relative to the RL
	Grade 2	>50% overlap of sesamoid relative to the RL
	Grade 3	Tibial sesamoid is displaced entirely lateral to the RL

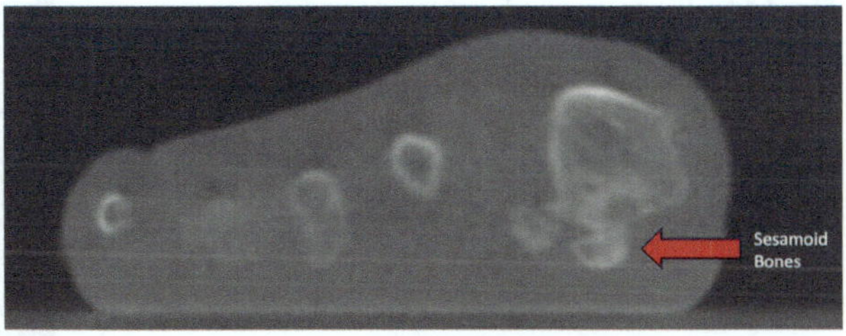

Fig. 14.7 Coronal section from a weight-bearing CT showing malalignment in sesamoid rotation

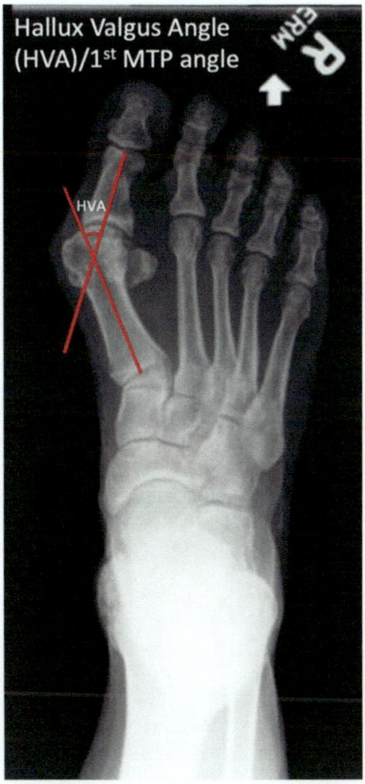

Fig. 14.8 Hallux valgus angle (first metatarsophalangeal angle)

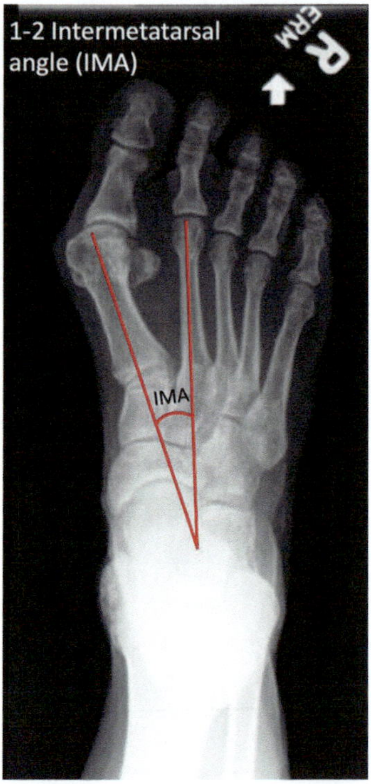

Fig. 14.9 First to second intermetatarsal angle

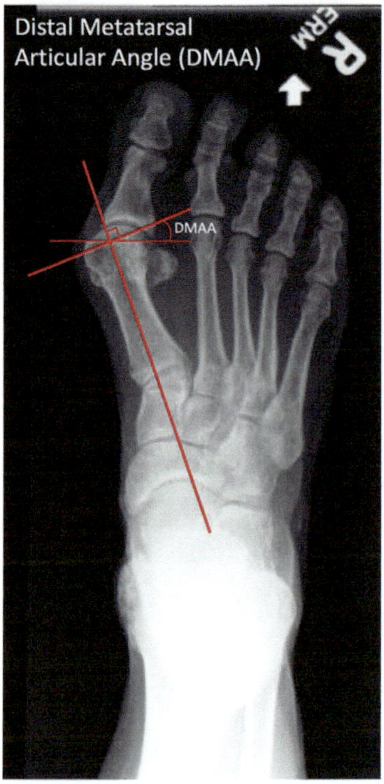

Fig. 14.10 Distal metatarsal articular angle (DMAA)

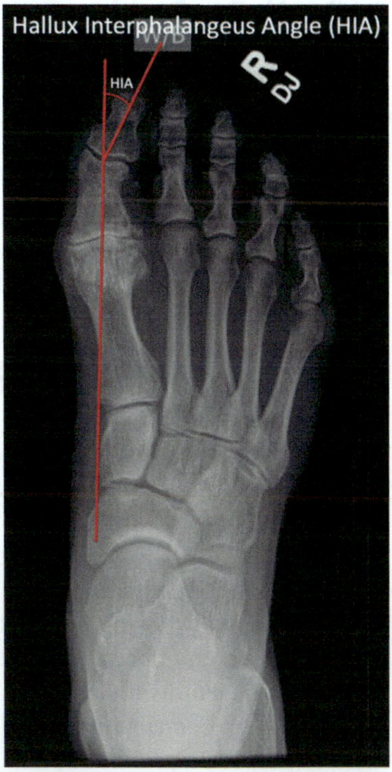

Fig. 14.11 Hallux interphalangeus angle (HIA)

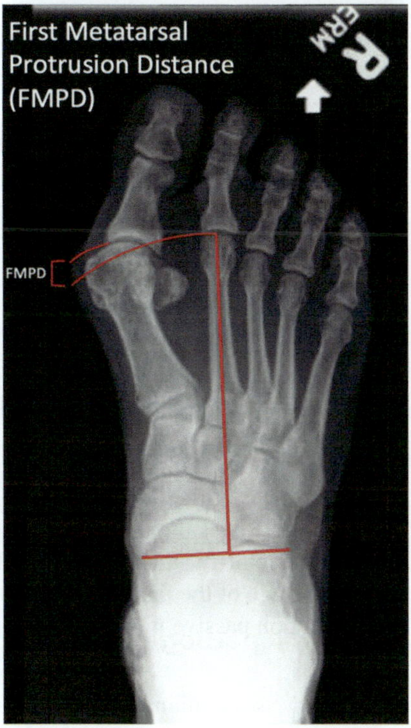

Fig. 14.12 First metatarsal protrusion distance

14.3 Hallux Rigidus

14.3.1 Introduction

Hallux rigidus (HR) is a disorder that entails degenerative osteoarthritis of the first MTP joint [31]. Initially, the term "hallux flexus" was introduced by Davies-Colley to describe the plantar flexion of the proximal phalanx in relation to the metatarsal head. However, Cotterill later coined the term "hallux rigidus" to better describe the condition, emphasizing its painful nature and the restricted motion experienced at the first MTP joint [31]. Similar to hallux valgus, females are more prone to developing HR, and the average age of onset is approximately 43 years. Family history also seems to play a significant role in this disease, as 67% of patients with HR had a parent with the same condition [32].

A wide variety of etiologies is described for the development of HR, with some instances being multifactorial. Repetitive stress caused by excessive walking or time standing is a leading cause of HR, as the first MTP joint can receive approximately 120–200% of the individual's body weight during the gait cycle [33]. Inflammatory and metabolic conditions (i.e., rheumatoid arthritis, gout) have been described as predisposing factors for the development of hallux rigidus, but conclusive studies have not been published confirming this link [34, 35]. Isolated tightness of the gastrocnemius muscle or Achilles tendon may also be associated with forefoot pathology contributing to hallux rigidus [32]. Both mechanical and structural factors (i.e., wide proximal phalanx/first MT, abnormally long sesamoids, a long second metatarsal) may contribute to hallucal osteoarthritis [36]. Additionally, first-ray pathologies, such as hallux valgus, metatarsus primus varus, or acute local trauma, could lead to HR in patients [31, 37–40].

14.3.2 History and Physical Examination

Patients with HR usually present with decreased ROM associated with pain at the level of the first MTP joint [41, 42]. Pain may be exacerbated with activities entailing dorsiflexion of the first MTP joint, such as walking up the stairs or running [43], and is usually worse in the heel rise lift-off phase of the gait cycle. Neuralgia often presents due to possible impingement of the superficial peroneal nerve caused by osteophytes [1, 44]. As the disease advances, the pain becomes more pronounced and typically arises in the middle of the range of motion with foot dorsiflexion and plantarflexion.

On physical examination, the patients often have a bony prominence representing osteophytes at the dorsal aspect of the joint that is usually painful to palpation (see Fig. 14.13). As the disease advances, there may be pain with passive compression of the joint by the clinician, which is called the "grind test" indicating an advanced stage of the disease. Additional examination may show redness, swelling, and pain upon passive motion [45, 46].

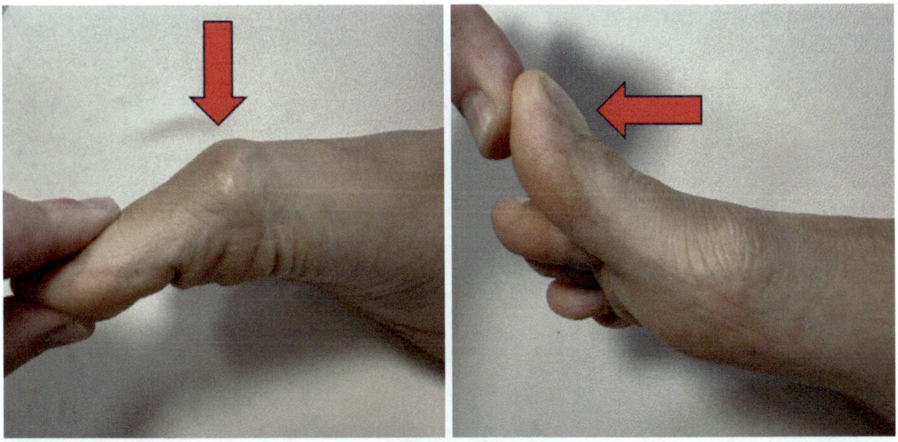

Fig. 14.13 Clinical photograph (left) shows plantar flexion and dorsal osteophyte of the hallux; clinical photograph (right) shows limited dorsiflexion associated with hallux rigidus

14.3.3 Radiological Examination

Plain radiographs are the gold-standard imaging technique used to diagnose this disorder [23], but other modalities such as magnetic resonance imaging (MRI) and computer tomography (CT) have been described as well [24]. Described findings in the AP view include narrowing of the joint space, and osteophytes at the margins of the first MT head indicating the severity of the disease. Likewise, the lateral view is very useful to evaluate the presence of osteophytes dorsally, and in advanced stages, those osteophytes may break off, creating intra-articular bony fragments called dorsal ossicles.

Other imaging modalities like MRI, CT, and US can be used to assist in the diagnosis of HR; however, the diagnosis of this disease is mainly based on clinical and plain film findings. MRI can be used to evaluate additional parameters associated with HR. The most common findings include dorsal and lateral osteophytes, marrow edema, geodes, and subchondral sclerosis [24]. CT can further confirm and characterize osteoarthritis of the first MTP joint leading to HR and is usually only indicated if clinicians suspect osteochondral cysts based on previous radiographic findings [47, 48]. US is rarely used but may be useful in determining possible concomitant injuries, such as tears in local tendons and ligaments [49]. Figure 14.14 displays the AP, lateral, and oblique views for hallux rigidus on plain radiographs.

There are four main classification systems described combining clinical and radiologic findings to determine the severity of the disease. The most common system used is the one described by Coughlin and Shurnass (Table 14.3).

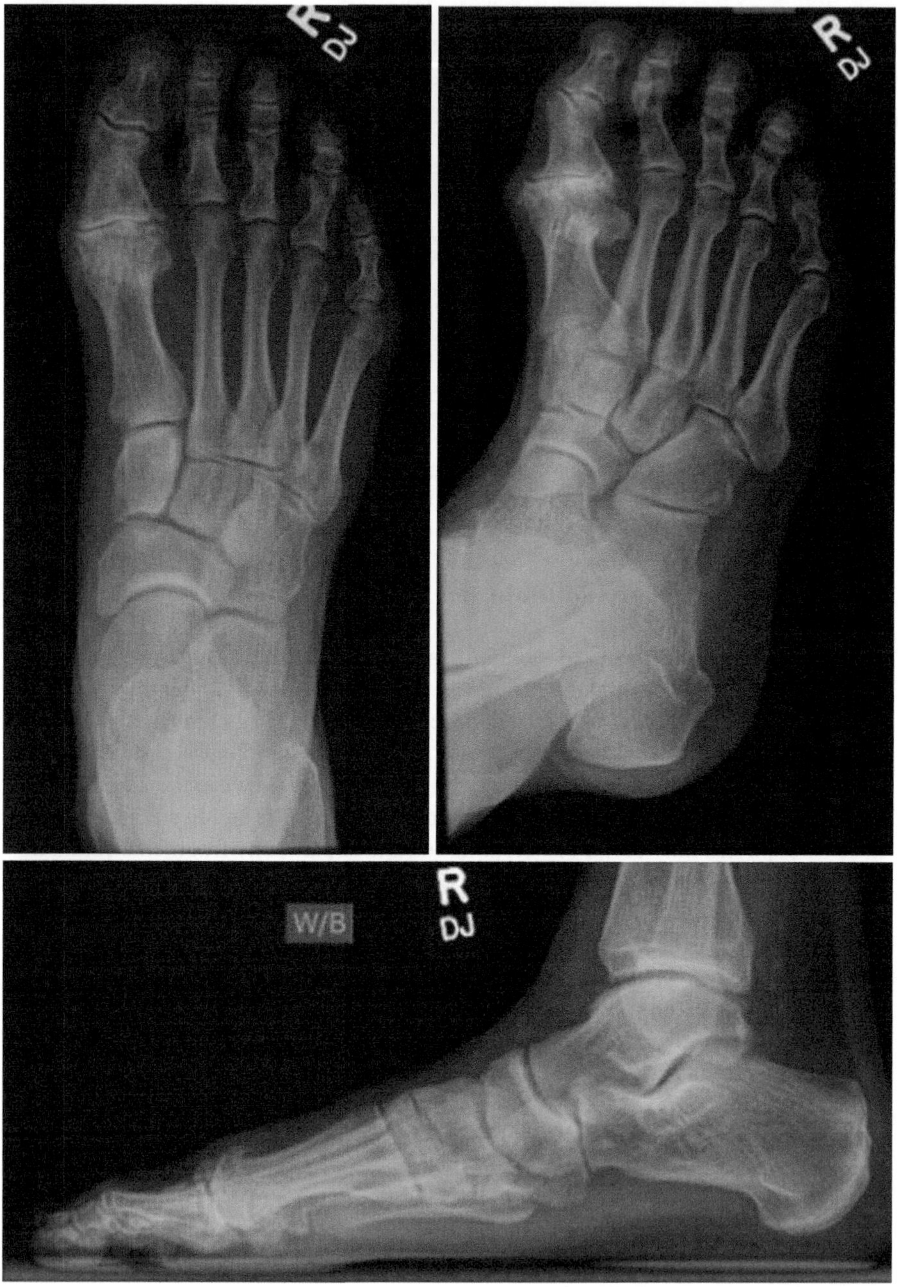

Fig. 14.14 Weight-bearing AP, lateral, and non-weight-bearing oblique views for hallux rigidus on plain radiographs

Table 14.3 Described classification systems for the diagnosis and severity of hallux rigidus

Name of classification system	Grade	Associated features
Regnauld	I	Functional hallux limitus
	II	The joint has adapted, pain at end ROM, and flat first MT head
	III	Arthrosis, osteophytes, erosions, asymmetric joint space narrowing, and severe flattening of the first MT head
Hattrup and Johnson	I	Mild-moderate osteophyte formation, no joint space narrowing/involvement
	II	Moderate osteophyte formation, joint space narrowing, subchondral sclerosis
	III	Greater than moderate osteophyte formation, loss of joint space
Coughlin and Shurnass	0	Dorsiflexion 40–60°, normal radiography, no pain
	1	Dorsiflexion 30–40°, dorsal osteophytes, minimal/no other joint changes
	2	Dorsiflexion 10–30°, mild to moderate joint narrowing or sclerosis, osteophytes
	3	Dorsiflexion less than 10°, severe radiographic changes, constant moderate to severe pain at extremities
	4	Stiff joint, severe changes with loose bodies, and osteochondritis dissecans
Roukis	1	Metatarsus primus elevates, periarticular subchondral sclerosis, minimal dorsal exostosis, and minimal flattening of the metatarsal head
	2	Moderate dorsal exostosis, flattening metatarsal head, minimal joint space narrowing, sesamoid hypertrophy
	3	Severe dorsal exostosis, focal joint space narrowing, cyst formation, loose bodies
	4	Excessive exostosis of the metatarsal head and proximal phalanx base, absent joint space, ankylosis

14.4 Hallux Varus

14.4.1 Introduction

Hallux varus deformity is the medial deviation of the first toe at the first MTP joint toward the midline of the foot [50]. The etiology of this disorder is highly variable. Hallux varus can present in children either as a primary developmental, congenital abnormality due to a weakened abductor hallucis tendon, or, more commonly, as a consequence of the fetus' position within the uterus. In the majority of cases, the primary pediatric deformity tends to self-correct when the child starts walking [51].

Table 14.4 Pathophysiologic changes associated with hallux varus after correction surgery for hallux valgus

Procedure	Pathophysiologic change
Chevron osteotomy	The capital fragment exhibits significant lateral displacement
Proximal osteotomy	The distal segment demonstrates excessive lateral translation
McBride procedure	Following the excision of the fibular sesamoid, the result is MTP joint hyperextension, IP joint flexion, and medial deviation of the hallux

Secondary hallux varus may be seen in association with metatarsus adductus or polydactyly of the hallux [50]. Likewise, in older patients, it can develop as a result of a surgical procedure. Most commonly, hallux varus occurs after hallux valgus correction surgery, possibly due to accidental overcorrection, the medial eminence being excessively resected, or the great toe being immobilized [52]. Similar to HV and HR, inflammatory conditions and trauma may also cause varus of the great toe.

Large-scale epidemiological studies are yet to be conducted, but prevalence data from postoperative hallux valgus surgery indicates that roughly 2–17% of patients may develop hallux varus after correction surgery [53]. Up to 10% of iatrogenic hallux varus cases have been related to crescentic osteotomies. The pathophysiology of hallux varus after correction surgery for hallux varus deformity is described in Table 14.4.

Similar to other hallux disorders, the gait cycle can become affected in hallux varus, leading to progressive changes in great toe position. The symptoms associated with hallux varus are non-specific and can be generalized to most hallux disorders; these include soreness, impaired ambulation/standing, weakness, ingrown toenails, limited ROM, and pain/swelling/redness. Improper footwear aggravates the pain, especially narrow-toe box shoes, as they place excessive pressure on the medial aspect of the great toe and can lead to skin irritation [54]. The aesthetic problem associated with hallux varus is usually a wide toe gap between the first and second rays.

14.4.2 History and Physical Examination

Symptoms can range from being asymptotic to pain, crepitus, swelling, redness, or ulceration of the hallux; these symptoms may be aggravated by narrow shoe wear, difficulty standing or walking, ingrown toenails, and limited MTP joint range of movement. During the physical examination, the clinician should assess if the deformity is rigid or passively correctible. Likewise, the patient should be evaluated sitting and standing to determine if there are any changes in the severity of the deformity. In some instances, there may also be an associated supination of the toe. The mobility of the first MTP and IP joints should be evaluated as this should be considered for treatment options. Alternatively, it is important to examine the integrity of the plantar hallucis longus and carefully inspect the plantar surface of the foot for any calluses or ulcerations [50, 55].

14.4.3 Radiological Examination

Weight-bearing plain radiographs of the foot are indicated to confirm the diagnosis. Anteroposterior, lateral, and sesamoid views are commonly obtained. Patients with hallux varus deformity have a negative hallux valgus angle with a medial deviation of the proximal phalanx. Additional findings include reduced IMA, osteotomies with overcorrection of the HVA, and medial subluxation of the tibia sesamoid [50]. The use of CT or MRI is not usually indicated unless osteonecrosis of the first MT is suspected. There is no established classification system for this deformity. Clinicians need to determine if the deformity is flexible or rigid and the degree of arthritis of the first MTP joint, as these parameters will determine the type of surgical intervention [56]. Figure 14.15 shows the AP, lateral, and oblique views for hallux varus on plain radiographs.

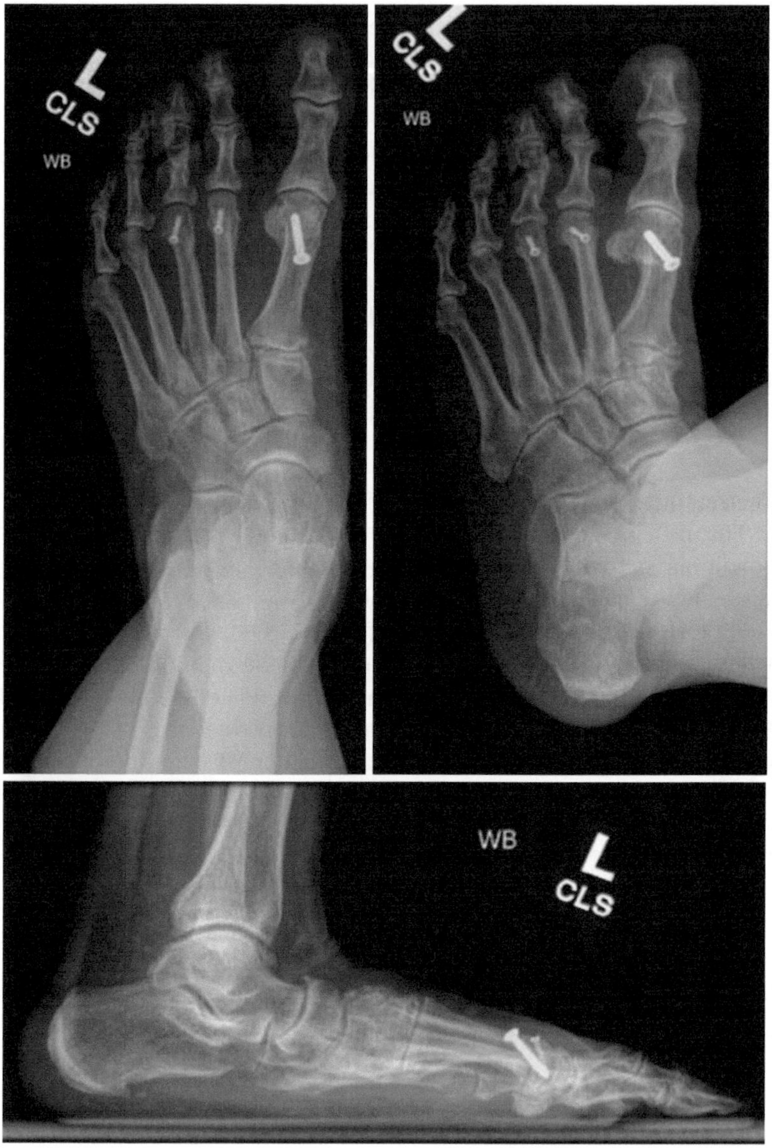

Fig. 14.15 Weight-bearing AP, lateral, and non-weight-bearing oblique views for hallux varus on plain radiographs

14.5 Sesamoiditis

14.5.1 Introduction

Sesamoids are bones that are not connected to any other bones at a joint but rather embedded in the muscle or connected to tendons (e.g., the largest sesamoid bone is the patella). Sesamoid bones of the foot are usually in pairs and commonly found on the plantar aspect of the distal first metatarsal but may also be found at all the MTP joints. Sesamoiditis is inflammation of the surrounding sesamoidal tendons, where pain is usually localized to the plantar aspect of the forefoot [57]. The prevalence of a unilateral hallucal sesamoid is less than 2%, while bilateral sesamoids can be found in over 98% of the general population [58]; additionally, the incidence of bipartite sesamoids is between 7% and 30% [59].

Repetitive movements in certain sports, like dancing and running, create excessive stress on the plantar aspect of the forefoot, which can lead to tendinopathy and sesamoiditis [59–61]. Individuals with pes cavus may develop sesamoiditis secondary to excessive load over the forefoot due to excessive plantarflexion of the first ray. Similarly, a significant correlation exists between pes planus and hallux valgus in patients; this highlights the connection between the abnormal biomechanical forces in a pronated, planus foot and the increasing severity of hallux valgus [62, 63]. Additionally, the increased load forces placed on the forefoot in patients with pes planus during activity may develop into excessive strain and local tendonitis of the tendons around the sesamoid bones of the foot [64, 65].

Additional associated factors include impaired peripheral blood flow, gout, regular use of high-arched footwear, and direct trauma to the bones themselves.

14.5.2 History and Physical Examination

The pain is usually localized to the medial and plantar aspect of the forefoot and is non-radiating. The severity can range from dull and aching to a sharp, throbbing pain. Symptoms are usually progressive and exacerbated by strenuous activity. Activity can be limited as pain can present when plantarflexing or dorsiflexing at the first MTP joint. An altered gait may be present in patients with sesamoiditis, as they may favor the lateral column of the foot when walking to reduce pressure on the medial, hallucal sesamoids [59]. The "push-off" phase and "toe-off" phase of ambulation may be reduced or modified as the patient intends to decrease the amount of pain on the medial portion of the foot, resulting in a biomechanical adjustment in their gait [66]. Sesamoiditis is not usually associated with erythematous skin or bruising, but some swelling may develop as inflammation is present. Sesamoiditis often stems from overuse or repetitive insult.

During a thorough history-taking, it's essential to inquire about activities that place repeated stress on the feet, like running, dancing, jumping, and wearing high heels. Additionally, it is crucial to ask about smoking habits and any previous injections around this area [59]. Physical examination findings consist of tenderness upon palpation in the region beneath the metatarsal head. Additionally, restricted range of motion may be observed in the metatarsophalangeal joint [59]. Patients also tend to avoid bearing weight on the affected area due to pain, which can be noticed while visually assessing their gait. As sesamoiditis is a diagnosis of exclusion, it is crucial to take into account other potential diagnoses and rule them out, including nerve compression, intractable plantar keratosis infection, fracture of the sesamoid bone, avascular necrosis, or subluxation of the sesamoids (see Fig. 14.16) [59].

14.5.3 Radiological Examination

Initial radiological evaluation includes plain radiographs of the foot, with changes like sclerosis and fragmentation of the sesamoids seen; however, these are usually associated with late-stage sesamoiditis. Advanced imaging, such as MRI or CT, is helpful in determining the degree of inflammation and the presence of osteonecrosis of the affected sesamoid. MRI can provide additional information about the surrounding structures that can contribute to the pain, such as tenosynovitis, plantar plate injuries, and osteochondral lesions of the first metatarsal head [67] (Table 14.5).

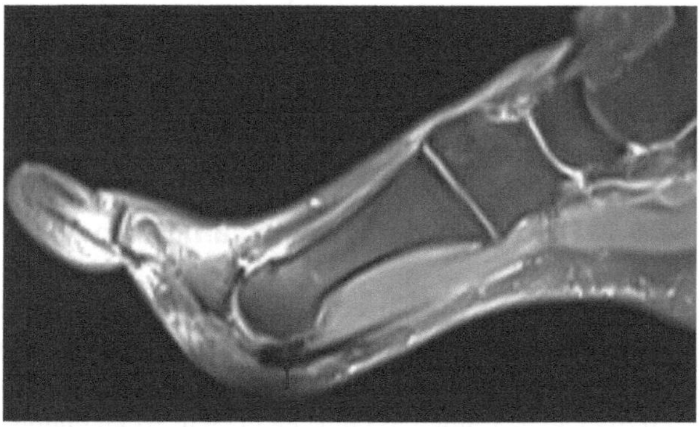

Fig. 14.16 MRI of sesamoid bone fracture

Table 14.5 Classification systems have been constructed for sesamoiditis

Name of classification system	Grades	Associated features
Katzer-Hoek-Yazhari	1	Tibial sesamoid position (TSP) 4 or 5 Tibial metatarsal-sesamoid joint space (TMSJS) ≥2 mm No radiographic indications
	2	TSP 4 or 5 TMSJS 1 mm < x <2 mm
	3	TSP ≥5 TMSJS 0 mm < x <1 mm
	4	TSP ≥5 No visible joint space (TMSJS = 0 mm)

14.6 Turf Toe

14.6.1 Introduction

Turf toe is a sprain of the plantar capsule ligament at the first MTP joint due to forceful hyperextension [68]. Athletes who compete in high-impact sports, such as gymnastics, soccer, basketball, rugby, and American football, are more likely to present this injury. The first MTP joint is both a hinge and sliding joint. There is limited bony stability at the first MTP joint due to its inherent anatomy with a convex articular surface at the first metatarsal and concave surface at the proximal phalanx base making the joint shallow. Consequently, the stability of the joint is maintained via a network of supporting tissues consisting of the joint capsule, plantar plate ligaments (i.e., inter-sesamoid, metatarso-sesamoid/suspensory, sesamoid-phalangeal), musculotendinous insertions (i.e., flexor hallucis longus and brevis, adductor, and abductor hallucis), an extensor apparatus (i.e., extensor hallucis longus, brevis, and hood), and the medial and lateral collateral/metatarsophalangeal ligaments [68, 69]. The incidence of turf toe has been relatively low among college football players, with a rate of 0.062 cases per thousand. Currently, this rate is likely lower as the properties of newer grass in the football field behave like natural grass, as opposed to the hard artificial grass or AstroTurf on older fields [68].

14.6.2 History and Physical Examination

Patients often present with a history of acute onset of sharp pain and swelling, which usually occurs following a forced hyperextension of the great toe upon impact with the ground. Physical examination includes evaluation of the first MTP joint for gross deformity, swelling, and ecchymosis. The range of motion and stability of the hallux MTP joint should be assessed. To assess the adjacent collateral ligaments, varus and valgus stress should be applied, and a dorsoplantar/anterior drawer test should be performed to assess the stability of the plantar ligament complex, associated pain, or displacement that may indicate a positive test. Active flexion and extension of the MTP and interphalangeal joints should also be evaluated for a potential disruption of the flexor hallucis brevis or, in rare cases, disruption of the flexor hallucis longus [70].

14.6.3 Radiological Examination

Plain radiographs are usually performed as the initial evaluation to rule out a bony injury; however advanced imaging is recommended. Although MRI is considered the test of choice for the diagnosis of turf toe injuries, US may also provide useful information and visualize the extent of the injury. Trained technicians and clinicians may be able to robustly maneuver a portable US to assess whether there is any disruption or discontinuity in the plantar plate (Fig. 14.17) [71]. Furthermore, MRI has

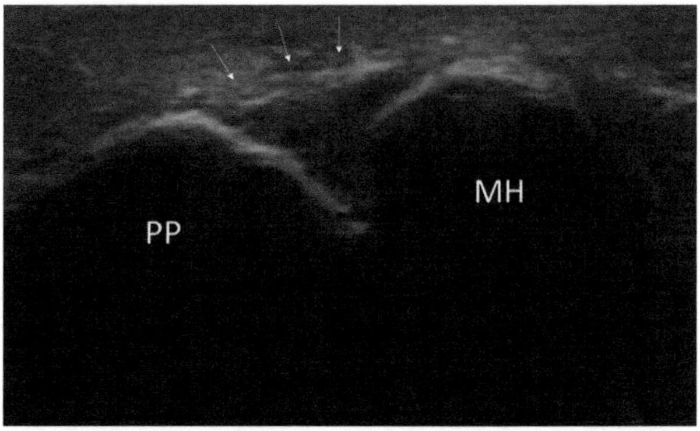

Fig. 14.17 Ultrasound assessment of an intact plantar plate

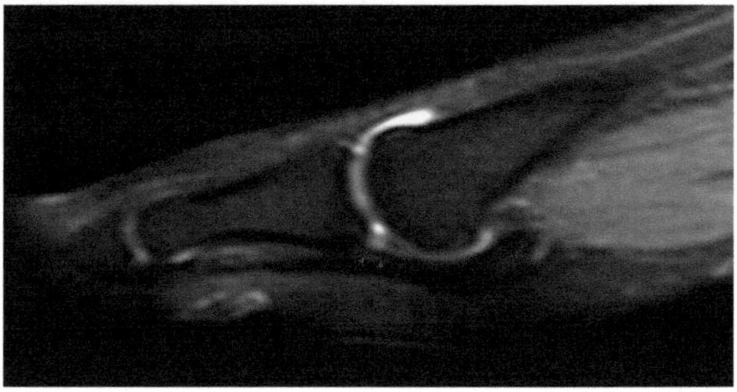

Fig. 14.18 MRI of an intact plantar plate

the ability to assess the degree of ligamentous tear along with other aspects of the plantar plate (Fig. 14.18). Partial-thickness tears may present with focal edema in the plantar plate, indicating a low-grade tear or sprain without complete discontinuity. The triad associated with full-thickness tears in turf toe includes proximal retraction, persistent hyperextension of the proximal phalanx or deviation of the great toe, and discontinuity [71–73].

Key Points

- Hallux valgus
  - Hallux valgus is a common forefoot deformity characterized by medial deviation of the first MT and lateral deviation of the proximal phalanx; it is often progressive with a strong genetic component.
  - History and physical examination findings commonly show pain over the medial eminence, skin changes, and restricted first MTP joint motion.
  - While radiographs remain the gold standard for the evaluation of hallux valgus through measurements like HVA and IMA, other advanced imaging modalities, such as WBCT, are increasingly gaining prominence in diagnosing this pathology.
  - Conservative treatment may be indicated for mild hallux valgus deformities to provide temporary pain relief; however, surgical intervention is often necessary to facilitate a patient's return to normal activity with minimal to no pain.

- Hallux rigidus
  - Hallux rigidus is a degenerative disease of the first MTP joint, characterized by restricted motion and pain. Its etiology is often multifactorial, including repetitive stress, inflammatory conditions, and mechanical/structural abnormalities.
  - Commonly, there is decreased ROM and pain at the first MTP joint. Symptoms usually worsen by activities involving dorsiflexion, such as walking upstairs or running; physical examination often shows dorsal/lateral osteophytes, tenderness at the dorsal MTP joints during palpation, and multiple signs of inflammation.
  - Multiple classification systems have been used to assess severity, with weight-bearing radiographs being the main imaging modality.
  - Conservative treatment modalities involve footwear modification, lifestyle changes, physical therapy, interarticular corticosteroid injections, and OTC pain relievers. Nevertheless, surgical intervention may be necessary for advanced cases, including corrective osteotomies, cheilectomy, or arthrodesis/arthroplasty of the first MTP joint.
- Hallux varus
  - This occurs when the first toe deviates medially at the first MTP joint. The etiology of this condition includes developmental abnormalities, iatrogenic hallux valgus overcorrection, inflammatory conditions, or trauma.
  - Symptoms of hallux varus include pain with shoe wear at the medial and distal aspect of the toe, impaired ambulation, limited ROM, and aesthetic issues with a wide toe-gap. Diagnosis relies on clinical assessment, as symptoms may greatly range; physical examination involves the determination of flexibility, observing any hallucal medialization or supination, and evaluating joint flexibility and integrity.
  - Radiologic assessment includes weight-bearing plain films of the foot. There is no specific classification system to determine the severity of this pathology.
  - Conservative treatment includes footwear modification and taping if detected early. Surgical intervention is based in the flexibility of the deformity and the presence of arthritis at the MTP joint.
- Sesamoiditis
  - Sesamoids are small bones connected to muscles or tendons, commonly found in pairs on the plantar aspect of the distal first MT. Inflammation of the surrounding tissue or bone edema of these structures can cause localized pain at the plantar aspect of the forefoot and is termed sesamoiditis. There is a higher incidence of sesamoiditis in subjects with bipartite sesamoids.
  - Patients often report progressive pain/redness/swelling, limited ROM, and difficulty WB on the affected region, often related to excessive running, dancing, or wearing high heels.
  - Plain radiographs are indicated as part of the initial assessment; however advanced imaging such CT can help distinguish between osteonecrosis and sesamoiditis, and MRI may be useful to evaluate soft tissue pathology in the area.

  - Sesamoiditis is initially treated conservatively with NSAIDs, corticosteroid injections, reduced WB with orthotics, and toe taping; if symptoms persist surgical, options include sesamoidectomy.
- Turf toe
  - Turf toe is defined as a sprain of the plantar capsule ligament due to forceful hyperextension of the first MTP joint. It commonly affects athletes in high-impact sports due to the limited bony stability in the joint. Luckily its incidence has decreased with the use of better field surfaces for athletes.
  - Patients with turf toe typically present with a sudden onset of sharp pain and swelling; physical examination involves evaluating the presence of any deformity and the stability of the first MTP joint. A dorsoplantar/anterior drawer test should be performed to assess the stability of the plantar ligament complex as associated pain or displacement that may indicate a positive test.
  - Advanced imaging such as MRI and US are helpful for diagnosing turf toe injuries, as radiographs are ineffective in visualizing soft tissue injuries. MRI allows accurate assessment of ligamentous tears and additional plantar plate injuries.
  - The treatment for turf toe is based on the severity of the injury, with most of them being treated conservatively.

References

1. Lezak B, Summers S. Anatomy, bony pelvis and lower limb: extensor hallucis longus muscle. In: StatPearls. Treasure Island: StatPearls Publishing; 2024. http://www.ncbi.nlm.nih.gov/books/NBK539875/. Accessed 24 January 2024.
2. Murdock CJ, Munjal A, Agyeman K. Anatomy, bony pelvis and lower limb: calf flexor hallucis longus muscle. In: StatPearls. Treasure Island: StatPearls Publishing; 2024. http://www.ncbi.nlm.nih.gov/books/NBK539776/. Accessed 24 January 2024.
3. Jenkins DB. Hollinshead's functional anatomy of the limbs and back. 9th ed. Amsterdam: Elsevier; 2008. https://shop.elsevier.com/books/hollinsheads-functional-anatomy-of-the-limbs-and-back/jenkins/978-1-4160-4980-7. Accessed 24 January 2024.
4. Coughlin MJ, Jones CP. Hallux valgus: demographics, etiology, and radiographic assessment. Foot Ankle Int. 2007;28(7):759–77. https://doi.org/10.3113/FAI.2007.0759.
5. Kuhn J, Alvi F. Hallux valgus. In: StatPearls. Treasure Island: StatPearls Publishing; 2023. http://www.ncbi.nlm.nih.gov/books/NBK553092/. Accessed 27 June 2023.
6. Nix S, Smith M, Vicenzino B. Prevalence of hallux valgus in the general population: a systematic review and meta-analysis. J Foot Ankle Res. 2010;3:21. https://doi.org/10.1186/1757-1146-3-21.
7. Menz HB, Roddy E, Marshall M, et al. Epidemiology of shoe wearing patterns over time in older women: associations with foot pain and hallux valgus. J Gerontol A Biol Sci Med Sci. 2016;71(12):1682–7. https://doi.org/10.1093/gerona/glw004.
8. Easley ME, Reach JS, Trnka HJ. Chapter 73 - What is the best treatment for hallux valgus? In: Wright JG, Ahn H, Graham B, et al., editors. Evidence-based orthopaedics. Philadelphia: W.B. Saunders; 2009. p. 479–91. https://doi.org/10.1016/B978-141604444-4.50076-X.
9. Louwerens JWK, Schrier JCM. Rheumatoid forefoot deformity: pathophysiology, evaluation and operative treatment options. Int Orthop. 2013;37(9):1719–29. https://doi.org/10.1007/s00264-013-2014-2.

10. Mosca M, Caravelli S, Vocale E, et al. Hallux valgus associated to osteoarthritis: clinical-radiological outcomes of modified SERI technique at mid- to long-term follow-up. A retrospective analysis. Foot Ankle Surg. 2022;28(1):49–55. https://doi.org/10.1016/j.fas.2021.01.012.
11. Golightly YM, Hannan MT, Dufour AB, Renner JB, Jordan JM. Factors associated with hallux valgus in a community-based cross-sectional study of adults with and without osteoarthritis. Arthritis Care Res. 2015;67(6):791–8. https://doi.org/10.1002/acr.22517.
12. Wong DWC, Wang Y, Chen TLW, et al. Finite element analysis of generalized ligament laxity on the deterioration of hallux valgus deformity (bunion). Front Bioeng Biotechnol. 2020;8:571192. https://doi.org/10.3389/fbioe.2020.571192.
13. Belanger L, Burt D, Callaghan J, Clifton S, Gleberzon BJ. Anterior cruciate ligament laxity related to the menstrual cycle: an updated systematic review of the literature. J Can Chiropr Assoc. 2013;57(1):76–86.
14. Goldsmith LT, Weiss G. Relaxin in human pregnancy. Ann N Y Acad Sci. 2009;1160:130. https://doi.org/10.1111/j.1749-6632.2008.03800.x.
15. Jia J, Li J, Qu H, et al. New insights into hallux valgus by whole exome sequencing study. Exp Biol Med. 2021;246(14):1607–16. https://doi.org/10.1177/15353702211008641.
16. Wülker N, Mittag F. The treatment of hallux valgus. Dtsch Ärztebl Int. 2012;109(49):857–68. https://doi.org/10.3238/arztebl.2012.0857.
17. Sferra J, Arndt S. The crossover toe and valgus toe deformity. Foot Ankle Clin. 2011;16(4):609–20. https://doi.org/10.1016/j.fcl.2011.09.001.
18. Rajan L, Kim J, An T, et al. The influence of concomitant hammertoe correction on post-operative outcomes in patients undergoing hallux valgus correction. Foot Ankle Surg. 2022;28(7):1100–5. https://doi.org/10.1016/j.fas.2022.03.008.
19. Silfverskiöld JP. Common foot problems. Relieving the pain of bunions, keratoses, corns, and calluses. Postgrad Med. 1991;89(5):183–8. https://doi.org/10.1080/00325481.1991.11700901.
20. Dittmar JM, Mitchell PD, Cessford C, Inskip SA, Robb JE. Fancy shoes and painful feet: hallux valgus and fracture risk in medieval Cambridge, England. Int J Paleopathol. 2021;35:90–100. https://doi.org/10.1016/j.ijpp.2021.04.012.
21. Ray JJ, Friedmann AJ, Hanselman AE, et al. Hallux valgus. Foot Ankle Orthop. 2019;4(2):2473011419838500. https://doi.org/10.1177/2473011419838500.
22. Cobden A, Camurcu Y, Sofu H, Ucpunar H, Duman S, Kocabiyik A. Evaluation of the association between plantar fasciitis and hallux valgus. J Am Podiatr Med Assoc. 2020;110(2):2. https://doi.org/10.7547/17-150.
23. Canella C. Imaging evaluation of hallux valgus. Radiol Bras. 2020;53(3):7. https://doi.org/10.1590/0100-3984.2020.53.3e2.
24. Schweitzer ME, Maheshwari S, Shabshin N. Hallux valgus and hallux rigidus: MRI findings. Clin Imaging. 1999;23(6):397–402. https://doi.org/10.1016/s0899-7071(00)00167-4.
25. Heineman N, Liu G, Pacicco T, Dessouky R, Wukich DK, Chhabra A. Clinical and imaging assessment and treatment of hallux valgus. Acta Radiol. 2020;61(1):56–66. https://doi.org/10.1177/0284185119847675.
26. Skweres J, Chhabra A, Hummel J, et al. Sesamoid malalignment in hallux valgus: radiographic and MRI measurements and their correlation with internal derangement findings of the first metatarsophalangeal joint. Br J Radiol. 2019;92(1100):20190038. https://doi.org/10.1259/bjr.20190038.
27. Mahmoud K, Metikala S, Mehta SD, Fryhofer GW, Farber DC, Prat D. The role of weightbearing computed tomography scan in hallux valgus. Foot Ankle Int. 2021;42(3):287–93. https://doi.org/10.1177/1071100720962398.
28. Wagner E, Wagner P. Metatarsal pronation in hallux valgus deformity: a review. JAAOS Glob Res Rev. 2020;4(6):e20.00091. https://doi.org/10.5435/JAAOSGlobal-D-20-00091.
29. Kim Y, Kim JS, Young KW, Naraghi R, Cho HK, Lee SY. A new measure of tibial sesamoid position in hallux valgus in relation to the coronal rotation of the first metatarsal in CT scans. Foot Ankle Int. 2015;36(8):944–52.

30. Stewart S, Ellis R, Heath M, Rome K. Ultrasonic evaluation of the abductor hallucis muscle in hallux valgus: a cross-sectional observational study. BMC Musculoskelet Disord. 2023;14:45. https://doi.org/10.1186/1471-2474-14-45.
31. Patel J, Swords M. Hallux rigidus. In: StatPearls. Treasure Island: StatPearls Publishing; 2023. http://www.ncbi.nlm.nih.gov/books/NBK556019/. Accessed 6 July 2023.
32. Coughlin MJ, Shurnas PS. Hallux rigidus: demographics, etiology, and radiographic assessment. Foot Ankle Int. 2003;24(10):731–43. https://doi.org/10.1177/107110070302401002.
33. Lam A, Chan JJ, Surace MF, Vulcano E. Hallux rigidus: how do I approach it? World J Orthop. 2017;8(5):364–71. https://doi.org/10.5312/wjo.v8.i5.364.
34. Di Caprio F, Mosca M, Ceccarelli F, et al. Hallux rigidus: current concepts review and treatment algorithm with special focus on interposition arthroplasty. Acta Bio Med Atenei Parm. 2022;93(5):e2022218. https://doi.org/10.23750/abm.v93i5.12811.
35. Clayton ML, Ries MD. Functional hallux rigidus in the rheumatoid foot. Clin Orthop. 1991;271:233–8.
36. Zammit GV, Menz HB, Munteanu SE. Structural factors associated with hallux limitus/rigidus: a systematic review of case control studies. J Orthop Sports Phys Ther. 2009;39(10):733–42. https://doi.org/10.2519/jospt.2009.3003.
37. Van der Griend R. Correction of Hallux valgus interphalangeus with an osteotomy of the distal end of the proximal phalanx (distal akin osteotomy). Foot Ankle Int. 2017;38(2):153–8. https://doi.org/10.1177/1071100716670389.
38. Roukis TS. Metatarsus primus elevatus in hallux rigidus: fact or fiction? J Am Podiatr Med Assoc. 2005;95(3):221–8. https://doi.org/10.7547/0950221.
39. Gilheany MF, Landorf KB, Robinson P. Hallux valgus and hallux rigidus: a comparison of impact on health-related quality of life in patients presenting to foot surgeons in Australia. J Foot Ankle Res. 2008;1:14. https://doi.org/10.1186/1757-1146-1-14.
40. Parvizi J, Kim GK. Chapter 104 - Hallux rigidus. In: Parvizi J, Kim GK, editors. High yield orthopaedics. Philadelphia: W.B. Saunders; 2010. p. 215–6. https://doi.org/10.1016/B978-1-4160-0236-9.00115-2.
41. Brantingham JW, Wood TG. Hallux rigidus. J Chiropr Med. 2002;1(1):31–7. https://doi.org/10.1016/S0899-3467(07)60025-1.
42. Anderson MR, Ho BS, Baumhauer JF. Current concepts review: hallux rigidus. Foot Ankle Orthopaed. 2018;3:2. https://doi.org/10.1177/2473011418764461.
43. Ho B, Baumhauer J. Hallux rigidus. EFORT Open Rev. 2017;2(1):13–20. https://doi.org/10.1302/2058-5241.2.160031.
44. Fortier LM, Markel M, Thomas BG, Sherman WF, Thomas BH, Kaye AD. An update on peroneal nerve entrapment and neuropathy. Orthop Rev. 2021;13(2):24937. https://doi.org/10.52965/001c.24937.
45. McNeil DS, Baumhauer JF, Glazebrook MA. Evidence-based analysis of the efficacy for operative treatment of hallux rigidus. Foot Ankle Int. 2013;34(1):15–32. https://doi.org/10.1177/1071100712460220.
46. Polzer H, Polzer S, Brumann M, Mutschler W, Regauer M. Hallux rigidus: Joint preserving alternatives to arthrodesis - a review of the literature. World J Orthop. 2014;5(1):6–13. https://doi.org/10.5312/wjo.v5.i1.6.
47. Cheung ZB, Myerson MS, Tracey J, Vulcano E. Weightbearing CT scan assessment of foot alignment in patients with hallux rigidus. Foot Ankle Int. 2018;39(1):67–74. https://doi.org/10.1177/1071100717732549.
48. Lee HY, Mansur NS, Lalevee M, et al. Weightbearing CT analysis of hallux rigidus: does metatarsus primus elevatus really exist? Foot Ankle Orthop. 2022;7(1):2473011421S00306. https://doi.org/10.1177/2473011421S00306.
49. Paczesny ŁM, Kruczyński J. Ultrasound-guided arthroscopic management of hallux rigidus. Videosurg Miniinvasive Tech. 2016;11(3):144–8. https://doi.org/10.5114/wiitm.2016.62811.

50. Munir U, Mabrouk A, Morgan S. Hallux varus. In: StatPearls. Treasure Island: StatPearls Publishing; 2023. http://www.ncbi.nlm.nih.gov/books/NBK470261/. Accessed 11 July 2023.
51. Zirngibl B, Grifka J, Baier C, Götz J. Hallux valgus. Orthopedics. 2017;46(3):283–96. https://doi.org/10.1007/s00132-017-3397-3.
52. Post-operative hallux varus: a review of treatment methods. SpringerLink. https://link.springer.com/article/10.1007/s00264-021-05143-w. Accessed 12 July 2023.
53. Izturov BZ, Abilmazhinov MT, Tsoi OG. Acquired hallux varus. Etiology, pathology, classification, and treatment algorithm. Literature review. Arch Razi Inst. 2022;77(6):2037–48. https://doi.org/10.22092/ARI.2022.360055.2537.
54. Hallux varus | Flushing, NY | Iris E. Cohn, DPM. https://www.iriscohnflushingpodiatrist.com/services/diseases-and-deformities/deformities/hallux-varus/. Accessed 24 July 2023.
55. Hallux varus: classification and treatment - ClinicalKey. https://www-clinicalkey-com.treadwell.idm.oclc.org/#!/content/playContent/1-s2.0-S1083751508000983?returnurl=null&referrer=null. Accessed 23 July 2023.
56. Devos Bevernage B, Leemrijse T. Hallux varus: classification and treatment. Foot Ankle Clin. 2009;14(1):51–65. https://doi.org/10.1016/j.fcl.2008.11.007.
57. Yeung AY, Arbor TC, Garg R. Anatomy, sesamoid bones. In: StatPearls. Treasure Island: StatPearls Publishing; 2023. http://www.ncbi.nlm.nih.gov/books/NBK578171/. Accessed 6 October 2023.
58. Sun T, Wang L, Zhao H, Wu W, Hu W. Prevalence, morphological variation and ossification of sesamoid bones of the forefoot: a retrospective radiographic study of 8,716 Chinese subjects. J Clin Transl Res. 2016;2(3):91–6.
59. Sims AL, Kurup HV. Painful sesamoid of the great toe. World J Orthop. 2014;5(2):146–50. https://doi.org/10.5312/wjo.v5.i2.146.
60. Kaur P, Carroll MR, Stewart S. The assessment and management of sesamoiditis: a focus group study of podiatrists in Aotearoa New Zealand. J Foot Ankle Res. 2023;16:29. https://doi.org/10.1186/s13047-023-00628-w.
61. Benslima N, Mniai EM, Kassimi M, Mahi M. Hallucal sesamoiditis on a bipartite sesamoid bone: an uncommon cause of chronic great toe pain. Radiol Case Rep. 2023;18(9):3357–60. https://doi.org/10.1016/j.radcr.2023.05.045.
62. Heyes G, Swanton E, Mason L, Molloy AP. Does the presence of pes planus deformity increase recurrence of hallux valgus deformity following surgery? Foot Ankle Orthop. 2019;4(4):2473011419S00211. https://doi.org/10.1177/2473011419S00211.
63. Atbaşı Z, Erdem Y, Kose O, Demiralp B, Ilkbahar S, Tekin HO. Relationship between hallux valgus and pes planus: real or fiction? J Foot Ankle Surg. 2020;59(3):513–7. https://doi.org/10.1053/j.jfas.2019.09.037.
64. Knadmin. Pes Cavus. PM&R KnowledgeNow. 2019. https://now.aapmr.org/pes-cavus/. Accessed 24 July 2023.
65. LeBlanc C, Houghton K. Chapter 47: Noninflammatory musculoskeletal pain conditions. In: Cassidy JT, Laxer RM, Petty RE, Lindsley CB, editors. Textbook of pediatric rheumatology. 6th ed. Philadelphia: W.B. Saunders; 2011. p. 697–717. https://doi.org/10.1016/B978-1-4160-6581-4.10047-0.
66. Waldman SD. Chapter 120 - sesamoiditis. In: Waldman SD, editor. Atlas of uncommon pain syndromes. 3rd ed. Philadelphia: W.B. Saunders; 2014. p. 350–2. https://doi.org/10.1016/B978-1-4557-0999-1.00120-9.
67. Sheikh Z. Sesamoiditis. Radiology. Radiopaedia. https://doi.org/10.53347/rID-58858.
68. Aran F, Ponnarasu S, Scott AT. Turf toe. In: StatPearls. Treasure Island: StatPearls Publishing; 2023. http://www.ncbi.nlm.nih.gov/books/NBK507810/. Accessed 24 July 2023.
69. Hallinan JTPD, Statum SM, Huang BK, et al. High-resolution MRI of the first metatarsophalangeal joint: gross anatomy and injury characterization. RadioGraphics. 2020;40(4):1107–24. https://doi.org/10.1148/rg.2020190145.

70. McCormick JJ, Anderson RB. Turf toe. Sports Health. 2010;2(6):487–94. https://doi.org/10.1177/1941738110386681.
71. Filippi J, Bastias GF, Escobar F, et al. Dynamic imaging evaluation in turf toe injuries: a prospective protocol in 18 patients. Foot Ankle Orthop. 2020;5(4):2473011420S00208. https://doi.org/10.1177/2473011420S00208.
72. Tewes DP, Fischer DA, Fritts HM, Guanche CA. MRI findings of acute turf toe. A case report and review of anatomy. Clin Orthop. 1994;304:200–3.
73. Wang JE, Jie BR, Li ZH, et al. High-resolution 3T magnetic resonance imaging and histological analysis of capsuloligamentous complex of the first metatarsophalangeal joint. J Orthop Surg. 2021;16(1):638. https://doi.org/10.1186/s13018-021-02795-7.

Lesser Toe Disorders

15

Jiyong Ahn

15.1 Introduction

Hammer toe and claw toe are the most common lesser toe diseases [1].

A hammer toe is a flexion deformity of the proximal interphalangeal (PIP) joint [1, 2]. In contrast, a claw toe combines a phalangeal hammer toe and an extension deformity of the metatarsophalangeal (MTP) joint [1]. In the case of a subsequent hyperextension of the MTP joint, the definition of the claw toe is not distinguishable from the description of the hammer toe [1, 3]. When it just affects one or two toes and has no other underlying disorders, the term "hammer toe" is commonly used [4]. The term "claw toe" is frequently used when it occurs along with neurological conditions or cavus foot deformity [2, 5]. A cock-up deformity is a different term for when the first toe develops a claw toe [6]. The severity of these lesser toe deformities varies from minor and flexible deformities to stiff and immovable deformities [7].

The proximal and MTP joints' deformities can be addressed to treat the hammer toe [4]. However, it is frequently necessary to correct the cavus foot deformity or muscle imbalance that is the underlying cause of the claw toe in addition to the deformities of the MTP joint and PIP joint [8].

The crossover deformity is a condition when an inward displacement follows the metatarsal joint instability [3]. Metatarsal joint instability refers to the instability of the metatarsal joint being translocated to the dorsum of the foot [7]. It usually occurs in one and two MTP joints accompanied by the hammer toe [9].

J. Ahn (✉)
Foot and Ankle Research and Innovation Lab, Harvard Medical School, Massachusetts General Hospital, Boston, MA, USA

Department of Orthopaedic Surgery, College of Medicine, The Catholic University of Korea, Seoul, Republic of Korea
e-mail: jahn14@mgh.harvard.edu

S. Sharma et al. (eds.), *Clinical and Radiological Examination of the Foot and Ankle*, https://doi.org/10.1007/978-981-97-4202-8_15

15.2 Mallet Toe

Uncertain factors result in flexion deformity of the distal interphalangeal (DIP) joint in mallet toes [4].

The second, third, and fourth toes are equally susceptible to mallet toes, although more frequently, the implicated toe is longer than the nearby toes [10]. Patients with a mallet toe deformity can exhibit tightness of the flexor digitorum longus tendon [11]. Early children may exhibit aberrant flexion of the PIP and DIP joint due to a stiff flexor tendon [12]. This childhood anomaly is described as a "curly toe" [12]. The main symptoms requiring surgical intervention in the adult population include discomfort from pressure on the toe tip, callus development, or dorsal pain over the DIP joint [13] (Figs. 15.1 and 15.2).

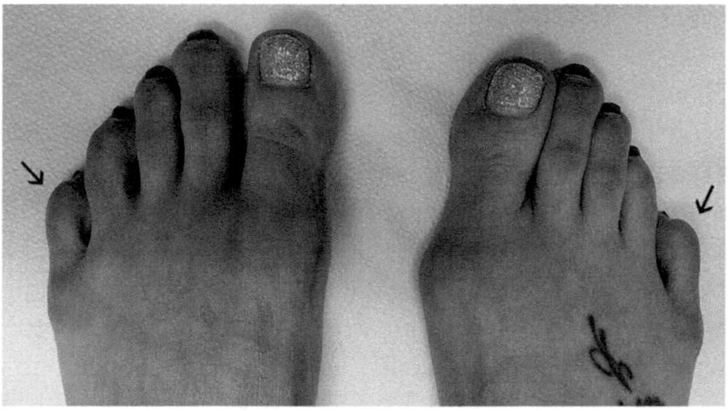

Fig. 15.1 Both fifth mallet toes (black arrow) in a 50-year-old female patient after left hallux valgus correction

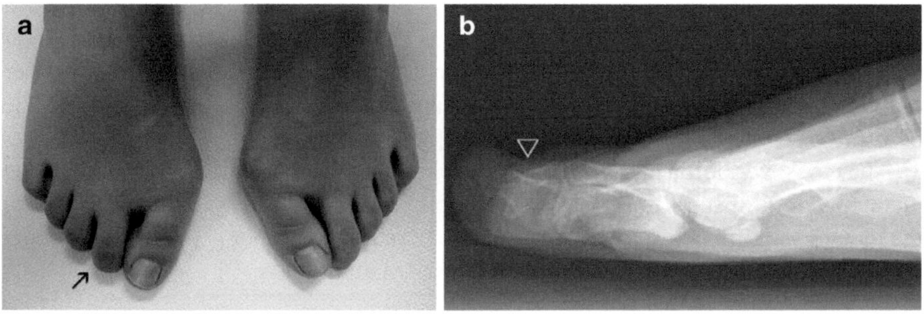

Fig. 15.2 (**a**) Multiple second, third, fourth, and fifth mallet toes (arrow) with hallux valgus in both. (**b**) Distal interphalangeal joint contracture of the second mallet toe (arrow head) on lateral plain radiography

15.3 Hammer Toe

Although the exact cause of the hammer toe is unknown, it is most likely brought on by poorly fitting shoes or a second metatarsal bone that is longer than the other metatarsal bones [14]. When the toes are tightly attached and bent to a narrow-toe box while wearing shoes, deformation of the MTP joint and the toe joint occurs in that state, and this is repeated and gradually proceeds to a fixed deformation [4, 14]. Additionally, hallux valgus, which happens while wearing narrow shoes, can push the second toe over the first toe, cause deformity, and can occur [15]. Generally, the intrinsic muscle of the foot is not working while standing. However, hammer toe deformity is not directly related to the functional anomaly of the intrinsic muscle because it is deteriorated while standing [16].

Hammer toe deformity can be divided into flexible deformation, which can passively be corrected to the neutral level, and fixed deformation, which cannot be corrected [16]. If the flexible deformation of the PIP joint is severe and prolonged, the elongated deformation of the MTP joint appears [14]. DIP joints are usually flexible but can also cause flexion or extension deformity [1]. The base of the metatarsal head, the dorsum of the PIP joint, and the tip of the toe may all be painful in the case of the hammer toe [17]. When the dorsum of the PIP joint comes into contact with the shoe, a callus is created [18]. This is known as the "corn" phenomenon and occurs when the tip of the toe presses the bottom of the shoe [18]. Ulcers and infections can occur in these areas if the senses are impaired, such as in diabetes or myelomeningoceles [19] (Fig. 15.3).

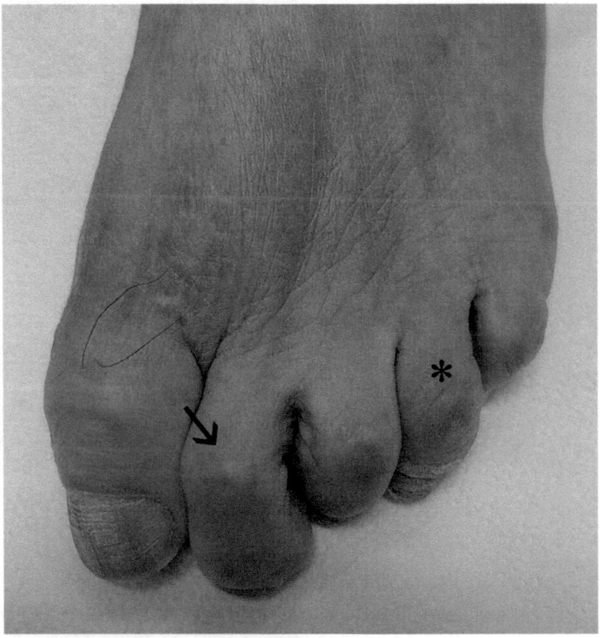

Fig. 15.3 Hammer toe deformity (arrow) with corn on the fourth proximal phalangeal joint (asterisk)

15.4 Claw Toe

There are numerous causes of claw toe deformity (Figs. 15.4, 15.5, and 15.6).

Claw toes are linked to arthritic disorders, cavus foot deformity, neurologic conditions with muscular imbalance, and the aftereffects of trauma to the lower extremities [20].

Additionally, during walking, the equinus deformity of the ankle can result in hyperextension of the MTP joint, which results in the claw toe deformity [21].

In 9% of cases, the third peroneus tertius, which attaches to the center of the fifth metatarsal bone on the outside of the extensor digitorum longus (EDL), is absent. If this is the case, the inner anterior tibial tendon is used to restore balance [22].

The absence of the third peroneus tertius causes the foot to invert, and theoretically, EDL causes hyperextension of the PIP joint, which results in claw toe [23].

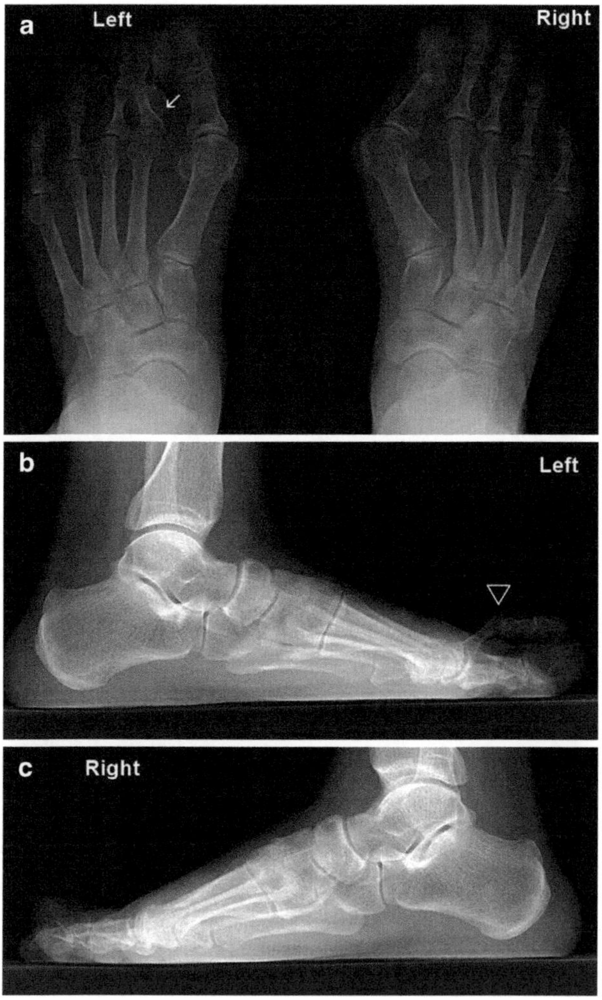

Fig. 15.4 Second claw toe deformity. (**a**) Second metatarsophalangeal (MTP) joint subluxation (arrow) with hallux valgus on an anteroposterior (AP) view of plain radiography. (**b**) Hyperextension of the left second MTP joint on a lateral view of weight-bearing X-ray. (**c**) Well-maintained right second MTP joint on a lateral view of weight-bearing X-ray

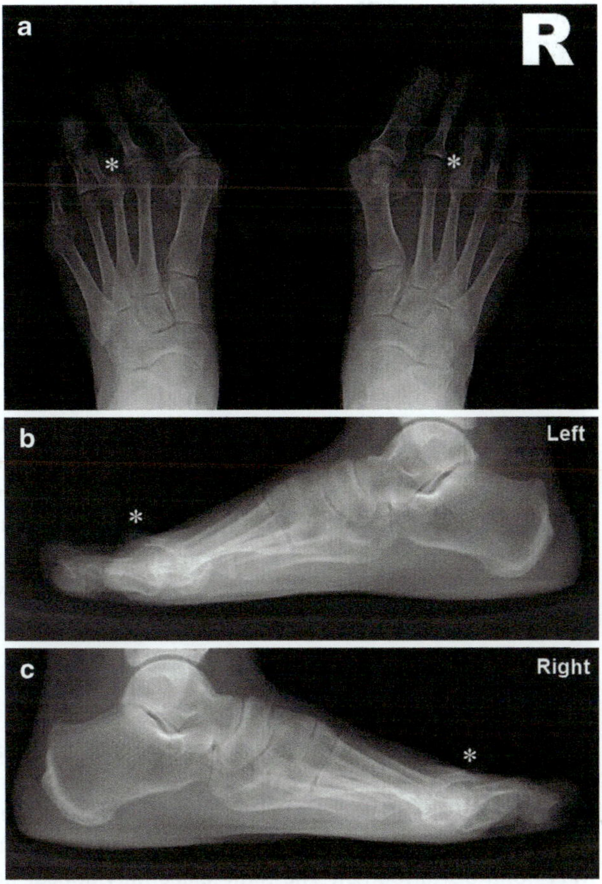

Fig. 15.5 Both third claw toes in a 61-year-old female patient. (**a**) Third MTP joint instability with dislocation of both MTP joints on AP plain radiography. (**b**, **c**) Contracture of PIP joint (asterisk) with hyperextension of MTP joint on lateral plain radiography

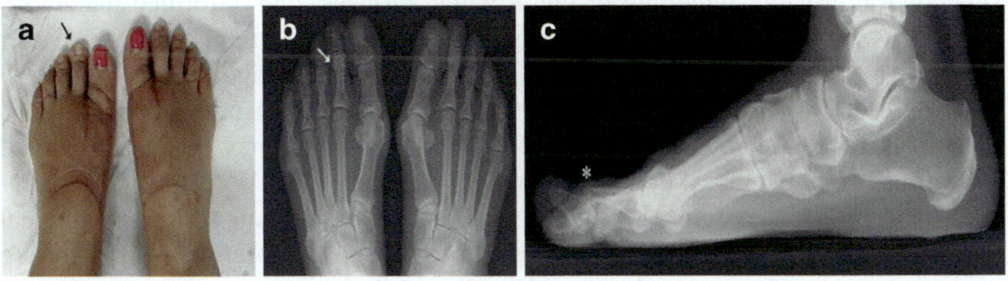

Fig. 15.6 Second hammer toes with both mild hallux valgus of a 61-year-old female patient. (**a**) Photograph of the left second hammer toe (arrow). (**b**) Left second hammer toe with blunting PIP joint on AP plain radiography. (**c**) PIP joint flexion (asterisk) without hyperextension of MTP joint on lateral plain radiography

15.5 General Physical Examination

It's crucial to assess the circulation of the lower extremities while inspecting feet with lesser toe disorders [8]. It's important to distinguish between patients' pain and neuropathic pain brought on by systemic illnesses including diabetes mellitus (DM) and spinal abnormalities [24, 25]. Similar to interdigital neuritis, inflammation of the digital nerve caused by compression requires thorough evaluation [13]. Precise attention should be taken from the toe tips to the medial and lateral portions of the joints if the toes demonstrate callosity [9]. Location of toe callosities is crucial when discussing toe flexion deformities like the hammer because deformity caused by joint contracture can result in callosities or calluses in any region of the toes with toe malalignment, such as deviation of the toe [3]. To differentiate between a claw toe and a hammer toe, the MTP joint should be carefully examined for hyperextension [26]. Analyzing the variables influencing foot posture when bearing weight is crucial. It is important to check to determine if the MTP joint and contraction change as patients bear weight [5]. Due to localized forefoot pressure concentration, tight Achilles tendons can also cause this [27]. The evaluation of the flexor digitorum tendon should be conducted to examine flexion deformity to predict postoperative alterations since it is generally significantly associated with callus or callosity [27].

15.6 General Radiographic Examination

To assess the severity of the lesser toe deformity, plain radiography is mainly necessary. Since they overlap, it can be challenging to determine whether the proximal interphalangeal (PIP) and MTP joints are deformed in plain radiography—severe deformity results in improper joint space, shorter-looking toes, or medial and lateral MTP joint deviation [2]. The joint space may appear narrow, especially if the subluxation or dislocation of the joint is chronic [28] (Fig. 15.6). In plain radiography, subchondral erosion, flattening of the articular surfaces, and deformed metatarsal heads caused by Freiberg infraction may all be present [29]. In general, it is easier for a lateral view to observe the deformation of an interphalangeal joint than an AP (anteroposterior) view. Depending on the type of lesion, ultrasound or MRI can be used to assess the state of soft tissue [30]. Bony structures may be evaluated with CT; in rare circumstances, bone scans may be necessary [8]. With the aid of pedobarography, the foot pressure can be assessed to help identify sites that need surgical intervention if the callosity is widespread throughout the forefoot or if there are many locations with varying degrees [31] (Fig. 15.7).

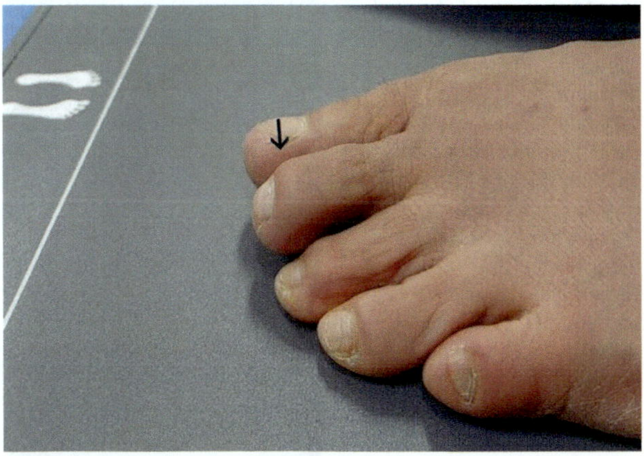

Fig. 15.7 Photograph of pedobarography to assess forefoot pressure on the second toe (arrow)

15.7 Metatarsophalangeal Joint Instability

There may be only instability in one direction toward the dorsum of the MTP joint, and there may be a crossover deformation in which the second toe is deformed medially and overlaps the first toe [3, 7, 26] (Fig. 15.8). This symptom is similar to severe deformation of the hammer toe because the over-stretching of the MTP joint is the leading cause [2]. Still, it differs in that the stability of the MTP joint is often restored only by shortening the second metatarsal bone during treatment [32]. It can be said that the severe deformation of the hammer toe has progressed, but the instability of the MTP joint is the most crucial point, whether there is a deformation of the PIP joint [7, 17].

It is common in the second MTP joint, especially in the second MTP joint, because (1) the second metatarsal is often longer than other metatarsal bones, (2) the first metatarsal bone forms a transverse axis, and (3) the second to fifth metatarsal bone forms an oblique axis so that a strong force can be applied at the intersection of the two axes during the push-off stage of the gait cycle [1, 7, 17, 26, 33]. Due to instability, active inflammation of the synovium and joint occurs at first, and it is an ordinary course to progress to subluxation and dislocations [34].

15.7.1 Etiology

The most common causes of this disease are (1) hyperextension of the MTP joint and (2) weakening and rupture of static structures such as collateral ligaments and plantar plates. In addition, (3) ruptured joint capsule and ligaments due to acute trauma, resulting in acute instability, and (4) chronic active inflammation caused by systemic arthritis may cause instability [1, 3, 7, 17, 33].

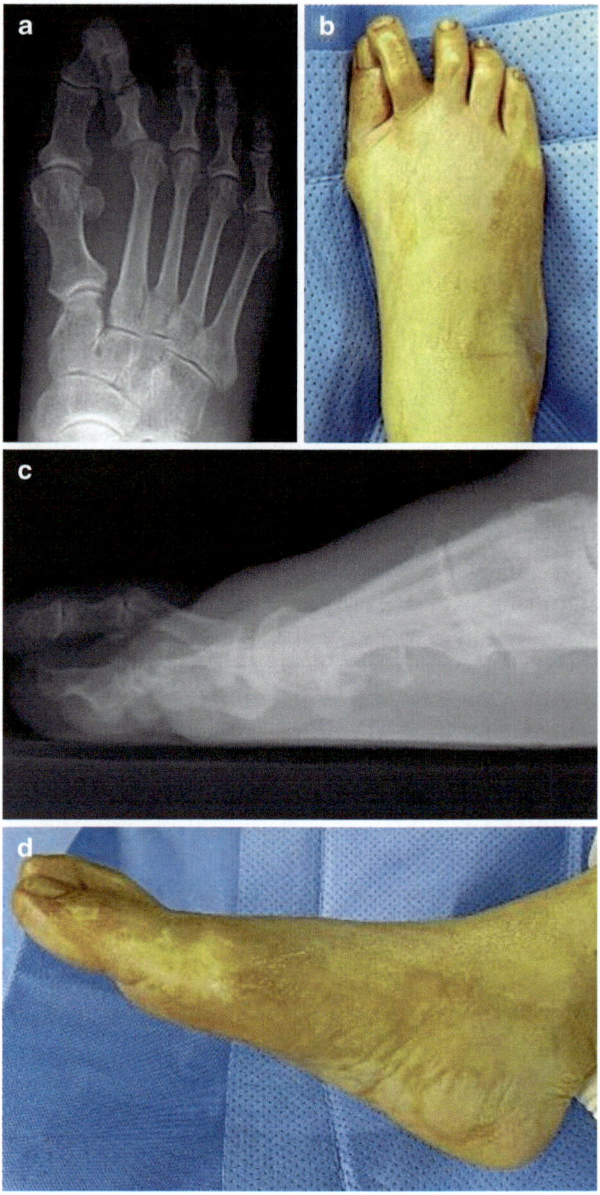

Fig. 15.8 Second claw toe deformity. (**a**) Second toe subluxation with medial deviation of metatarsophalangeal (MTP) joint on the anteroposterior (AP) view of plain radiography. (**b**) Second toe overriding on the first toe. (**c**) Hyperextension of the second MTP joint on a lateral view of weight-bearing X-ray. (**d**) Reduced hyperextension of the second MTP join on a non-weight-bearing state

15.7.1.1 Hyperextension of the MTP Joint

Flexor digitorum longus (FDL) has the function of extending the proximal phalanges only when the MTP joint is neutral or flexed [35]. In a state where the MTP joint is extended, FDL increases the hyperextension of the MTP joint instead of the role of extending the proximal phalanges and serves as a force to subluxate or dislocate

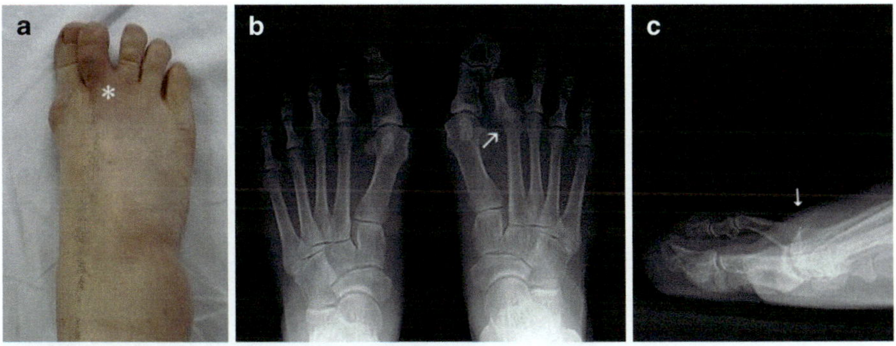

Fig. 15.9 Instability of the second MTP joint. (**a**) Photograph of the second MTP joint hyperextension (asterisk). (**b**, **c**) Second MTP joint dislocation (arrow) on plain radiography

the MTP joint [12, 35, 36]. Intrinsic muscles play an essential role as dynamic structures that resist hyperextension of these MTP joints, and static structures include plantar plates and collateral ligaments [27].

However, the interosseous muscle among the intrinsic muscles that prevent hyperextension of the MTP joint is placed in the dorsum of the flexion-extension axis when the MTP joint is hyperextended and loses the flexural function of the MTP joint [27]. Static structures prevent subluxation or dislocation, and if this state persists chronically, static structures expand and rupture, leading to subluxation and dislocation [36]. Flexor digitorum longus and brevis flexed the phalanges, and the action of flexion of the MTP joint is weak [12, 36]. However, when the MTP joint is hyperextended, the tension of the flexion tendons on the plantar side naturally increases [37]. The flexion increases, resulting in flexion deformation of the proximal phalanges [28].

For the above reasons, when the MTP joint is hyperextended, the FDL acts as a force to subluxate, and the intrinsic muscle loses the function of flexing the MTP joint; thus deformity gradually increases [11, 19, 35, 36] (Fig. 15.9).

15.7.1.2 The Wear and Tear of a Static Structure

The static structures are worn out due to the chronic progression of hyperextension in the MTP joint [28]. However, if the second metatarsal bone is long, the pressure under the second metatarsal head increases, and the plantar plate may gradually wear out [34, 38, 39] (Fig. 15.10).

Women usually wear shoes with narrow-toe boxes and high heels, but shoes with high heels cause the MTP joint to be hyperextended, causing the pathologic condition. In other words, if the second metatarsal bone is long, has high heels, and wears shoes with a narrow-toe boxes, such lesions are most likely to occur [5, 40]. However, it does not occur only in middle-aged women who have been wearing shoes with high heels and narrow-toe boxes for a long time, but it can also be caused by overuse in young people. Instability can result from the hallux valgus deformity, which is connected to the MTP joint's active inflammation [15, 41].

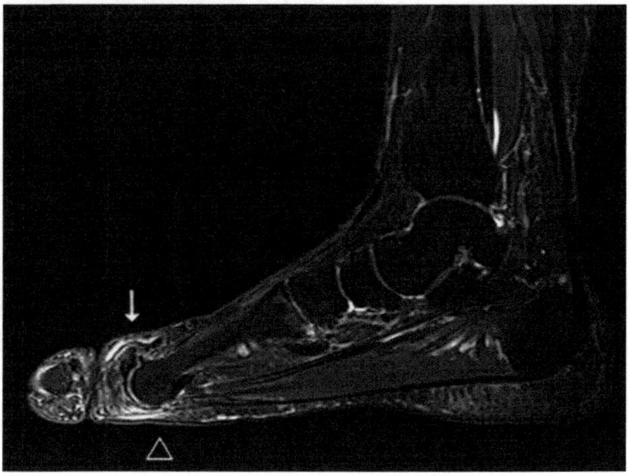

Fig. 15.10 Second MTP joint instability on MRI. Dorsal dislocation of the second proximal phalanges (arrow) and plantar plate rupture (arrow head) on a T2-weighted sagittal image

15.7.2 Clinical Symptoms

High-heeled shoes first cause pain, which most people report progressing slowly and being chronic. Additionally, the activation of the interdigital nerve in the second web space may cause pain in the second and third phalanges. In addition to women, men also have MTP joint instability that is unrelated to wearing high heels [42]. When the joint is subluxated or dislocated, there is acute pain and swelling at first, but over time, there is swelling of the second metatarsal joint, although the pain may not be severe [26].

15.7.3 Physical Examinations

It may be accompanied by a hammer toe deformity, although it may not be. Due to the swelling of the joint, the form of the extensor tendon in the center of the MTP joint is not clearly evident [1, 2]. There is extensive tenderness in the MTP joint, and the range of joint motion is less than normal [17].

Similar to the method of examining anterior cruciate ligament rupture of the knee joint, the degree of instability can be determined by holding the toe and performing a drawer test that moves toward the dorsum and plantar toe, and most of these are accompanied by severe pain [3] (Fig. 15.11).

The results of the drawer test are classified as follows: (1) the degree to which it is not dislocated at all by the dorsum of the foot is called stage 0, (2) the degree to which it is subluxated but not dislocated is called stage 1, (3) the degree to which it can be reduced is called stage 2, and (4) the state to which it is dislocated is called stage 3. Examine the degree to which the proximal phalanges dislocate from the MTP joint to the foot dorsum [3].

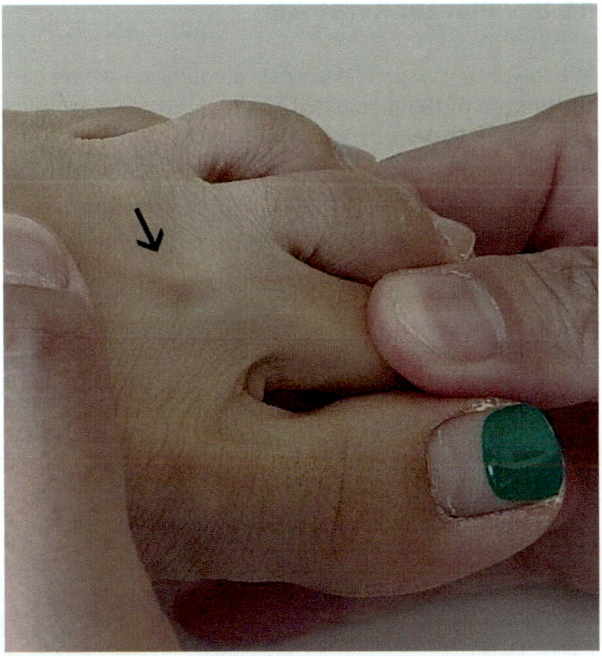

Fig. 15.11 Anterior drawer test determining how much the proximal phalanges have dislocated (arrow) from the metatarsophalangeal joint to the dorsum

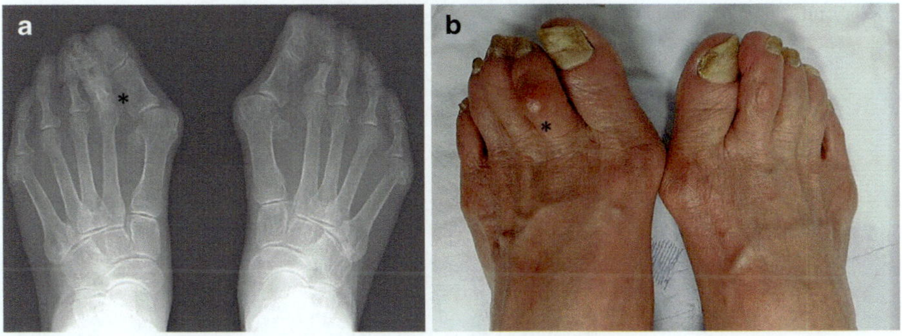

Fig. 15.12 Second claw toes with both severe hallux valgus of an 81-year-old patient. (**a**) Left second claw toe with subluxation on AP plain radiography. (**b**) Photograph of the left second claw toe with hard corn on the PIP joint

15.7.4 Radiographic Evaluation

Plain radiography should be used to assess joint congruency, the relative length of the metatarsal bone, the presence of degenerative arthritis, and the severity of the deformity [43]. The AP view of plain radiography revealed a narrowing of the joint space as well as an overlap of the proximal phalanges and the metatarsal head [43, 44] (Fig. 15.12).

15.8 Plantar Keratosis

Whatever the cause, it describes a condition where callosity and pain are present on the bottom of the metatarsal head, and it often involves a hammer or claw toe [45]. Mann and DuVries called intractable plantar keratosis (IPK) that continuously develops callosity in the plantar aspect of metatarsal head or sesamoid bone [46].

Skin keratosis develops in reaction to pressure and may be widespread or limited to small, localized areas, depending on the cause [46] (Fig. 15.13).

15.8.1 Anatomical Factors [47, 48]

(1) When the first metatarsal-cuneiform joint is hypermobility or (2) the first metatarsal bone is short, the weight load of the first metatarsal head is reduced than normal, so the load increases under the second, third, and fourth metatarsal head, resulting in wide calluses. (3) The length of the first metatarsal bone is normal, but if the second metatarsal bone is long, the weight load increases under the second

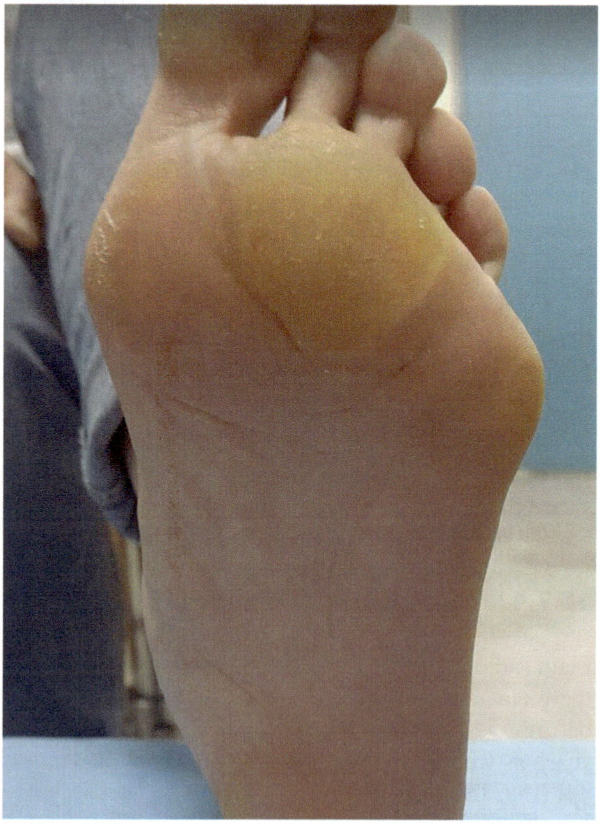

Fig. 15.13 Intractable plantar keratosis. Widespread callosity across the two and three MTP joints

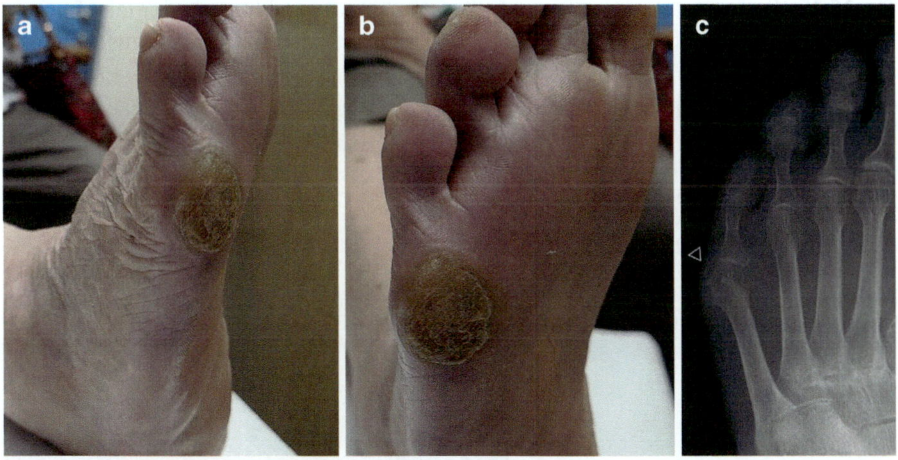

Fig. 15.14 IPK-related bunionette. (**a**, **b**) Plantar-lateral large callosity. (**c**) Fifth bunionette with subluxation of MTPJ (arrow head)

metatarsal head, resulting in calluses. (4) It also occurs in the case where the lateral condyle of the metatarsal head protrudes a lot toward the plantar aspect and (5) in the area where the bunionette is located. (6) If the first metatarsal bone is flexed or the second metatarsal bone is short, calluses occur at the bottom of the sesamoid bone under the first metatarsal head, especially the medial sesamoid bone. (7) Calluses also occur when a sesamoid bone is at the first interphalangeal joint (Fig. 15.14).

15.8.2 Secondary Keratosis [49]

(1) After a metatarsal fracture, the distal bone fragment may occasionally flex at the fracture site, or (2) following hallux valgus correction, transfer lesions may develop secondary to dorsal malunion of the first metatarsal bone. Additionally, tumors like epidermal inclusion cysts, which result in plantar keratosis, may develop beneath the metatarsal head (Fig. 15.15).

15.8.3 Physical Examination

Hypermobility of the first MTP joint, as well as a localized or expanding lesion, should be recognized during a physical examination [50, 51]. Localized lesions can develop in three situations: (1) when the medial sesamoid bone of the first metatarsal head protrudes markedly, (2) when the lateral condyle of the metatarsal head protrudes significantly, and (3) when the position of the sesamoid bone shifts following hallux valgus surgery [3, 7, 51].

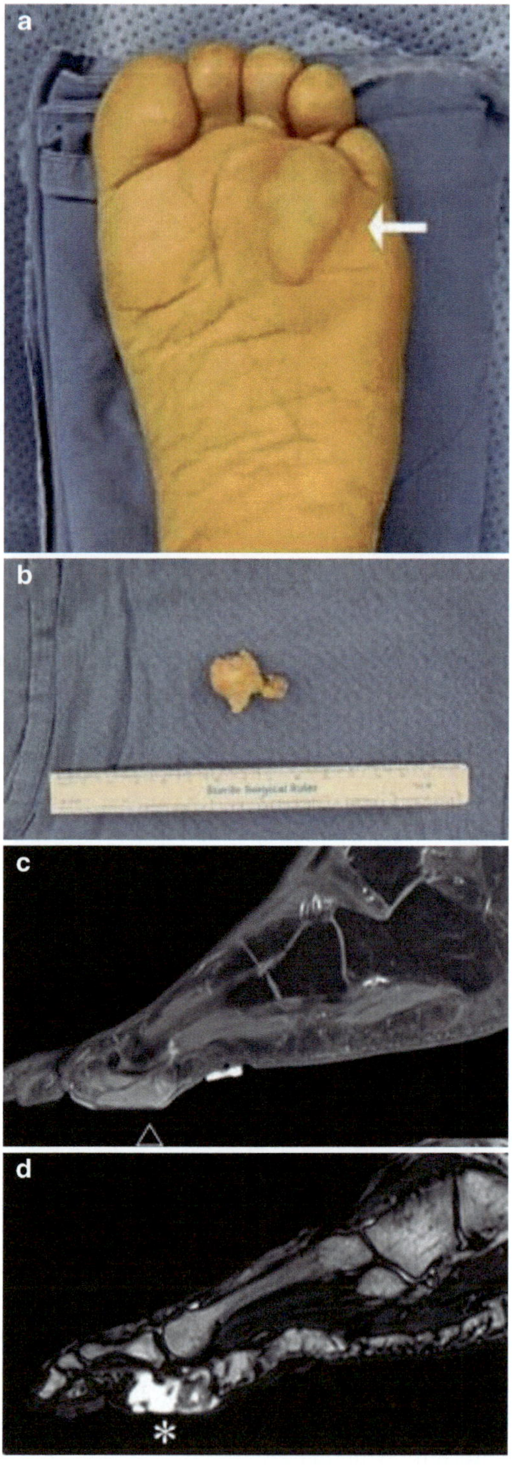

Fig. 15.15 Benign adnexal tumor-related IPK. (**a**) Photograph of large plantar callosity (arrow). (**b**) Intraoperative finding of excisional biopsy. (**c, d**) Space-occupying lesion on low signal intensity of (arrow head) T1-weighted MRI (**c**) and high signal intensity of (asterisk) T2-weighted MRI (**d**)

Attaching metallic objects to areas with calluses and taking weight-bearing radiographs and an axial view of the metatarsal head helps determine which areas are responsible for calluses.

It is vital to distinguish it from warts, which are normally held with the thumb and index finger. The pain is worse when pressure is applied to the side, intractable plantar keratosis is not severe when pressure is applied to the side, and when cut with a surgical blade, there is spot bleeding, but calluses do not bleed [45].

15.9 Corns

15.9.1 Causes and Clinical Symptoms

Hard and soft skin corns are two different types of corns, which are hyperkeratotic lesions that develop in the stratum corneum (horn layer) of the skin [52]. Due to the typical reaction of the skin appearing where a lot of pressure is applied, it is known as skin corn when it develops on the toe and callosity when it does so on the plantar aspect of the foot [45, 53].

15.9.1.1 Hard Corn

The contact of the condyle of the phalangeal bone with the toe box of the shoe causes a hard corn [54]. It mainly occurs at the lateral side of the proximal phalanges of the fifth toe and is hard and tender [55]. When acute and severe irritation is applied to the skin corn, redness and local fever may occur around it [45, 52, 55]. It needs to be differentiated from warts [45] (Fig. 15.16).

15.9.1.2 Soft Corn

The same factors that lead to hard corn can cause soft corn, though interdigital corn is a more appropriate term since it refers to those between the toes [45, 46].

It's known as soft corn. Because it is continuously humid between the toes, it is soft in comparison to hard corn [56] (Fig. 15.17).

There are two types of corn between the toes [54, 56]: (1) The most common occurrence is in the base of the distal bone of the short toe (in the case of the fourth to fifth toes) and the corresponding area of the proximal bone of the long toe. Sometimes, the toe nail grows strangely, which can cause corn. (2) More than this, what occurs between the toes of the proximal part occurs when the lateral side of the proximal bone of the fourth toe and the head of the proximal bone of the fifth toe contact. In this case, knowing the protruding part by touching or seeing it is difficult [53].

The corn occurs between the toes for various reasons in various parts, and there are relatively many occurrences between the first and second toes due to the hallux valgus deformity and between the third and fourth limbs in the curly toe [48, 49, 57].

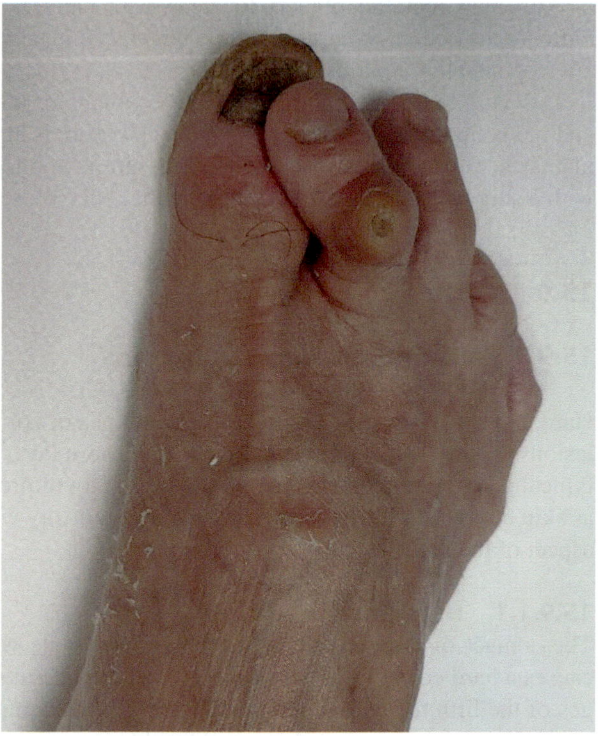

Fig. 15.16 Photograph of a hard corn on the second PIP joint

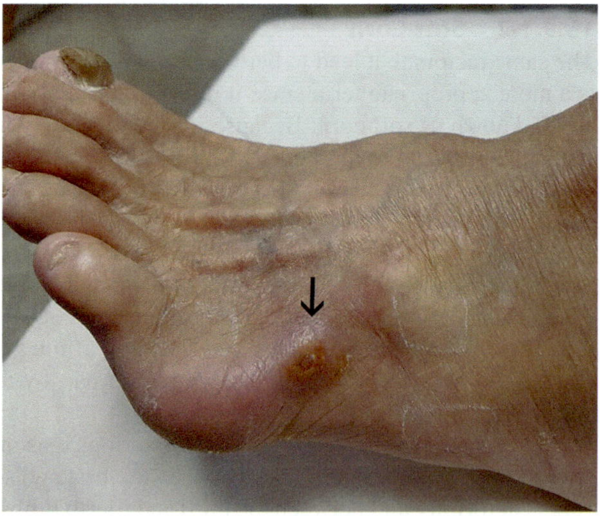

Fig. 15.17 Soft corn on the fifth MTP joint dorsolateral aspect

15.10 Bunionette

Since the lateral side of the fifth metatarsal head is in direct contact with the shoes, there is a bunionette, which is a disease similar to hallux valgus deformity, and it is treated in many ways similar to hallux valgus deformity [58–60]. The distinction from hallux valgus is that in hallux valgus deformity, symptoms develop on the medial first metatarsal joint, but in bunionette, symptoms appear on the lateral aspect and slightly on the plantar aspect [58, 61].

The fifth metatarsal head is also pressed against the floor and shoes because the width of the foot increases if there is hallux valgus deformity, which is often accompanied by hallux valgus deformity, and it occurs a lot in the splay foot with a wide foot width [61–63].

During a physical examination, paying attention to the hyperkeratotic lesions near the fifth metatarsal head is crucial [62]. However, the precise location should be verified by palpation. Symptoms typically manifest at the fifth metatarsal head's most projecting portion [60]. Laterally, dorsolaterally, and plantarly manifesting discomfort to the fifth metatarsal head is how patients typically describe their pain [58, 60, 64, 65]. Frequently, this area will also exhibit erythema and edema. Determining surgical treatment options requires accurate localization of the tenderness [60–62, 66] (Fig. 15.18).

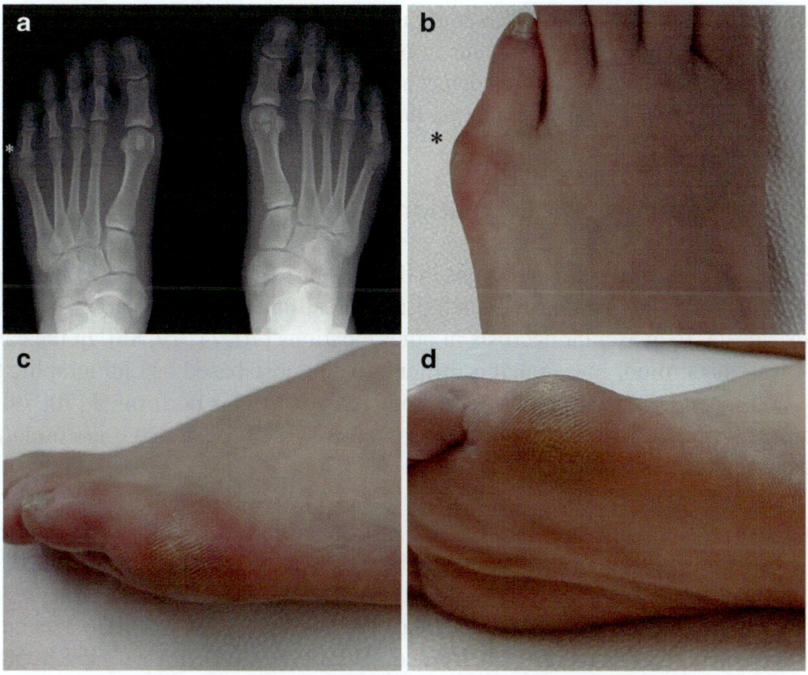

Fig. 15.18 Symptomatic bunionette. (**a**) Soft tissue swelling (asterisk) with congruent fifth MTP joint on AP plain radiography. (**b**) Callosity on lateral protrusion (asterisk) of the fifth MTP joint. (**c**, **d**) Callosity on plantar-lateral aspect of the fifth MTP joint

15.11 Freiberg Infraction

The ratio of female to male is 5:1, as only osteochondrosis in females is more than in males [67]. It is unclear whether there is true avascular necrosis in the second metatarsal head despite it being categorized as a kind of osteochondrosis [68–72]. Knowing the precise etiology of the disease is challenging because patients receiving treatment already have metatarsal head deformity, secondary degenerative arthritis, and loose bodies in the joint [70, 71, 73]. Even patients without symptoms may accidentally take radiographs to examine other symptoms or trauma, so it is difficult to know the exact frequency because severe deformation is found [68]. It is also difficult to know the exact time of onset because there are no symptoms in the early stages, but it occurs most frequently between the ages of 11 and 17 because symptoms of degenerative arthritis may appear later [74]. On the contrary, many patients experience symptoms prior to the onset of radiologic abnormalities [75–77]. Approximately 68–82% occur in the metatarsal head, and 6.6% are bilateral [78–80]. Symptoms include pain on the weight-bearing, local tenderness of the MTP joint, limited range of motion, and edema [68, 70].

15.11.1 Physical Examination

Palpation can result in discomfort and tenderness, most frequently over the second or third metatarsal head [68, 73]. Symptoms can be mistaken for just the usual metatarsalgia because they initially occur in one metatarsal bone [70, 81, 82]. Extensor tendon shape becomes difficult to detect as the symptoms worsen [78]. Stiffness and edema in the joints may occur, making degenerative changes visible [83].

15.11.2 Radiographic Evaluation

There are sclerosing cystic alterations in the metatarsal head, loose body and flattening of the metatarsal head, subchondral fractures, flattening, and larger metatarsal heads, similar to avascular necrosis in other locations [70, 71, 79, 84]. In the case of early lesions or minor invasion, it is difficult to diagnose based on radiographic findings, and it can be known only when a bone scan or MRI is performed [70, 79, 84]. Compared to radiographic findings, more severe cartilage damage is common in surgical findings [75, 82, 85].

Key Points

- Lesser toe disorder-related metatarsalgia should be thoroughly evaluated with physical examination and radiographic evaluation.
- Understanding the etiology of metatarsalgia in lesser toe disorder depends on where the callus or callosity is located. The possibilities for surgical treatment are directly related to the precise location of tenderness.

- Systemic disorder, Achilles tendon tightness, and lumbar problems should be ruled out to diagnose lesser toe disorders correctly.
- Clinically, instability of the metatarsophalangeal joint should be evaluated in the coronal and sagittal plane in lesser toe disorders.
- The location of the callosity does not always coincide with the symptoms in lesser toe disorder. Before making a diagnosis, the location of tenderness must be accurately identified and checked to see if it matches radiographic findings.

References

1. Coughlin MJ. Lesser toe abnormalities. Instr Course Lect. 2003;52:421–44.
2. Harrasser N, Toepfer A, Lenze F, et al. Lesser toe deformities: what the general practitioner should know. MMW Fortschr Med. 2015;157:43–4. https://doi.org/10.1007/s15006-015-3217-9.
3. Kinter CW, Hodgkins CW. Lesser metatarsophalangeal instability: diagnosis and conservative management of a common cause of metatarsalgia. Sports Health. 2020;12:390–4. https://doi.org/10.1177/1941738120904944.
4. Smith BW, Coughlin MJ. Disorders of the lesser toes. Sports Med Arthrosc Rev. 2009;17:167–74. https://doi.org/10.1097/JSA.0b013e3181a5cd26.
5. Doty JF, Coughlin MJ, Weil L, Nery C. Etiology and management of lesser toe metatarsophalangeal joint instability. Foot Ankle Clin. 2014;19:385–405. https://doi.org/10.1016/j.fcl.2014.06.013.
6. Lui TH. Arthroscopy and endoscopy of the foot and ankle: indications for new techniques. Arthroscopy. 2007;23:889–902. https://doi.org/10.1016/j.arthro.2007.03.003.
7. Peck CN, Macleod A, Barrie J. Lesser metatarsophalangeal instability: presentation, management, and outcomes. Foot Ankle Int. 2012;33:565–70. https://doi.org/10.3113/FAI.2012.0565.
8. Reddy SC. State of the art in lesser metatarsophalangeal instability. Foot Ankle Clin. 2019;24:627–40. https://doi.org/10.1016/j.fcl.2019.08.007.
9. Park CH, Chang MC. Forefoot disorders and conservative treatment. Yeungnam Univ J Med. 2019;36:92–8. https://doi.org/10.12701/yujm.2019.00185.
10. Molloy A, Shariff R. Mallet toe deformity. Foot Ankle Clin. 2011;16:537–46. https://doi.org/10.1016/j.fcl.2011.08.004.
11. Schmitz P, Scheffer R, De Gier S, et al. The effect of percutaneous flexor tenotomy on healing and prevention of foot ulcers in patients with claw deformity of the toe. J Foot Ankle Surg. 2019;58:1134–7. https://doi.org/10.1053/j.jfas.2019.03.004.
12. Jacobs R, Vandeputte G. Flexor tendon lengthening for hammer toes and curly toes in paediatric patients. Acta Orthop Belg. 2007;73:373–6.
13. DiPreta JA. Metatarsalgia, lesser toe deformities, and associated disorders of the forefoot. Med Clin North Am. 2014;98:233–51. https://doi.org/10.1016/j.mcna.2013.10.003.
14. Scheck M. Etiology of acquired hammertoe deformity. Clin Orthop Relat Res. 1977;123:63–9.
15. Roan L-Y, Tanaka Y, Taniguchi A, et al. Why do lesser toes deviate laterally in hallux valgus? A radiographic study. Foot Ankle Int. 2015;36:664–72. https://doi.org/10.1177/1071100715573051.
16. Schrier JCM, Verheyen CCPM, Louwerens JW. Definitions of hammer toe and claw toe: an evaluation of the literature. J Am Podiatr Med Assoc. 2009;99:194–7. https://doi.org/10.7547/0980194.
17. Mizel MS, Yodlowski ML. Disorders of the lesser metatarsophalangeal joints. J Am Acad Orthop Surg. 1995;3:166–73. https://doi.org/10.5435/00124635-199505000-00006.

18. Oliver TP, Armstrong DG, Harkless LB, Krych SM. The combined hammer toe-mallet toe deformity with associated double corns: a retrospective review. Clin Podiatr Med Surg. 1996;13:263–8.
19. Scott JE, Hendry GJ, Locke J. Effectiveness of percutaneous flexor tenotomies for the management and prevention of recurrence of diabetic toe ulcers: a systematic review. J Foot Ankle Res. 2016;9:25. https://doi.org/10.1186/s13047-016-0159-0.
20. Kaz AJ, Coughlin MJ. Crossover second toe: demographics, etiology, and radiographic assessment. Foot Ankle Int. 2007;28:1223–37. https://doi.org/10.3113/FAI.2007.1223.
21. Mandiroglu S, Firat M. Foot and ankle deformities in stroke: relationship with ambulation, balance, and daily living activities: a cross-sectional study. Somatosens Mot Res. 2022;39:106–10. https://doi.org/10.1080/08990220.2021.2018294.
22. Ellis H. Morphology of peroneus tertius muscle. Clin Anat. 2007;20:230. https://doi.org/10.1002/ca.20449.
23. Brody PJ, Grumbine N. Peroneus tertius reconstruction for flexible clawtoes associated with cavus deformity: a preliminary report. J Foot Surg. 1984;23:357–61.
24. Bellemans M, de Somerhausen N, Lê PQ. Osteoid osteoma: a unique presentation in a child's lesser toe. Case Rep Orthop. 2021;2021:8876584. https://doi.org/10.1155/2021/8876584.
25. Roukis TS. Lesser toe salvage with external fixation and autogenous bone grafting: a case series. Foot Ankle Spec. 2010;3:108–11. https://doi.org/10.1177/1938640010363322.
26. Hsu RY, Barg A, Nickisch F. Lesser metatarsophalangeal joint instability: advancements in plantar plate reconstruction. Foot Ankle Clin. 2018;23:127–43. https://doi.org/10.1016/j.fcl.2017.09.009.
27. Keenan MA, Gorai AP, Smith CW, Garland DE. Intrinsic toe flexion deformity following correction of spastic equinovarus deformity in adults. Foot Ankle. 1987;7:333–7. https://doi.org/10.1177/107110078700700604.
28. Hulstaert T, Shahabpour M, Provyn S, et al. Forefoot pain in the lesser toes: anatomical considerations and magnetic resonance imaging findings. Can Assoc Radiol J. 2019;70:408–15. https://doi.org/10.1016/j.carj.2019.06.010.
29. Seybold JD, Zide JR. Treatment of Freiberg disease. Foot Ankle Clin. 2018;23:157–69. https://doi.org/10.1016/j.fcl.2017.09.011.
30. Nery C, Baumfeld D, Umans H, Yamada AF. MR imaging of the plantar plate: normal anatomy, turf toe, and other injuries. Magn Reson Imaging Clin N Am. 2017;25:127–44. https://doi.org/10.1016/j.mric.2016.08.007.
31. Khurana A, Kadamabande S, James S, et al. Weil osteotomy: assessment of medium term results and predictive factors in recurrent metatarsalgia. Foot Ankle Surg. 2011;17:150–7. https://doi.org/10.1016/j.fas.2010.04.003.
32. Vandeputte G, Dereymaeker G, Steenwerckx A, Peeraer L. The Weil osteotomy of the lesser metatarsals: a clinical and pedobarographic follow-up study. Foot Ankle Int. 2000;21:370–4. https://doi.org/10.1177/107110070002100502.
33. Hyer CF, Consul D, McAlister JE, Cottom JM. Lesser metatarsophalangeal plantar plate repair. Clin Podiatr Med Surg. 2022;39:167–85. https://doi.org/10.1016/j.cpm.2021.11.008.
34. Doty JF, Coughlin MJ. Metatarsophalangeal joint instability of the lesser toes and plantar plate deficiency. J Am Acad Orthop Surg. 2014;22:235–45. https://doi.org/10.5435/JAAOS-22-04-235.
35. Tonogai I. Flexor tenotomy for mallet toe with penetration of the middle phalanx head by dual-component intramedullary implant following proximal interphalangeal arthrodesis. Int J Surg Case Rep. 2022;91:106703. https://doi.org/10.1016/j.ijscr.2021.106703.
36. Finney FT, Cata E, Holmes JR, Talusan PG. Anatomy and physiology of the lesser metatarsophalangeal joints. Foot Ankle Clin. 2018;23:1–7. https://doi.org/10.1016/j.fcl.2017.09.002.
37. Tafur M, Iwasaki K, Statum S, et al. Magnetic resonance imaging of the pulleys of the flexor tendons of the toes at 11.7 T. Skeletal Radiol. 2015;44:87–95. https://doi.org/10.1007/s00256-014-2019-y.
38. Blazek CD, Brandão RA, Manway JM, Burns PR. Multiplanar correction of the lesser digital deviation and indirect plantar plate rupture repair using a braided polyethyl-

ene nylon suture: a technique guide. Foot Ankle Spec. 2017;10:551–4. https://doi.org/10.1177/1938640017724544.
39. Finney FT, McPheters A, Singer NV, et al. Microvasculature of the plantar plate using nano-computed tomography. Foot Ankle Int. 2019;40:457–64. https://doi.org/10.1177/1071100718816292.
40. Mickle KJ, Munro BJ, Lord SR, et al. Gait, balance and plantar pressures in older people with toe deformities. Gait Posture. 2011;34:347–51. https://doi.org/10.1016/j.gaitpost.2011.05.023.
41. DiDomenico LA, Rizkalla J, Cartman J, Abdelfattah S. Hallux and lesser digits deformities associated with cavus foot. Clin Podiatr Med Surg. 2021;38:343–60. https://doi.org/10.1016/j.cpm.2020.12.013.
42. Wang B, Guss A, Chalayon O, et al. Deep transverse metatarsal ligament and static stability of lesser metatarsophalangeal joints: a cadaveric study. Foot Ankle Int. 2015;36:573–8. https://doi.org/10.1177/1071100714563310.
43. Tobimatsu H, Ikari K, Yano K, Okazaki K. Radiographic factors associated with painful callosities after forefoot surgery in patients with rheumatoid arthritis. Mod Rheumatol. 2023;33:104–10. https://doi.org/10.1093/mr/roab131.
44. Christman RA. Radiographic anatomy of the foot and ankle-part 5. The phalanges. J Am Podiatr Med Assoc. 2015;105:141–9. https://doi.org/10.7547/0003-0538-105.2.141.
45. Bae JM, Kang H, Kim HO, Park YM. Differential diagnosis of plantar wart from corn, callus and healed wart with the aid of dermoscopy. Br J Dermatol. 2009;160:220–2. https://doi.org/10.1111/j.1365-2133.2008.08937.x.
46. Mann RA, DuVries HL. Intractable plantar keratosis. Orthop Clin North Am. 1973;4:67–73.
47. Booth J, McInnes A. The aetiology and management of plantar callus formation. J Wound Care. 1997;6:427–30.
48. Mann RA, Mann JA. Keratotic disorders of the plantar skin. Instr Course Lect. 2004;53:287–302.
49. DeLauro TM. Keratotic lesions of the foot. Clin Podiatr Med Surg. 1996;13:73–84.
50. Biz C, Maso G, Malgarini E, et al. Hypermobility of the first ray: the Cinderella of the measurements conventionally assessed for correction of hallux valgus. Acta Biomed. 2020;91:47–59. https://doi.org/10.23750/abm.v91i4-S.9769.
51. Myerson MS, Badekas A. Hypermobility of the first ray. Foot Ankle Clin. 2000;5:469–84.
52. Gibbs RC, Boxer MC. Abnormal biomechanics of feet and their cause of hyperkeratoses. J Am Acad Dermatol. 1982;6:1061–9. https://doi.org/10.1016/s0190-9622(82)70091-x.
53. Day RD, Reyzelman AM, Harkless LB. Evaluation and management of the interdigital corn: a literature review. Clin Podiatr Med Surg. 1996;13:201–6.
54. Lee WJ, Lee SM, Won CH, et al. Efficacy of intralesional bleomycin for the treatment of plantar hard corns. Int J Dermatol. 2014;53:572–7. https://doi.org/10.1111/ijd.12586.
55. Roukis TS, Landsman AS. Adipofasciocutaneous flap coverage of a fifth toe cicatrix following medicated pad treatment of a hard corn. Plast Reconstr Surg. 2004;113:789–91. https://doi.org/10.1097/01.PRS.0000104535.48597.A5.
56. Gillet HG. Interdigital clavus: predisposition is the key factor of soft corns. Clin Orthop Relat Res. 1979;142:103–9.
57. Okuda R, Kinoshita M, Morikawa J, et al. Surgical treatment for hallux valgus with painful plantar callosities. Foot Ankle Int. 2001;22:203–8. https://doi.org/10.1177/107110070102200306.
58. Martijn HA, Sierevelt IN, Wassink S, Nolte PA. Fifth metatarsal osteotomies for treatment of bunionette deformity: a meta-analysis of angle correction and clinical condition. J Foot Ankle Surg. 2018;57:140–8. https://doi.org/10.1053/j.jfas.2017.08.006.
59. Moran MM, Claridge RJ. Chevron osteotomy for bunionette. Foot Ankle Int. 1994;15:684–8. https://doi.org/10.1177/107110079401501212.
60. Shen Y-T, Huang P-J, Chen S-J, Chang S-M. Symptomatic bunionette treated with sliding oblique metatarsal osteotomy-case series. J Clin Med. 2022;11:3962. https://doi.org/10.3390/jcm11143962.
61. Shi GG, Humayun A, Whalen JL, Kitaoka HB. Management of bunionette deformity. J Am Acad Orthop Surg. 2018;26:e396–404. https://doi.org/10.5435/JAAOS-D-17-00345.

62. Coughlin MJ. Etiology and treatment of the bunionette deformity. Instr Course Lect. 1990;39:37–48.
63. Koti M, Maffulli N. Bunionette. J Bone Joint Surg Am. 2001;83:1076–82. https://doi.org/10.2106/00004623-200107000-00016.
64. Cohen BE, Nicholson CW. Bunionette deformity. J Am Acad Orthop Surg. 2007;15:300–7. https://doi.org/10.5435/00124635-200705000-00008.
65. Shimobayashi M, Tanaka Y, Taniguchi A, et al. Radiographic morphologic characteristics of bunionette deformity. Foot Ankle Int. 2016;37:320–6. https://doi.org/10.1177/1071100715614923.
66. Weil L, Weil LS. Osteotomies for bunionette deformity. Foot Ankle Clin. 2011;16:689–712. https://doi.org/10.1016/j.fcl.2011.08.012.
67. Katcherian DA. Treatment of Freiberg's disease. Orthop Clin N Am. 1994;25:69–81.
68. Ary KR, Turnbo M. Freiberg's infraction: an osteochondritis of the metatarsal head. J Am Podiatry Assoc. 1979;69:131–2. https://doi.org/10.7547/87507315-69-2-131.
69. Johnson H. Osteochondritis of the second metatarsal head; Freiberg's infraction. J Natl Assoc Chirop. 1953;43:39–40.
70. Talusan PG, Diaz-Collado PJ, Reach JS. Freiberg's infraction: diagnosis and treatment. Foot Ankle Spec. 2014;7:52–6. https://doi.org/10.1177/1938640013510314.
71. Viana SL, Machado BB, Mendlovitz PS. MRI of subchondral fractures: a review. Skeletal Radiol. 2014;43:1515–27. https://doi.org/10.1007/s00256-014-1946-y.
72. Young MC, Fornasier VL, Cameron HU. Osteochondral disruption of the second metatarsal: a variant of Freiberg's infraction? Foot Ankle. 1987;8:103–9. https://doi.org/10.1177/107110078700800206.
73. Scartozzi G, Schram A, Janigian J. Freiberg's infraction of the second metatarsal head with formation of multiple loose bodies. J Foot Surg. 1989;28:195–9.
74. Gillespie H. Osteochondroses and apophyseal injuries of the foot in the young athlete. Curr Sports Med Rep. 2010;9:265–8. https://doi.org/10.1249/JSR.0b013e3181f19488.
75. Arnold AI. Freiberg's infraction: its sequelae and treatment with ultrasound therapy. J Natl Assoc Chirop. 1957;47:240–1.
76. Carter KR, Chambers AR, Dreyer MA. Freiberg infraction. In: StatPearls. Treasure Island: StatPearls Publishing; 2023.
77. Smillie IS. Treatment of Freiberg's infraction. Proc R Soc Med. 1967;60:29–31.
78. Ashman CJ, Klecker RJ, Yu JS. Forefoot pain involving the metatarsal region: differential diagnosis with MR imaging. Radiographics. 2001;21:1425–40. https://doi.org/10.1148/radiographics.21.6.g01nv071425.
79. Couturier S, Gold G. Imaging features of avascular necrosis of the foot and ankle. Foot Ankle Clin. 2019;24:17–33. https://doi.org/10.1016/j.fcl.2018.10.002.
80. Maresca G, Adriani E, Falez F, Mariani PP. Arthroscopic treatment of bilateral Freiberg's infraction. Arthroscopy. 1996;12:103–8. https://doi.org/10.1016/s0749-8063(96)90229-7.
81. Hodes A, Umans H. Metatarsalgia. Radiol Clin North Am. 2018;56:877–92. https://doi.org/10.1016/j.rcl.2018.06.004.
82. Shane A, Reeves C, Wobst G, Thurston P. Second metatarsophalangeal joint pathology and Freiberg disease. Clin Podiatr Med Surg. 2013;30:313–25. https://doi.org/10.1016/j.cpm.2013.04.009.
83. Ozkan Y, Oztürk A, Ozdemir R, et al. Interpositional arthroplasty with extensor digitorum brevis tendon in Freiberg's disease: a new surgical technique. Foot Ankle Int. 2008;29:488–92. https://doi.org/10.3113/FAI-2008-0488.
84. Lee S, Saifuddin A. Magnetic resonance imaging of subchondral insufficiency fractures of the lower limb. Skeletal Radiol. 2019;48:1011–21. https://doi.org/10.1007/s00256-019-3160-4.
85. Trnka H-J, Lara JS. Freiberg's infraction: surgical options. Foot Ankle Clin. 2019;24:669–76. https://doi.org/10.1016/j.fcl.2019.08.004.

Foot and Ankle Trauma

16

Cumhur Deniz Davulcu, Ece Davutluoglu, and Abdisalam Mutaj Shafaj Nur

Foot and ankle traumas are among the common reasons for admission to the emergency department. Patients present with a wide spectrum of injuries, ranging from osseous pathologies to ligament injuries. Treatment options may include surgery or conservative methods, depending on the nature and severity of the injury. Injury occurs as a result of twisting mechanism or direct trauma with a bimodal distribution occurring mostly in young active males between the ages of 15 and 24 and elderly females at the ages of between 75 and 84 years [1, 2].

16.1 Symptoms

Depending on the severity of the injury, ankle and foot trauma patients can present with a wide range of symptoms.

1. Pain: Acute injuries of the foot and ankle such as fractures and ligament injuries can result in sudden-onset severe pain that makes it difficult or impossible to bear weight on the affected limb.
2. Instability: Ankle and foot trauma can result in joint instability often manifested by the inability to bear weight and the sensation of the joint "giving out" during movement. Ligament tears are the most common causes of joint instability.
3. Swelling and ecchymosis: Accumulation of fluids as edema in the affected tissues, which may be mild or pronounced, is a common response to injury in ankle and foot trauma depending on the severity of the injury. Blood leakage from blood vessels in the affected region to the surrounding tissues can result in ecchymosis. This can be apparent immediately or may develop days later. The

C. D. Davulcu · E. Davutluoglu (✉) · A. M. S. Nur
Department of Orthopaedics, Istanbul University-Cerrahpasa, Istanbul, Turkey

S. Sharma et al. (eds.), *Clinical and Radiological Examination of the Foot and Ankle*, https://doi.org/10.1007/978-981-97-4202-8_16

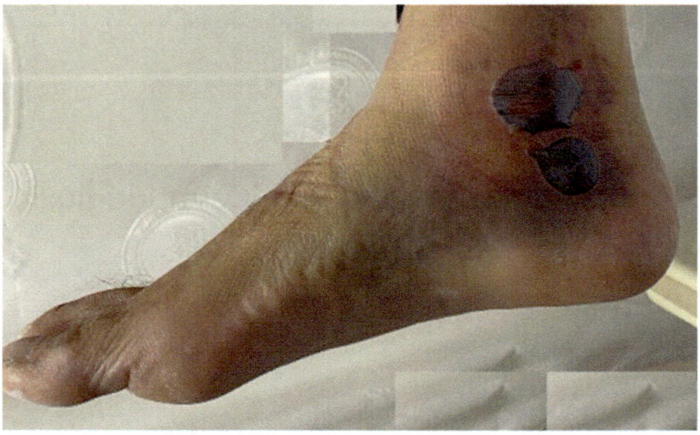

Fig. 16.1 Ecchymosis on the medial aspect of foot and ankle, hemorrhagic blisters

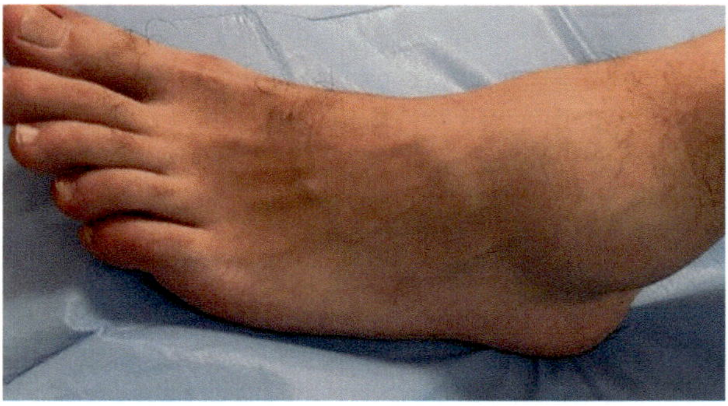

Fig. 16.2 Swelling around the ankle

color of the ecchymosis can change over time (Fig. 16.1). Pain, swelling, and instability following foot and ankle trauma can result in limited range of motion of the joints (Fig. 16.2).

16.2 Inspection

During the foot and ankle examination, the patient's shoes and socks should be removed, and their clothing rolled up to the knees for a bilateral comparison. The examiner carefully inspects the foot and ankle noting for obvious visible abnormalities while comparing to the contralateral side. The foot and ankle should be observed during walking, standing, and while bearing weight if the patient is able to [3]. The toes should be assessed for deformities, calluses, nail infections, and circulatory problems. Open wounds, deformities, joint malalignment, ecchymosis, fracture blisters, and skin tenting should be noted (Fig. 16.3). Ecchymosis at the medial plantar aspect of the foot and swelling at the midfoot area is highly suggestive of Lisfranc injury (Fig. 16.4). Fracture blisters can be categorized as fluid-filled and blood-filled and should be evaluated with respect to the anticipated surgical incisions (Figs. 16.5 and 16.6). Fluid-filled blisters represent a lesser injury [4].

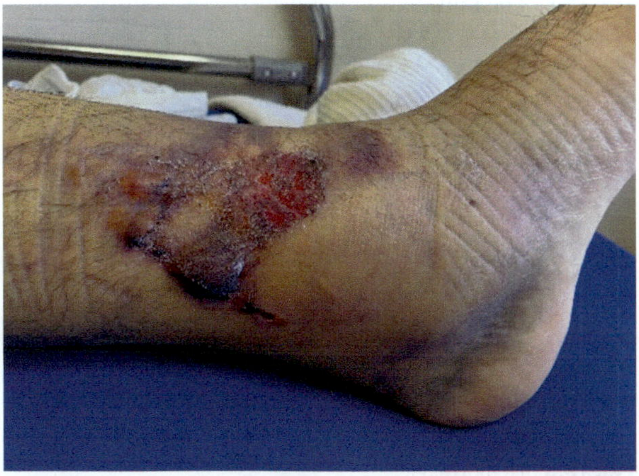

Fig. 16.3 Dermabrasion, ecchymosis, and blisters

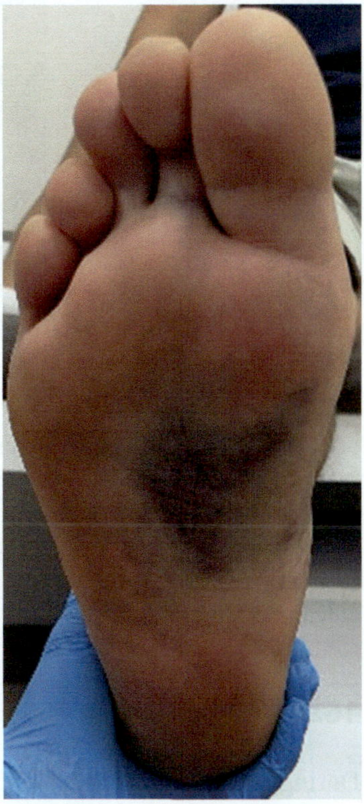

Fig. 16.4 Plantar ecchymosis suggestive of Lisfranc injury

The walking pattern, heel walking, and tiptoes can give a quick evaluation of neurovascular status, joint range of motion, and muscle strength. The presence of calluses and ulcers should also be observed. Additionally, evaluating the contact

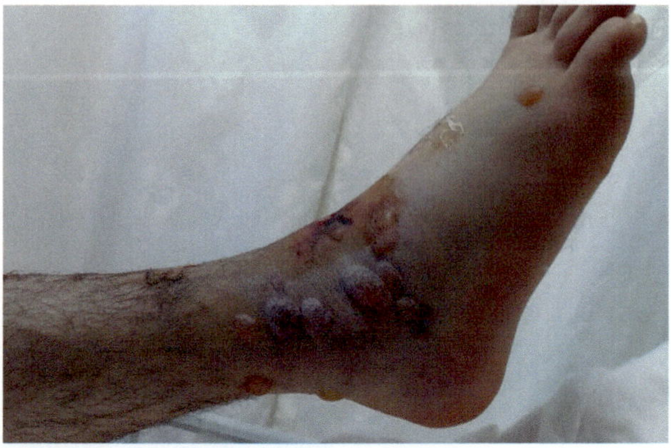

Fig. 16.5 Fluid-filled blisters

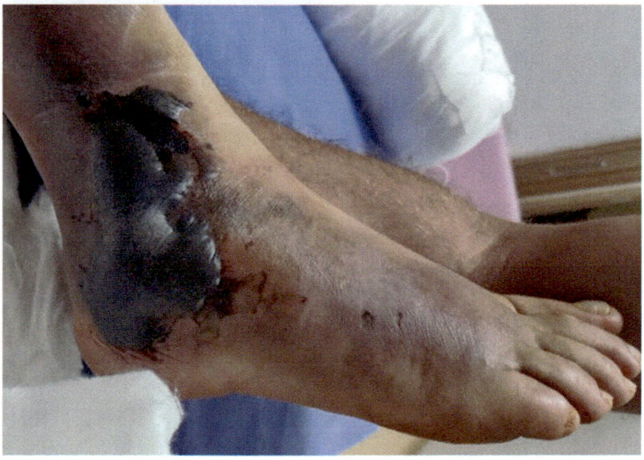

Fig. 16.6 Hemorrhagic blister

area and inner surface of frequently worn shoes can reveal areas of increased deformation, indicating excessive pressure during regular use.

Grossly, alignment evaluation should be done from the front, back, medial, and lateral aspects during inspection. The arch of the foot should be observed for flat feet (pes planus) or high-arched feet (pes cavus). The angulation of the tibia and calcaneus should be assessed for valgus-varus [4].

16.3 Palpation

During palpation the patient is asked to point with their finger in the area that is most tender. The examiner starts palpating the affected area starting away from the pain site and moving toward the most tender region along bony landmarks, ligaments, tendons, and joint line noting for crepitus, pain, or pathologic bony movement. Osseous prominences should be palpated with moderate pressure. The ideal pressure is when the clinician's finger depresses the skin until the nail bed blanches. Pain upon pressure over the lateral and medial malleolus may indicate a fracture [5]. The palpation of the proximal fibula should not be forgotten as syndesmotic injuries of the ankle can result in Maisonneuve fracture [6]. Pain on the anterior joint surface of the ankle during forced plantar flexion may suggest a talus osteochondral defect. Additionally, the tarsometatarsal joint should be palpated for tenderness to evaluate for Lisfranc injury [7]. The metatarsals, especially the second and third metatarsals, should be palpated along their longitudinal axes in athletes with metatarsalgia and in obese individuals who are at risk for stress fracture [8]. In patients with acute trauma, the base of the fifth metatarsal should be evaluated for the Jones fracture.

Soft tissue assessment should be conducted alongside the osseous evaluation. Tendon integrity should be evaluated by palpating the tendons with active and passive ankle movement. Thickening of the tendon may indicate tendinosis. Pain over the tendons on palpation could indicate tendinitis [9].

Pathologies in ligamentous structures are commonly observed due to acute or chronic trauma. The anterior talofibular ligament, where the most frequent injury is observed, should be palpated with pressure on the anteroinferior lateral malleolus in every patient [10]. The deltoid ligament, which significantly contributes to ankle stability, should also be assessed with pressure on the anteromedial medial malleolus [11].

During the examination of the plantar surface, particular attention should be given to plantar fascia for tenderness. The fascia can be stretched and accentuated through forceful passive toe extension, making it more prominent. Thickening and tenderness in the fascia should be evaluated, and an increase in pain with pressure on the calcaneal insertion may indicate plantar fasciitis [12].

16.4 Evaluation of Joint Range of Motion

The assessment of joint range of motion should be done with bilateral comparison. The examiner assesses for the extents of the range of movement both active and passive noting for pain and limitations. Range of motion could be limited due to pain and assessment of tendon function could be difficult. Each joint should be individually evaluated during this stage of the examination.

Dorsiflexion and plantar flexion of the ankle should be evaluated actively and then passively. Limited dorsiflexion may indicate pathologies such as Achilles tendon disorders, joint capsule contractures, arthritis, or tendon interposition. Similarly, limited plantar flexion may indicate capsule contracture and posterior impingement [13].

16.5 Neurovascular Evaluation

The neurovascular examination begins with the assessment of the perfusion noting for signs of pallor and cyanosis and the capillary refill. The examiner notes for the pulses and compares with the contralateral side as this gives an idea on the general vascular status of the patient. When it is difficult to get a pulse due to swelling, Doppler ultrasound examination can be used. The sensation of the foot to light touch should be assessed. The first web space is supplied by the deep peroneal nerve, the rest of the dorsum of the foot by the superficial peroneal nerve, the lateral border of the foot by the sural nerve, and the plantar aspect of the foot by the lateral and medial plantar nerve. The saphenous nerve supplies the medial aspect of the foot and ankle. The examiner compares the findings with the contralateral side.

After trauma, acute injuries, or compression syndromes, paresthesia and hyperalgesia may develop [14]. Nerve injuries or compression, such as superficial peroneal nerve entrapment, may result in sensory loss on the dorsal side of the foot and tenderness after percussion over the proximal part of the lateral malleolus [15]. Anterior tarsal tunnel syndrome may cause burning pain between the first and second toes on the dorsal side of the foot after percussion over the intermetatarsal space. Tarsal tunnel syndrome may produce pain and paresthesia in the plantar region after percussion over the posterior part of the medial malleolus [16].

16.6 Special Tests

16.6.1 Stability Evaluation

Stability tests have been described, especially for the two most commonly injured ligaments, the anterior talofibular ligament and the calcaneofibular ligament. These tests are used for both acute and chronic injury evaluations, but in acute injuries, pain may hinder the complete application of the tests, leading to suboptimal assessment [17].

1. Anterior drawer test: The anterior drawer test is used to evaluate for the integrity of the anterior talofibular ligament (ATFL) in case of ligamentous injury. The test can be performed with the patient seated or lying supine. The knee is held at 90° flexion and the ankle plantar flexed to 10–20°. One hand of the examiner stabilizes the lower leg above the ankle while grasping the other hand with the heel. An anterior force is applied to the heel. In a normal ankle, there should be minimal anterior translation of the talus. However, if the ATFL is injured or

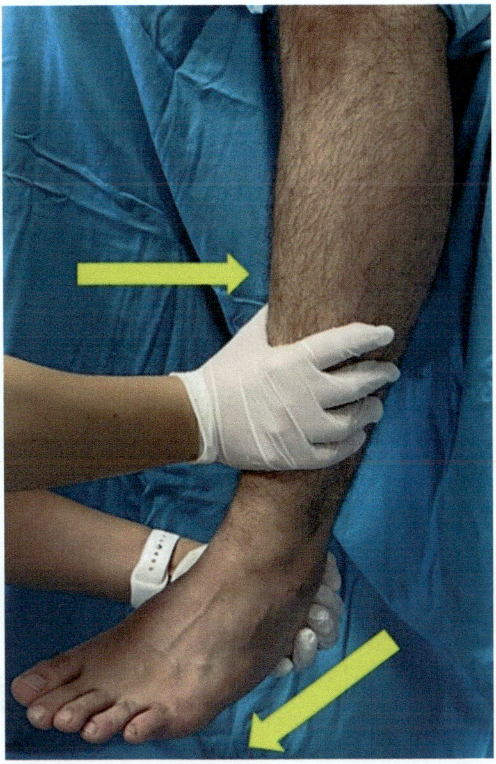

Fig. 16.7 Anterior drawer test performed. One hand of the examiner stabilizes the lower leg. The other hand of the examiner applies anterior force to the heel

unstable, excessive anterior movement or a visible "drawer-like" motion may be observed in the affected side compared to the uninjured side. This abnormal anterior displacement indicates a positive anterior drawer test and suggests an injury to the ATFL. Ankle injuries that cause elongation of the ligament by 4 mm or 20% of its length might result in failure in the ligament and positive anterior drawer test (Fig. 16.7).

2. Inversion stress test: The inversion stress test, also called the varus stress, or talar tilt test is used to assess the integrity of the calcaneofibular ligament (CFL). The test is performed with the patient seated with the knee flexed to 90°. The lower leg is stabilized, and the foot held in neutral position. A varus moment is applied to the calcaneus, while the talus and the calcaneus must be held as a unit to limit the subtalar motion. Excessive or increased inversion compared to the uninjured side indicates ligamentous laxity or injury. A difference of more than 10° compared to the normal side is considered abnormal (Fig. 16.8).

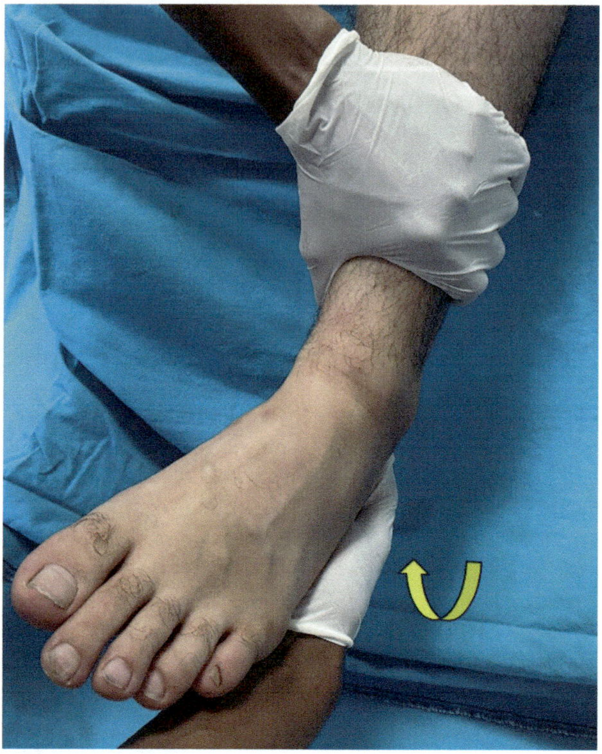

Fig. 16.8 Inversion stress test performed. One hand of the examiner stabilizes the cruris, while the other hand of the examiner applies inversion force to the calcaneus

16.7 Syndesmosis Injury Evaluation

The ankle syndesmosis is stabilized by four ligaments: the anterior inferior tibiofibular ligament (AITFL), posterior-inferior tibiofibular ligament (PITFL), the interosseous ligament (IOL), and the inferior transverse ligament (ITL). The AITFL is the most superficial and originates from the anterolateral tubercle of the tibia, takes an oblique course, and inserts on the anterior tubercle of the fibula. The PITFL is the strongest component of the ankle syndesmosis, and the interosseous ligament (IOL) is the main restraint to the proximal migration of the talus [18]. The most common mechanism of injury of the ankle syndesmosis is the external rotation injuries [19].

1. Squeeze test: The squeeze test is used to assess for syndesmotic injury or high ankle sprain. The test is performed by applying lateromedial pressure at the middle and distal-third part of the leg. A positive test result indicates tenderness at

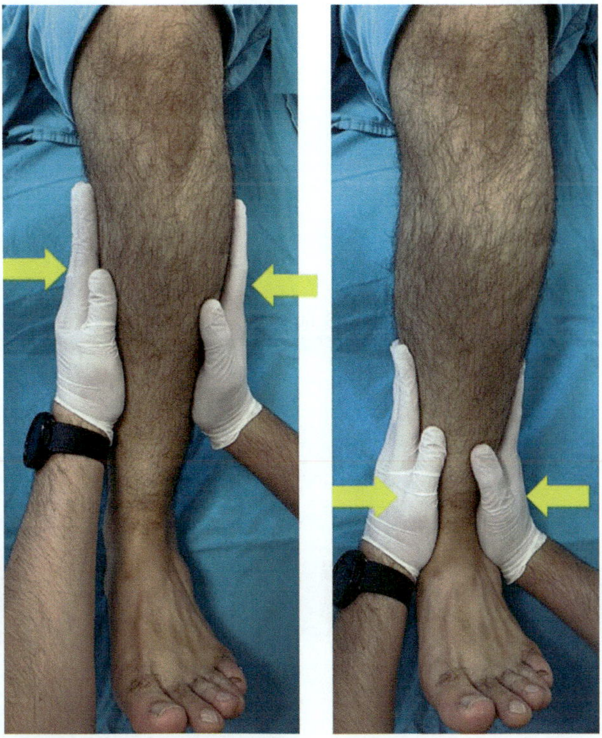

Fig. 16.9 Squeeze test performed. The examiner applies lateromedial pressure at the middle (**a**) and distal-third (**b**) part of the leg

the distal syndesmosis. A positive squeeze test has the highest specificity test of about 95% (Fig. 16.9) [20].

2. External rotation test: This test is performed with the patient seated and the knee flexed to 90°. The leg is stabilized, the ankle is placed in a neutral position, and the foot is externally rotated. If pain is elicited with the external rotation maneuver at the anterolateral side of the ankle, the test is considered positive and can indicate deltoid ligament or distal tibiofibular syndesmosis injury (Fig. 16.10) [21].
3. The Cotton test: This test is performed while stabilizing the distal tibia and fibula with one hand and applying lateral translation force to the foot with the other hand of the examiner. The test is positive when the translation of more than 3 mm or if a clunk is felt, indicating syndesmotic injury (Fig. 16.11) [22].

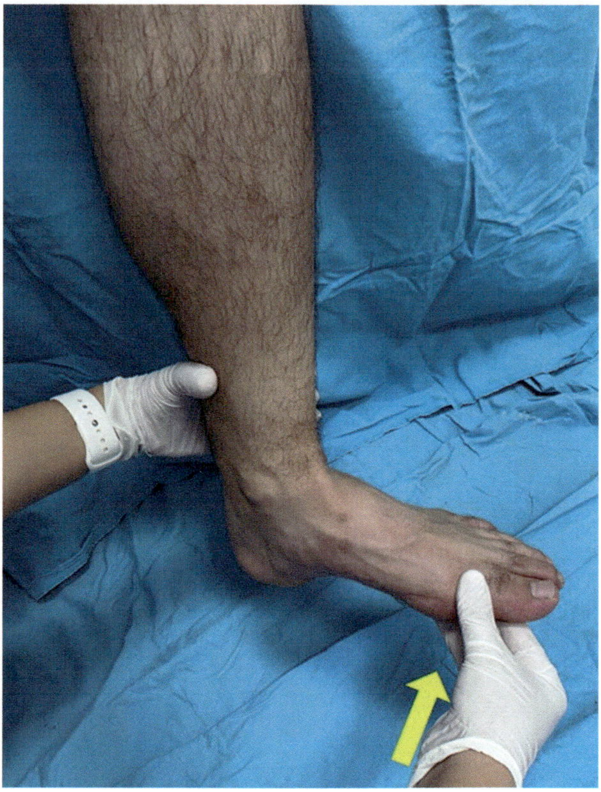

Fig. 16.10 External rotation test performed. The right hand of the examiner applies external rotation force to the forefoot

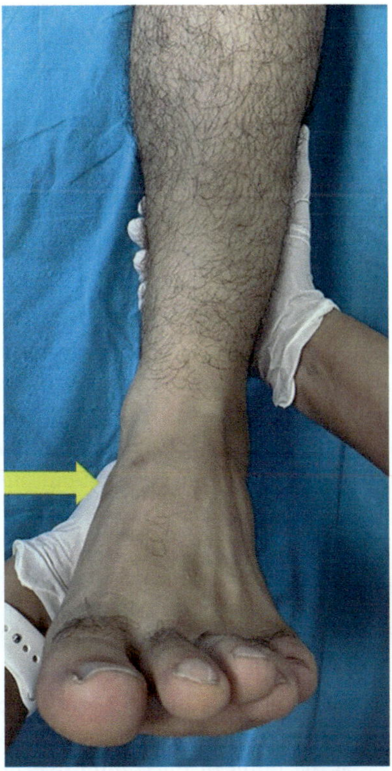

Fig. 16.11 The Cotton test performed. The examiner stabilizes the distal tibia and fibula with the right hand. The left hand of the examiner applies lateral translation force to the foot

16.8 Imaging

The Ottawa Ankle Rule (OAR) is employed as a clinical decision tool to determine the necessity of ankle and foot X-rays in patients with acute ankle and foot injuries. It assesses several criteria, including the ability to bear weight, defined as the capacity to take four steps in the emergency department or immediately following the trauma, tenderness at the medial and lateral malleoli, tenderness at the base of the fifth metatarsal, and navicular tenderness. An ankle X-ray series is only required if there is pain at the mentioned locations or an inability to bear weight either immediately after the injury or in the emergency department. The Ottawa criteria has been extensively validated and has a high sensitivity (96–99%) and a <1% chance of false-negative findings [23]. Patients who do not meet the Ottawa criteria are considered low risk for ankle fractures and do not require immediate X-ray.

16.9 Plain Radiography

The standard X-ray views of ankle and foot trauma are the anteroposterior (AP) and lateral views of the ankle and foot, with a mortise view of ankle and oblique view of the foot. AP and lateral view of the entire length of the tibia and fibula have to be taken as well, to rule out proximal fibula fracture and associated syndesmotic injury. Dynamic stress views are done to assess stability or subtle injuries.

16.10 Common Radiologic Findings in Foot and Ankle Trauma

1. Increased medial clear space: The medial clear space is the distance between the lateral border of the medial malleolus and the medial border of the talus at the level of the talar dome. The medial clear space is used to evaluate the integrity of deltoid ligament and medial ankle instability. The function of the deltoid ligament is to provide medial stability to the tibiotalar joint by preventing valgus shifting of the talus. The normal value for medial clear space is 5 mm on an external rotation stress view (gravity stress view) which is indicative for deltoid rupture and medial ankle instability (Fig. 16.12a, b).
2. Decreased tibiofibular overlap: The tibiofibular overlap is the overlap of the lateral malleolus and the anterior tibial tubercle measured 1 cm proximal to the plafond. The tibiofibular overlap is assessed on the AP and mortise view radiographs. The normal value which is >1 mm on the mortise and >6 mm on the AP at the point of maximum overlap decreased, or no overlap is indicative of syndesmotic injury (Fig. 16.13a, b).

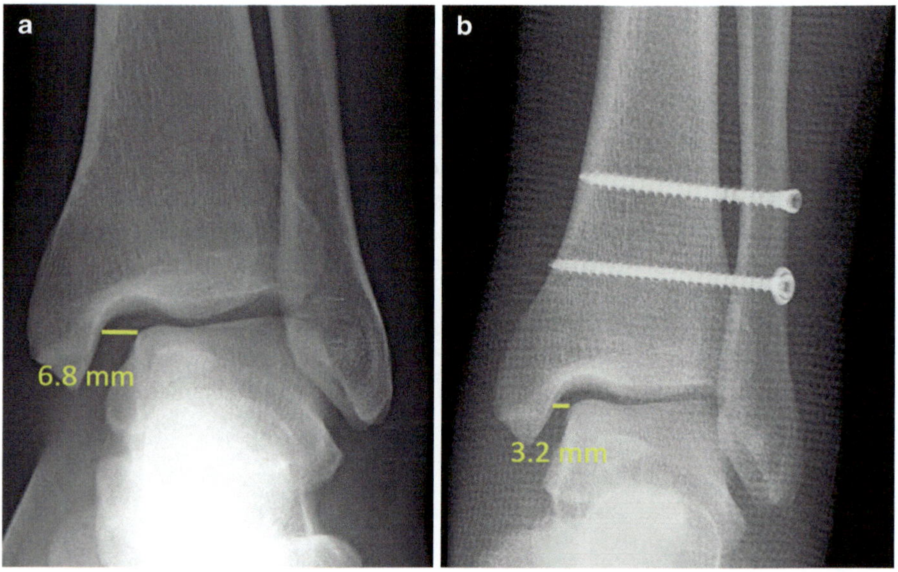

Fig. 16.12 (**a**) Increased medial clear space. (**b**) Postoperative decreased medial clear space

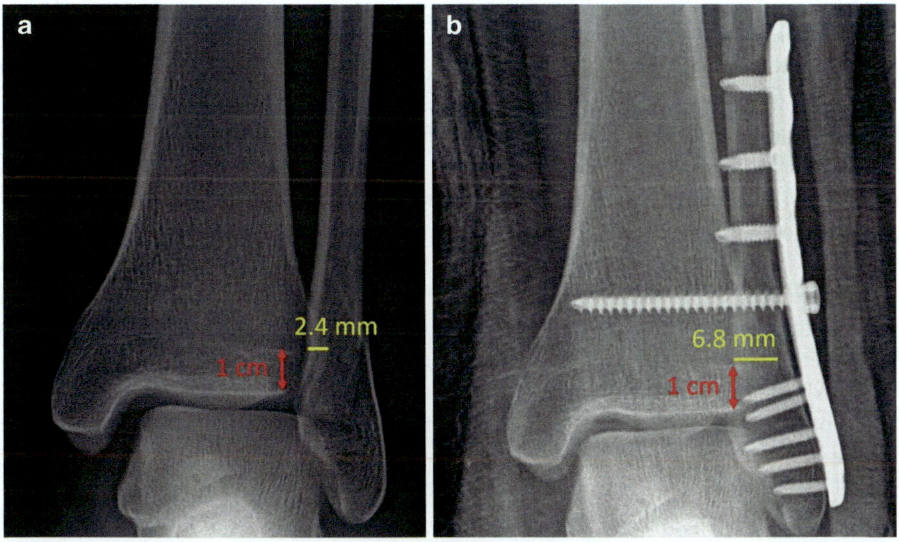

Fig. 16.13 (**a**) Decreased tibiofibular overlap. (**b**) Postoperative increased tibiofibular overlap

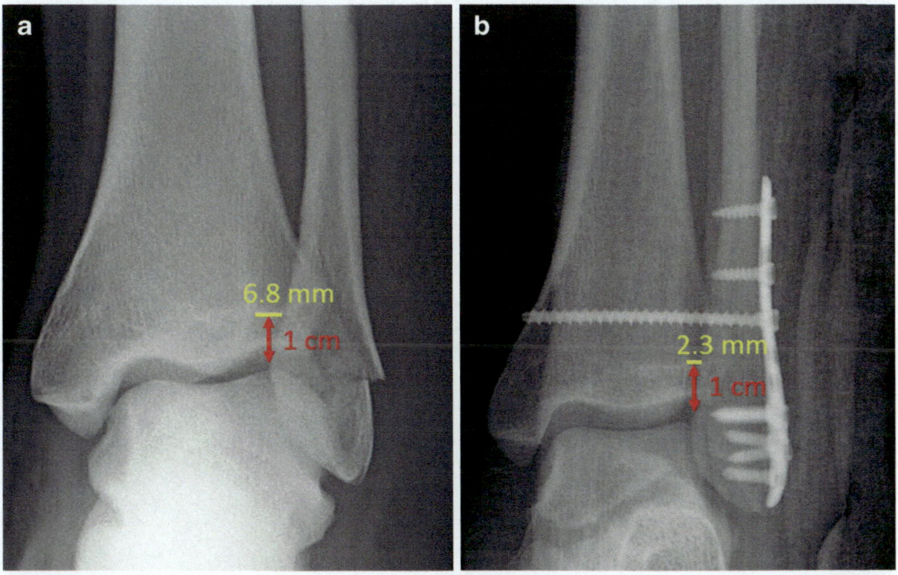

Fig. 16.14 (**a**) Increased tibiofibular clear space. (**b**) Postoperative decreased tibiofibular clear space

3. Increased tibiofibular clear space: It is measured on both AP and mortise views 1 cm above the joint line. An increased tibiofibular clear space above 6 mm is predictive of syndesmotic injury (Fig. 16.14a, b).

Midfoot Instability: The Lisfranc ligament spans across the tarsometatarsal joint originating from the medial aspect of the medial cuneiform and inserting to the plantar aspect of the base of the second metatarsal. It maintains the alignment and stability of the midfoot. Injury to the Lisfranc ligament is manifested radiologically as the widening of the space between the first and the second ray accompanied by avulsion of the Lisfranc ligament from the second metatarsal seen as fleck sign on the X-ray. Radiological evaluation could be done by measuring the shortest horizontal distance between the bases of the first and second metatarsal. Although patients mostly cannot be able to bear weight, the gold standard of the measurement has to be done in a weight-bearing AP view of the foot. The distance between the first and second metatarsals should be 1–2 mm normally. Approximately 2–5 mm distance is considered subtle injury, and the distance more than 5 mm indicates severe injury [24]. Alignment evaluation is also important in assessment of midfoot instability. Disruption of the line on the medial side of the second metatarsal and the medial site of the middle cuneiform seen on an AP view and the medial side of the fourth metatarsal and the medial side of cuboid on an oblique view demonstrate midfoot instability (Fig. 16.15a, b). Another finding seen with Lisfranc injury is the dorsal displacement of the first and the second metatarsal on the lateral view (Fig. 16.16) [25].

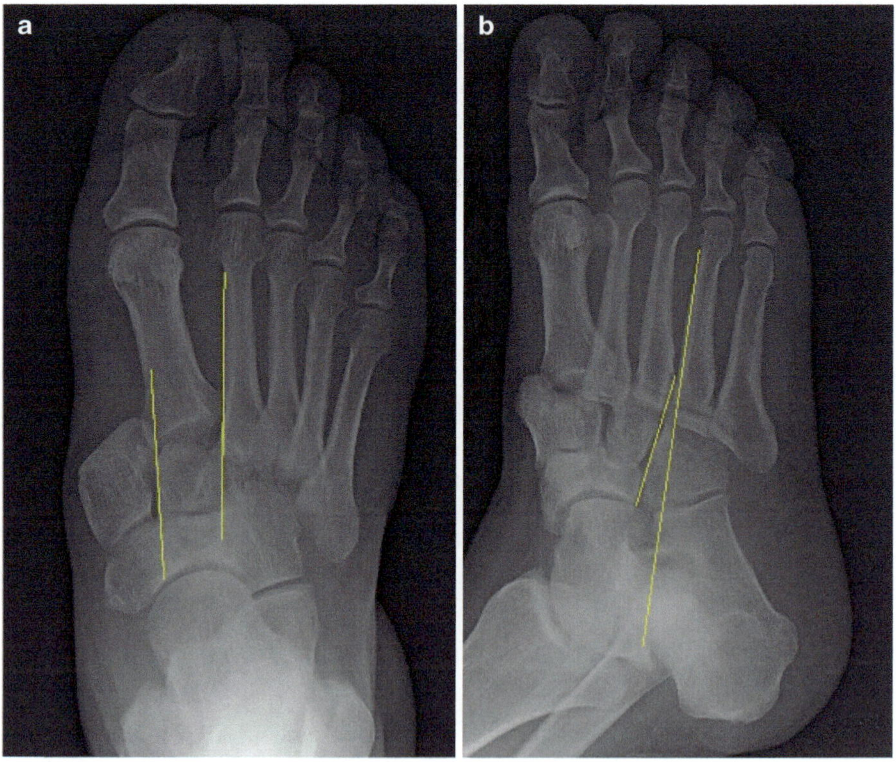

Fig. 16.15 (**a**) Disruption of the line on the medial side of the second metatarsal and the middle cuneiform on AP foot X-ray. (**b**) Disruption of the line on the fourth metatarsal and medial side of cuboid on oblique foot X-ray

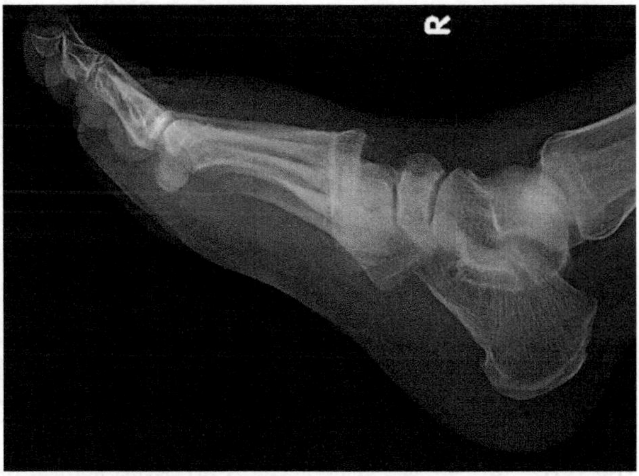

Fig. 16.16 Dorsal displacement of the first and the second metatarsal on lateral foot X-ray

16.11 Computed Tomography, Magnetic Resonance Imaging, and Ultrasonography

3D imaging methods are frequently used when plain radiographs are not sufficient. Patients with more suspicions of bone pathologies should be evaluated by computed tomography, whereas patients with more suspicions of soft tissue pathologies should be evaluated with MRI-advanced imaging.

Conventional CT is commonly used in assessing injuries in foot and ankle trauma for subtle injuries that might not be apparent in conventional x-rays, when intra-articular involvement is suspected and in surgical planning (Fig. 16.17a, b). Weight-bearing CT gives a cross-sectional images while the patient in a weight-bearing position. This allows viewing of structures in their anatomical positions with the utilization of the contralateral ankle view as an internal control. It is specifically useful in assessing syndesmotic instability, lateral ligament instability, and Lisfranc ligament injury [26]. CT allows for assessment of anteroposterior, mediolateral, and rotational changes of the syndesmosis on axial CT images [27].

MRI is excellent for detecting and characterizing soft tissue injuries, such as sprains, strains, and tears in ligaments like ATFL and syndesmosis, tendons like the Achilles tendon, and bone contusions or osteochondral lesions (Fig. 16.18). Additionally, MRI is also used in diagnosing stress fractures as a result of overuse activities before X-ray can detect significant changes on the bone.

Evaluation with ultrasonographic imagining is useful adjunct to the physical examination of foot and ankle trauma. Ultrasound is useful in assessing hematomas, ligamentous sprain, and tendon injuries like the Achilles tendon rupture (Fig. 16.19). Although USG is an operator-dependent technique and it provides limited assessment of the intra-articular structures, being non-invasive, economical, and readily available as well as having a significant advantage in providing dynamic assessment by providing real-time image during stress tests that helps in determining the extent of injury and instability, make it useful as a diagnostic imaging method [28].

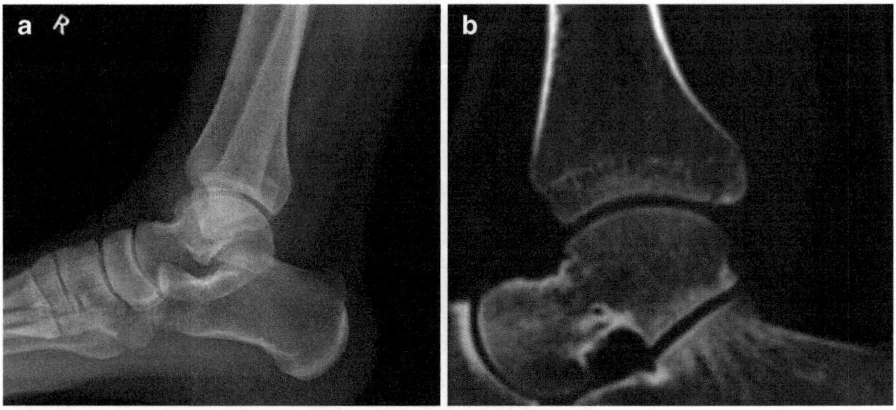

Fig. 16.17 (**a**) Foot and ankle lateral X-ray seems normal. (**b**) CT sagittal view of the same patient reveals posterior malleolus fracture

Fig. 16.18 Ruptured Achilles tendon on MRI

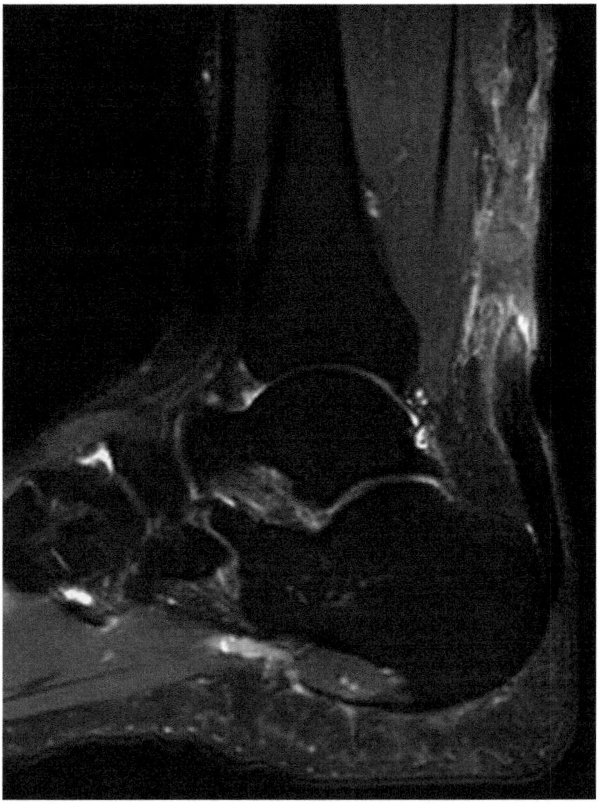

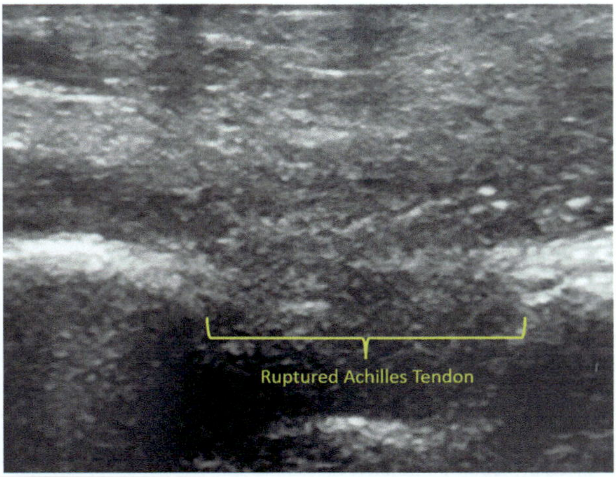

Fig. 16.19 Ruptured Achilles tendon on USG

Key Points

- Foot and ankle trauma is a common injury, occurring due to various mechanisms and affecting different age groups, with a distinct pattern in young males and elderly females.
- A systematic approach to physical examination and the appropriate use of imaging techniques are crucial for accurate diagnosis and guiding treatment options for the effective management of foot and ankle trauma.
- A thorough examination involves comparing the affected area to the uninjured side, looking for deformities, ecchymosis, and skin abnormalities and palpating along key bony landmarks to assess for tenderness, crepitus, and pathologic movements.
- Depending on the mechanism of injury, a high index of clinical suspicion must be maintained for possible ligament integrity loss and syndesmosis injury and should be evaluated with special provocative tests like anterior drawer test, inversion stress test, squeeze test, and external rotation test.
- The Ottawa Ankle Rule is a clinical decision tool used to determine the necessity of ankle and foot X-rays in patients with acute injuries, based on criteria like the ability to bear weight, tenderness at specific locations, and navicular tenderness.
- Radiologic evaluation should be started with plain X-rays, and where subtle or specific injuries are suspected, advanced imaging like CT and MRI is warranted. Advanced imaging is also useful in surgical planning. Ultrasound complements the physical examination by providing dynamic real-time assessment of soft tissue injuries.

References

1. Doherty C, Delahunt E, Caulfield B, et al. The incidence and prevalence of ankle sprain injury: a systematic review and meta-analysis of prospective epidemiological studies. Sports Med. 2014;44:123–40. https://doi.org/10.1007/s40279-013-0102-5.
2. Gribble PA, Bleakley CM, Caulfield BM, et al. Evidence review for the 2016 International Ankle Consortium consensus statement on the prevalence, impact and long-term consequences of lateral ankle sprains. Br J Sports Med. 2016;50:1496–505. https://doi.org/10.1136/bjsports-2016-096189.
3. Besse J. Metatarsalgia. Orthop Traumatol Surg Res. 2017;103:29–39. https://doi.org/10.1016/j.otsr.2016.06.020.
4. Leiß F, Schwarz T, Baier C, Grifka J. Clinical examination of the foot. Orthopade. 2019;48:541–52. https://doi.org/10.1007/s00132-019-03748-z.
5. Mcinnis KC, Ramey LN. High-risk stress fractures: diagnosis and management. PM R. 2016;8:S113–24. https://doi.org/10.1016/j.pmrj.2015.09.019.
6. Millen JC, Lindberg D. Maisonneuve fracture. JEMS. 2011;41:77–8. https://doi.org/10.1016/j.jemermed.2008.08.021.
7. Grewal US, Onubogu K, Southgate C, Dhinsa BS. Lisfranc injury: a review and simplified treatment algorithm. Foot. 2020;45:101719. https://doi.org/10.1016/j.foot.2020.101719.
8. Welck MJ, Hayes T, Pastides P, et al. Stress fractures of the foot and ankle. Injury. 2015;48(8):1722–6. https://doi.org/10.1016/j.injury.2015.06.015.
9. Ivanova V, Todd NW, Yurgelon J. Dance - related foot and ankle injuries and pathologies dance ballet sesamoiditis impingement stress fracture. Clin Podiatr Med Surg. 2023;40:94040.
10. Bonnel F, Toullec E, Mabit C, Tourné Y. Chronic ankle instability: biomechanics and pathomechanics of ligaments injury and associated lesions. Orthop Traumatol Surg Res. 2010;96:424–32. https://doi.org/10.1016/j.otsr.2010.04.003.
11. Cain JD. Anatomy of the deltoid - spring ligament complex. Foot Ankle Clin. 2021;26:237–47. https://doi.org/10.1016/j.fcl.2021.03.001.
12. Gutteck N, Schilde S, Delank K. Pain on the plantar surface of the foot. Dtsch Arztebl Int. 2019;116(6):83–8. https://doi.org/10.3238/arztebl.2019.0083.
13. Papaliodis DN, Vanushkina MA, Richardson NG, Dipreta JA. The foot and ankle examination. Med Clin. 2014;98:181–204. https://doi.org/10.1016/j.mcna.2013.10.001.
14. Ferkel E, Davis WH, Ellington JK. Entrapment neuropathies of the foot and ankle. Clin Sports Med. 2015;34:791–801. https://doi.org/10.1016/j.csm.2015.06.002.
15. Chari B. Nerve entrapment in ankle and foot: ultrasound imaging. Semin Musculoskelet Radiol. 2018;22:354–63.
16. Lenchik L, Brigido MK, Shahabpour M, Marcelis S. Normal anatomy and compression areas of nerves of the foot and ankle: US and MR imaging with anatomic correlation. Radiographics. 2015;35(5):1469–82.
17. Larkins LW, Baker RT, Baker JG. Physical examination of the ankle: a review of the original orthopedic special test description and scientific validity of common tests for ankle examination. Arch Rehabil Res Clin Transl. 2020;2:100072. https://doi.org/10.1016/j.arrct.2020.100072.
18. Hermans JJ, Beumer A, De Jong TAW, Kleinrensink G. Anatomy of the distal tibiofibular syndesmosis in adults: a pictorial essay with a multimodality approach. J Anat. 2010;217:633–45. https://doi.org/10.1111/j.1469-7580.2010.01302.x.
19. Sman AD, Hiller CE, Rae K, et al. Diagnostic accuracy of clinical tests for ankle syndesmosis injury. Br J Sports Med. 2013;49(5):323–9. https://doi.org/10.1136/bjsports-2013-092787.
20. Bleakley C. Physical therapy in sport diagnostic accuracy of clinical tests assessing ligamentous injury of the ankle syndesmosis: a systematic review with meta-analysis. Phys Ther Sport. 2021;49:214–26. https://doi.org/10.1016/j.ptsp.2021.03.005.
21. Porter DA, Jaggers RR, Barnes AF, Rund AM. Optimal management of ankle syndesmosis injuries. Open Access J Sports Med. 2014;5:173–82.

22. Sman AD, Hiller CE, Refshauge KM. Diagnostic accuracy of clinical tests for diagnosis of ankle syndesmosis injury: a systematic review. Br J Sports Med. 2013;47(10):620–8. https://doi.org/10.1136/bjsports-2012-091702.
23. Daş M, Temiz A, Çevik Y. Implementation of the Ottawa ankle rules by general practitioners in the emergency department of a Turkish district hospital. Semin Musculoskelet Radiol. 2016;22:2–5. https://doi.org/10.5505/tjtes.2016.72662.
24. Sugino Y, Yoshimura I, Hagio T, et al. Reliability of various diastasis measurement methods on weightbearing radiographs in patients with subtle Lisfranc injuries. Skeletal Radiol. 2022;51:801–6. https://doi.org/10.1007/s00256-021-03892-8.
25. Pflüger P, Harder F, Müller K, et al. Evaluation of ankle fracture classification systems in 193 trimalleolar ankle fractures. Eur J Trauma Emerg Surg. 2022;48:4181–8. https://doi.org/10.1007/s00068-022-01959-2.
26. Rojas EO, Mansur NS, Dibbern KN, et al. Weightbearing computed tomography for assessment of foot and ankle deformities: the Iowa experience. Foot Ankle Orthop. 2022;7:111–9. https://doi.org/10.1177/2473011421S00419.
27. Ebinger T, Goetz J, Dolan L, Phisitkul P. 3D model analysis of existing CT syndesmosis measurements. Iowa Orthop J. 2013;33:40–6.
28. Roberts CS, Beck DJ, Heinsen J, Seligson D. Review article diagnostic ultrasonography: applications in orthopaedic surgery. Clin Orthop Relat Res. 2002;401:248–64. https://doi.org/10.1097/00003086-200208000-00028.

The Diabetic Foot

17

Ali Seker, Goker Utku Deger, and Tolgahan Korkmaz

17.1 Introduction

The number of diabetic patients is increasing, and it is estimated to reach 783,000,000 by 2045. In addition to acute metabolic problems, it can induce neuropathy and vascular complications. These two complications are the important causes of diabetic foot problems. The International Working Group on the Diabetic Foot (IWGDF) defines diabetic foot as having one or more of the following conditions: peripheral neuropathy, peripheral artery disease, infection, ulcer(s), neuro-osteoarthropathy, gangrene, and amputation [1–3].

Foot ulcer is defined as a wound extending to the dermis. It is one of the most important diabetic complications, affecting 15–25% of the patients. Aside from decreasing the quality of life, it increases morbidity and mortality. Unfortunately, 14–24% of these individuals will require lower-extremity amputations. Diabetes is the major cause of amputations in the United States, accounting for 60–70% of all amputations performed each year [4]. The complication rates after diabetes-related amputations are also high. Local infections and soft tissue problems are seen in 9–59.3% of patients, with a reamputation risk of 16.5% in the first 3 months and 20–25% in the first year (Fig. 17.1) [5–7]. The mortality rates in diabetic amputees are reported as 13–49% in the first year and 50–65.3% after 5 years [7–9].

A. Seker (✉) · T. Korkmaz
Cerrahpasa Medical Faculty, Department of Orthopaedics and Traumatology, Istanbul University-Cerrahpasa, Istanbul, Turkey
e-mail: ali.seker@iuc.edu.tr

G. U. Deger
Beykoz State Hospital, Istanbul, Turkey

S. Sharma et al. (eds.), *Clinical and Radiological Examination of the Foot and Ankle*, https://doi.org/10.1007/978-981-97-4202-8_17

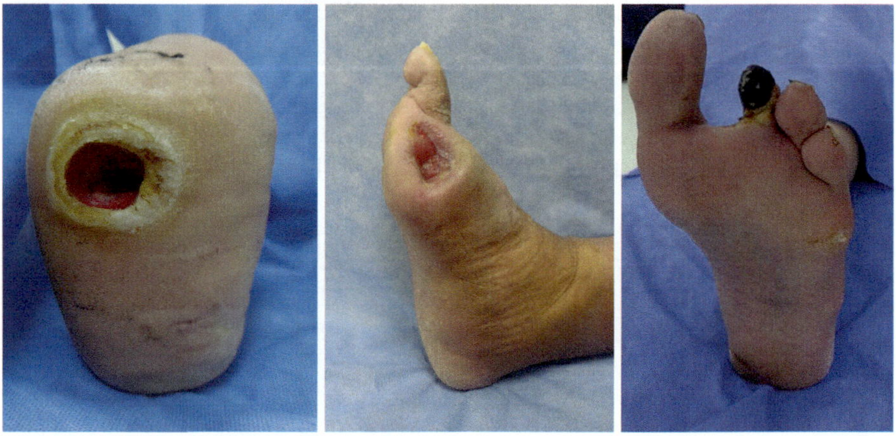

Fig. 17.1 The risk of ulceration and reamputation following diabetic foot amputation is considerable. The images show patients who had previous amputations but now have new ulcer/gangrene

Aside from their medical significance, diabetic foot problems raise healthcare expenses. Over a 10-year period, the estimated costs of diabetic foot care and lower-extremity amputations in the United States were $18 billion and $11.7 billion, respectively [7, 10]. According to Kerr et al., the United Kingdom spent more than £900 million on diabetic foot care in 2019 [11]. These findings highlight the significance of diabetic foot care in both medical and economic implications. That's why a routine foot examination is advised for all diabetics.

17.2 Etiopathogenesis

Neuropathy is the main underlying cause of diabetic foot ulcers. It is bilateral and symmetrical and starts from distal. Motor, sensory, and autonomic impairments may occur [3]:

- Motor neuropathy causes intrinsic muscle weakness. This can result in toe deformities and gait problems.
- Sensorial neuropathy causes loss of protective sensation (LOPS). Pain, temperature, and pressure sensations all decrease. Proprioception is also affected. Patients become oblivious to injuries and the resulting skin disorders. They continue to walk despite the ulcers and healing impairs afterwards [3, 12].
- Autonomic neuropathy promotes decreased perspiration and skin dryness, resulting in skin fissures. Bacterial invasion occurs through these pores, increasing the likelihood of infection. Autonomic neuropathy may potentially impair arteriolar regulation. Blood flow and tissue perfusion will be disrupted by arteriovenous shunting. Pulses may be felt, and veins may be swollen in this scenario [13].

Calluses and pressure ulcers are caused by bony prominences and LOPS, especially after wearing ill-fitting footwear. Calluses are skin thickenings that grow as a result of extreme pressure, particularly on bony prominences. When there is a deformity, the pressure distribution shifts, and calluses appear on the skin that has been subjected to excessive pressure. They are most noticeable on the plantar surface around the metatarsal heads, the lateral border of the fifth metatarsal head, the first metatarsal in the case of hallux valgus, and the dorsum of the lesser toes. Untreated calluses would result in further increased pressure. Increased foot loading causes subcutaneous hemorrhage and, eventually, ulceration.

Peripheral artery disease (PAD) also influences the pathogenesis of diabetic foot ulcers. Decreased blood flow in major arteries and occlusion of the minor arteries cause limb ischemia. This would result in decreased tissue healing and necrosis. It should be noted that PAD and pure ischemia without neuropathy account for a minor proportion of diabetic ulcers. The majority of ulcers are either neuropathic or neuro-ischemic [14].

17.3 History

A thorough history should be obtained prior to the physical examination. All previous diagnoses and treatments should be asked about and documented. Time since diabetes was first diagnosed, glycemic control and medications, prior history of foot problems, and treatments (e.g., wound care, amputation, vascular surgeries, and so on) must all be documented. Calf pain/claudication while walking is an indication of vascular insufficiency. Diabetic foot ulcer risk factors must be investigated (Table 17.1) [2, 12, 15–17].

Table 17.1 Risk factors for diabetic foot ulcers

History of previous foot ulcer
Symptoms related to peripheral neuropathy (numbness, decreased sensation, burning pain, tingling, and electrical sensation)
Symptoms related to peripheral vascular disease (pain at rest, claudication, delay in wound healing)
Foot deformity (hallux valgus, hammer toe, mallet toe, claw toe, Charcot arthropathy)
History of previous amputation
Visual impairment
Use of tobacco and alcohol
Poor glycemic control
Diabetic nephropathy (need for dialysis)
Inappropriate shoe use (wearing ill-fitting or narrow shoes, lack of footwear during the day)

17.4 Physical Examination and Basic Tests

The physical examination of the diabetic foot has to be comprehensive. The patient's general condition may be impacted, and the patient may appear unwell depending on the existence of deep or disseminated infection. As a result, all patients' vital signs should be evaluated, and a systemic examination should be undertaken. Following that, the foot is examined. Both lower limbs must be examined and compared:

1. *Inspection*: The examination may begin with visual gait analysis. The patient should be observed while entering the room. You can also ask the patient to walk in the room. The presence of a limp may be associated with pain or deformity. Shoes and socks must be removed. As previously stated, ill-fitting shoes play an important role in the mechanism of ulceration. As a result, it is critical to inspect the shoes. Shoes with narrow toe box, worn shoes, or shoes that are too small for the patient's feet are considered inappropriate.

   It is preferable to examine the entire lower extremity for any color changes (redness in cellulitis or pallor/darkness in ischemia), muscle atrophy, edema, decreased hair distribution, skin dryness, hemorrhages, fissures, and sores (Fig. 17.2). Neuropathy may result in diminished pain awareness, and the patient

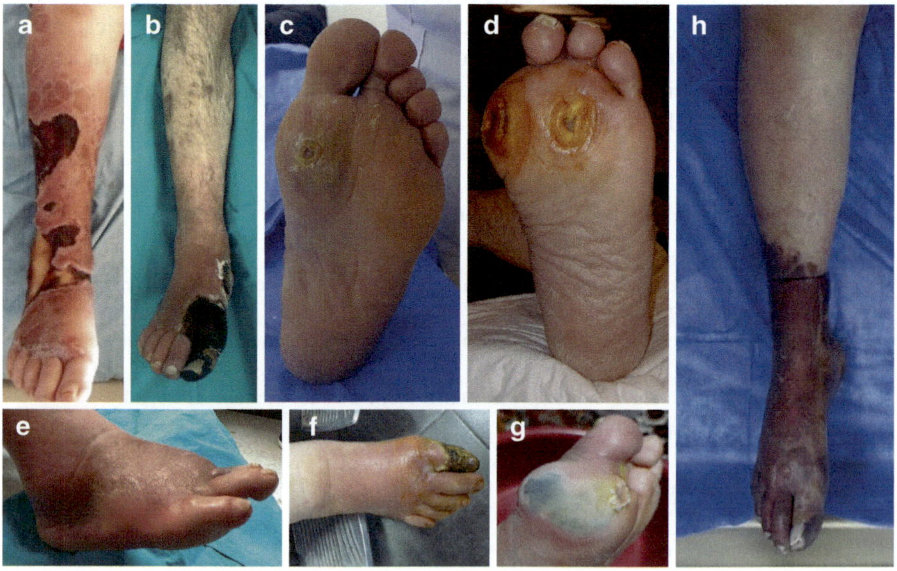

Fig. 17.2 (**a**) The dryness of the skin, change of color, and dispersed wounds on a diabetic foot. (**b**) Gangrene affecting the medial part of the foot. Change of color and decreased body hair indicate ischemia. (**c**) Callus on first metatarsal head. (**d**) Ulcers on metatarsal heads. (**e**) Redness and swelling in an infected diabetic foot. (**f**) The patient has gangrene on the hallux, but soft tissue infection is extending to the level of ankle. (**g**) Callus with intradermal hemorrhages. (**h**) Lower-extremity gangrene with demarcated border

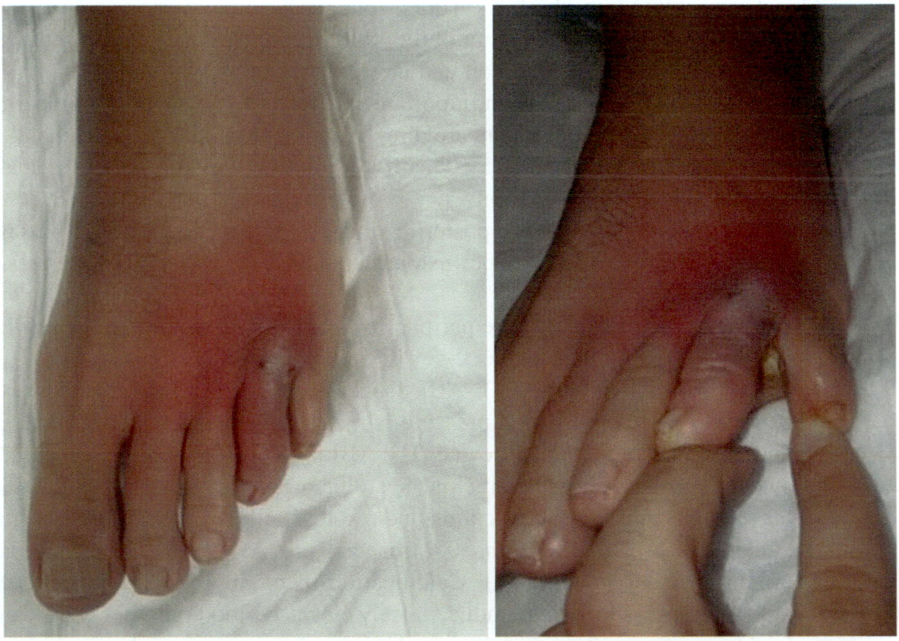

Fig. 17.3 It is crucial to check the web spaces in a diabetic foot for possible ulcers

may be oblivious of his ulcers. Fungal infections in the nails and web spaces should be checked (Fig. 17.3). Sinuses or presence of pus are important findings of infection. Toe deformities and rocker-bottom foot may indicate the presence of neuropathy [2, 3, 12].

2. *Palpation and detailed examination*: Any callus or ulcer on the foot should be evaluated carefully, and it should be kept in mind that small scars may mask underlying deep ulcers and abscesses [17]. The followings should be checked during detailed examination of the patient:
   - *Temperature*: An increase in temperature may be local or systemic. The former can be caused by infection or deep vein thrombosis. Warmth can be felt after arteriovenous shunting even in the presence of ischemia. Coldness of the extremity is an important sign of impaired circulation.
   - *Tenderness*: This should be done to look for foreign bodies, fractures, infection, or other soft tissue problems. The patient may not have pain because of neuropathy, but swelling, crepitation, or solid object beneath the skin can be noticed.
   - *Range of motion (ROM)*: Joint stiffness would result in aberrant loading and impaired shock absorption of the foot. This will raise pressure and may result in ulceration. Check the mobility of the ankle, subtalar, tarsal, and phalangeal joints. Pain and restriction of ROM must be recorded [17, 18].

- *Muscle strength*: Strength of ankle dorsi-/plantar flexors, invertor, and evertor muscles should be checked. Immobility and neuropathy may cause muscle wasting which will affect the gait and increase risk of ulceration [18].
- *Vascular evaluation*: PAD is an important risk factor for diabetic foot ulcers. One-third of diabetic foot ulcers are directly related to PAD (neuro-ischemic or pure ischemic ulcers) [12, 15]. Therefore, vascular assessment is a crucial part of the physical examination:
  Pulses: The dorsalis pedis, posterior tibial, popliteal, and femoral pulses should all be checked. Well perfusion is not always guaranteed by the presence of pulses.
  Capillary refilling test: If the peripheric circulation is disturbed, the refilling time could be longer than 2–3 s [2].
  Ankle brachial index (ABI): ABI can be used for evaluation of lower-extremity perfusion. Blood pressure is measured at the level of the ankle and arm by using Doppler device. The ABI is calculated by dividing ankle systolic pressure by the brachial systolic pressure. A normal value is between 0.9 and 1.3. Claudication and rest discomfort are both linked with values of lower than 0.8 and 0.4, respectively. ABI less than 0.5 needs urgent vascular imaging [3]. Calcification of blood vessels may reduce artery compressibility and induce falsely elevated blood pressure. This can lead to misinterpretation of ABI, and if it is >1.3, measuring the transcutaneous oxygen tension, toe-brachial index (use of toe pressure instead of ankle) or absolute toe systolic pressure would be preferable [3, 12]. Toe systolic pressure lesser than 45 mmHg is accepted as abnormal, and it should be greater than 30 mmHg for normal wound healing [19, 20].
- *Neurologic evaluation*: Peripheral neuropathy is the leading cause of diabetic foot ulcers. The first step of neurologic examination is to determine whether there is loss of protective sensation. The evaluation of neuropathy may need advanced tests [21]:
  Sensation: Cutaneous innervation should be checked by using cotton wool. Ask patient to close his eyes and move from distal to proximal. You can also lightly touch to the toe tips with your index finger for 1 or 2 s.
  Reflexes: Achilles tendon reflex should be checked. While the patient is lying on a couch, the ankle joint is kept in a neutral position, and Achilles tendon is struck by a reflex hammer. Decreased or absent reflex is a sign of peripheral neuropathy [17].
  10-g monofilament (Semmes-Weinstein) test: This test is used to check the loss of sensation. Ask patient to close his eyes. A nylon monofilament is reclined against the skin and sustained to 10-g force which will cause bending (Fig. 17.4). The patient must recognize the pressure and the application point. The inability to feel this pressure is accepted as abnormal. It is recommended to test metatarsal heads and the plantar surface of the hallux; in general however checking more sites increases sensitivity of the test. Avoid the skin with calluses, and before the test make a tutorial on upper extremities [12, 22].

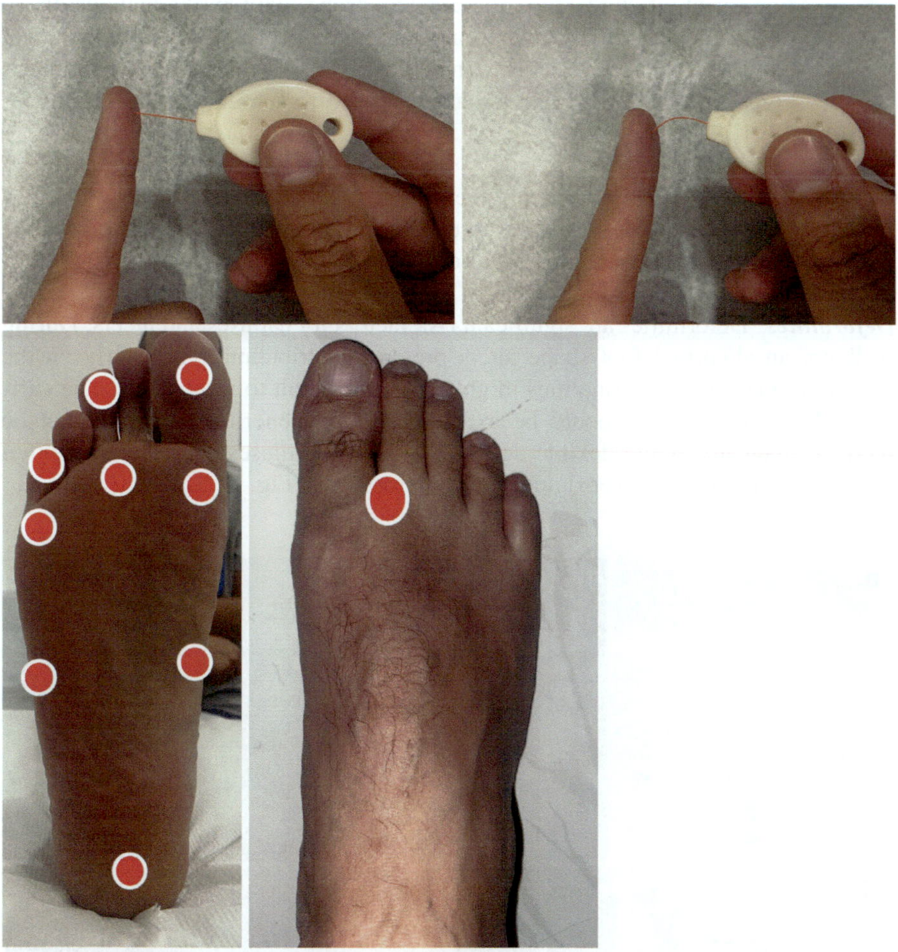

Fig. 17.4 Application of 10-g monofilament test and test sites

Ipswich touch test: This test is performed by touching to the first, third, and fifth toes bilaterally (total of six sites) by the index finger for 1–2 s. The existence of LOPS is defined by feeling four or fewer sites [17].

Pinprick test: A pin is used to distort the dorsal skin of the great toe just proximal to the nail. The test should be done bilaterally, and inability to perceive pinprick is accepted as abnormal [12, 22].

Tuning fork vibration test: This test is used to evaluate the sensation of vibration. A 128 Hz tuning fork is held on the tip of the great toe. It is considered abnormal if the patient loses vibration sensation, while the examiner continues to feel it [3].

Vibration perception threshold (VPT) test: This test is performed by using a special device (biothesiometer or neurothesiometer) that measures vibration sensitivity threshold. The device's probe is put on the hallux's dorsal or plantar surface. The vibration's amplitude is increased until the patient can feel it. That level is recognized as VPT. A value of 25 V or more is considered abnormal and is closely linked to subsequent ulcerations [2, 12, 17].

One of these tests can be used for the diagnosis of neuropathy. The American Diabetes Association (ADA) advises to use 10-g monofilament test and one of the other tests. This approach increases the sensitivity of the diagnosis by 87% [12, 16].

Deformities Deformities are significant risk factors for calluses and ulcers. During walking, an abnormal foot shape alters pressure distribution, particularly on the plantar surface. Toe abnormalities might make it difficult to wear shoes. The stiffness of the deformities should be examined. Rigid abnormalities could not be rectified manually, while flexible deformities could. Calluses and ulcers are often caused by prominences over rigid forefoot deformities (Fig. 17.5).

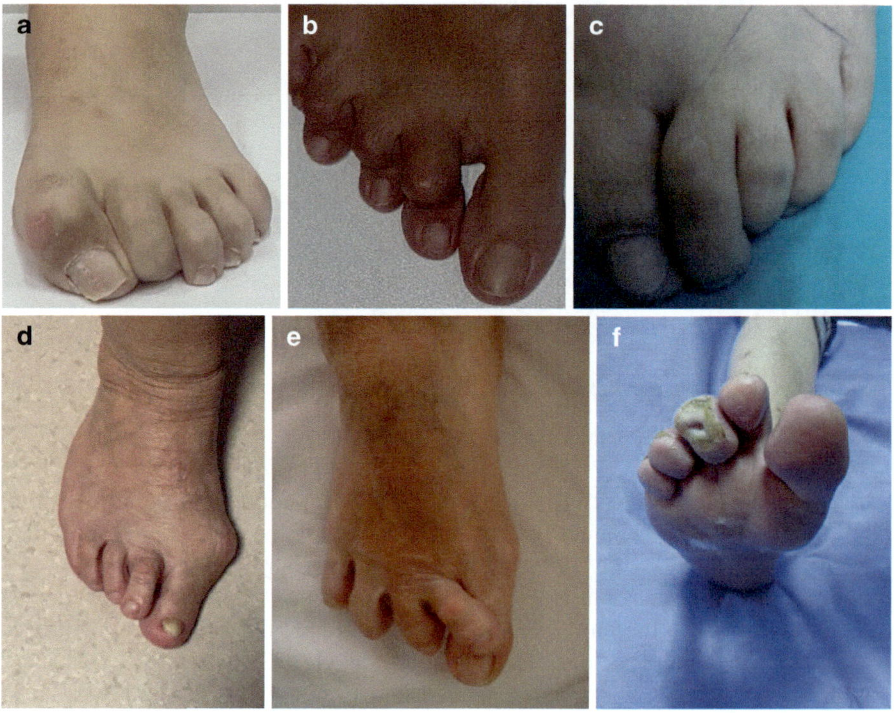

Fig. 17.5 (**a**) Mallet toe deformities of the hallux and third toe. Distal interphalangeal (DIP) joint has flexion deformity. (**b**) Hammer toe deformity in the second, third, and fourth toes. There is flexion deformity in proximal interphalangeal (PIP) joint. (**c**) Claw toe deformity in the second, third, and fourth toes. Note the hyperextension of metatarsophalangeal joints and flexion deformities of the PIP and DIP joints. (**d**) The patient has hallux valgus, hammer toe and claw toe deformities. (**e**) Hammer toe deformity and hallux valgus may cause overlapping of the toe. (**f**) Ulceration caused by claw toe

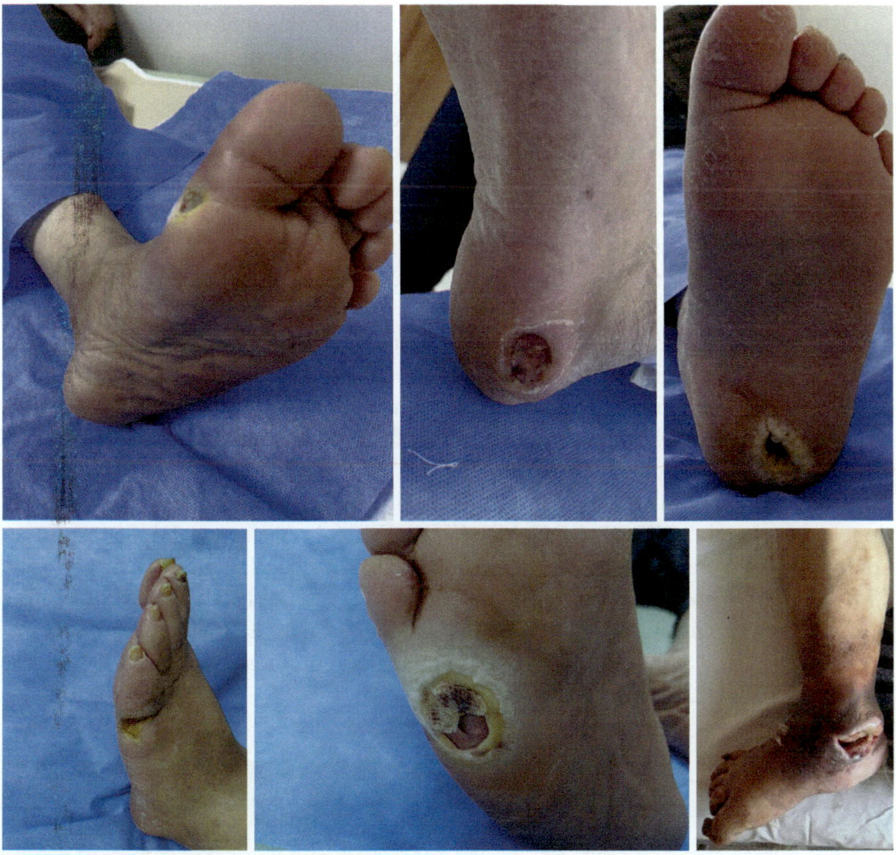

Fig. 17.6 Ulcers caused by ill-fitting shoes in diabetic patients

Examination of Ulcers This stage is critical in the treatment decision, and the ulcer must be carefully examined for success. The etiology (neuropathic, ischemic, or neuro-ischemic) and possible cause of the ulcer (trauma, ill-fitting shoes) must be determined (Fig. 17.6). The location of the ulcer may provide useful information. Neuropathic ulcers are most commonly found on the plantar surface of the foot, over bony prominences. Neuro-ischemic and ischemic ones are typically found on the toes and the lateral margins of the foot [12, 14].

The ulcer may be covered by a callus or necrotic tissue. Signs of infection should be checked and graded according to Infectious Diseases Society of America and International Working Group on the Diabetic Foot Classifications of Diabetic Foot Infection (ISDA/IWGDF). The ulcer is graded as 1 if there is no evidence of infection. Mild infection (superficial) is graded as 2, moderate infection (deep/extensive) as 3, and severe infection (associated systemic symptoms) as 4. The addition of "O" denotes concomitant osteomyelitis [3].

Debridement can be done in order to determine the real depth of the ulcer. Evaluate the ulcer with a sterile probe. If it touches the bone, it is more likely to be osteomyelitis. Cultures are required for effective antibiotherapy if the wound is infected. Tissue specimens should be obtained for culture in that case. Swabs and superficial samples are inappropriate for this purpose. Bone biopsy is the best choice for isolating pathogens in osteomyelitis. If there is a suspicion of local or systemic infection, blood samples can be taken. Complete blood count, C-reactive protein (CRP), erythrocyte sedimentation rate (ESR), procalcitonin, blood culture, serum glucose and electrolyte levels, HbA1c level, and kidney/liver function tests should be ordered [2, 3].

Following this thorough investigation, the ulcers should be categorized. The goal of this categorization is to assist healthcare providers in determining a treatment and follow-up plan. Examiners can choose from a variety of classification systems for diabetic foot ulcers. Wagner classification stratifies foot according to the depth of the ulcer and presence of necrosis (Fig. 17.7). The University of Texas Wound Classification System (UTWCS) uses depth of the ulcer and the presence of infection and ischemia. PEDIS, SINBAD, IWGDF/IDSA, WIfI, and DUSS are other commonly used classification systems (Tables 17.2, 17.3, 17.4, 17.5, 17.6, 17.7, and 17.8). Among these, IWGDF recommends to use IWGDF/ISDA criteria for infection assessment, the WIfI system to assess the risk of amputation and the benefit of revascularization, and the SINBAD system for communication among healthcare professionals [3, 23–25].

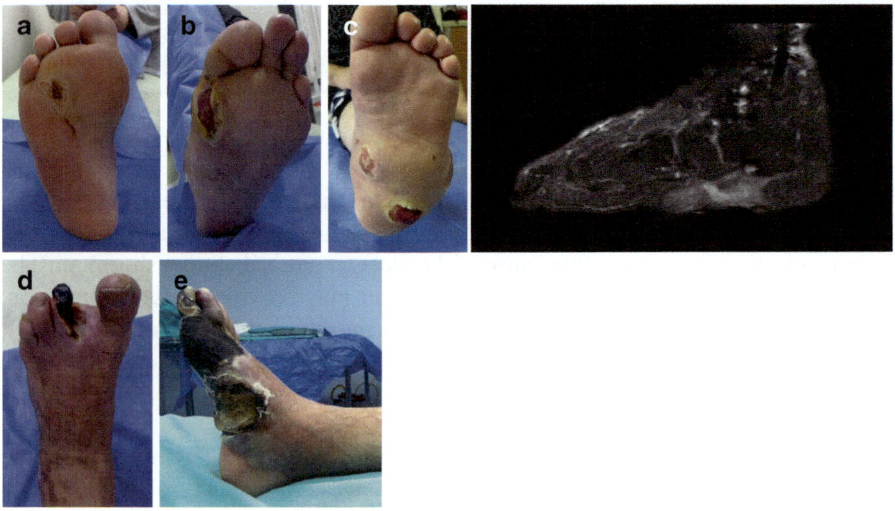

Fig. 17.7 Wagner classification for diabetic ulcer. (**a**) Grade 1 ulcer. (**b**) Grade 2 ulcer. (**c**) Grade 3 ulcer. (MRI demonstrating abscess) (**d**) Grade 4 ulcer. (**e**) Grade 5 ulcer

Table 17.2 Wagner classification

Grade	Lesion
0	No ulcer, high-risk foot (presence of callus, deformity, or sensorial deficit)
1	Superficial ulcer
2	Deep ulcer (tendon, ligament, or joint capsule is affected, no osteomyelitis or abscess)
3	Deep ulcer with abscess/osteomyelitis
4	Localized gangrene of the forefoot or heel
5	Gangrene of whole foot

Table 17.3 University of Texas Wound Classification System (UTWCS)

	Grade			
Stage	0	1	2	3
A	Pre- or post-ulcerative lesion completely epithelialized	Superficial wound, not involving tendon, capsule or bone	Wound penetrating to tendon or capsule	Wound penetrating to the bone or joint
B	With infection	With infection	With infection	With infection
C	With ischemia	With ischemia	With ischemia	With ischemia
D	With infection and ischemia	With infection and ischemia	With infection and ischemia	With infection and ischemia

Table 17.4 PEDIS (perfusion-extent-depth-infection-sensation) classification

Grade	Perfusion	Extent	Depth	Infection	Sensation
1	PVD−	Skin intact	Superficial ulcer (not deeper than dermis)	No symptom or sign of infection	No loss
2	PVD+ CLI−	<1 cm^2	Penetrating below dermis	Infection involving the skin and superficial subcutaneous tissue but no systemic signs	Loss of sensation
3	CLI−	1–3 cm^2	All layers of the foot including bone/joint	Infection involving deep tissues, presence of abscess or osteomyelitis, but no systemic signs	–
4	Any foot infection with signs of systemic inflammatory response syndrome				

PVD peripheral vascular disease, *CLI* critical limb ischemia

Table 17.5 SINBAD (site-ischemia-neuropathy-bacterial infection-area ulcer-depth) system

Category	Definition	Score
Site	Forefoot Midfoot and hindfoot	0 1
Ischemia	Pedal blood flow intact: at least one palpable pulse Clinical evidence of reduced pedal flow	0 1
Neuropathy	Protective sensation intact Protective sensation lost	0 1
Bacterial infection	None Present	0 1
Area ulcer	Ulcer <1 cm^2 Ulcer ≥ 1 cm^2	0 1
Depth	Ulcer confined to the skin and subcutaneous tissue Ulcer reaching the muscle, tendon, or deeper	0 1
Total possible score		0–6

Table 17.6 IWGDF/IDSA system

Clinical manifestation	Infection severity	PEDIS grade
Wound lacking purulence or any manifestations of inflammation	Uninfected	1
Presence of ≥ 2 manifestations of inflammation (purulence, erythema, tenderness, warmth, or induration), but any cellulitis/erythema extends ≤ 2 cm around the ulcer, and infection is limited to the skin or superficial subcutaneous tissues; no other local complications or systemic illness	Mild	2
Infection (as above) in a patient who is systemically well and metabolically stable but which has ≥ 1 of the following characteristics: Cellulitis extending >2 cm, lymphangitic streaking, spread beneath the superficial fascia, deep-tissue abscess, gangrene, and involvement of the muscle, tendon, joint, or bone Infection involving the bone (osteomyelitis)	Moderate	3 Add "O"
Infection in a patient with systemic toxicity or metabolic instability (e.g., fever, chills, tachycardia, hypotension, confusion, vomiting, leucocytosis, acidosis, severe hyperglycemia, or azotemia) Infection involving the bone (osteomyelitis)	Severe	4 Add "0"

Table 17.7 WIfI system

Wound		
Grade	DFU	Gangrene
0	No ulcer Clinical description: minor tissue loss. Salvageable with simple digital amputation (one or two digits) or skin coverage	No gangrene
1	Small shallow ulcer(s) on the distal leg or foot; no exposed bone, unless limited to the distal phalanx Clinical description: minor tissue loss. Salvageable with simple digital amputation (one or two digits) or skin coverage	No gangrene

Table 17.7 (continued)

Wound			
Grade	DFU		Gangrene
2	Deeper ulcer with exposed bone, joint or tendon; generally not involving the heel; shallow heel ulcer, without calcaneal involvement Clinical description: major tissue loss salvageable with multiple (≥3) digital amputations or standard TMA ± skin coverage		Gangrenous changes limited to digits
3	Extensive, deep ulcer involving the forefoot and/or midfoot; deep, full-thickness heel ulcer ± calcaneal involvement Clinical description: extensive tissue loss salvageable only with a complex foot reconstruction or non-traditional TMA (Chopart or Lisfranc); flap coverage or complex wound management needed for large soft tissue defect		Extensive gangrene involving the forefoot and/or midfoot; full-thickness heel necrosis ± calcaneal involvement
Ischemia			
Grade	ABI	Ankle systolic pressure (mmHg)	TP, $TcPO_2$ (mmHg)
0	≥0.80	>100	≥60
1	0.6–0.79	70–100	40–59
2	0.4–0.59	50–70	30–39
3	≤0.39	<50	<30
Foot infection			
Grade	Clinical manifestation		
0	No symptoms or signs of infection Infection present, as defined by the presence of at least two of the following items: • Local swelling or induration • Erythema >0.5 to ≤2 cm around the ulcer • Local tenderness or pain • Local warmth • Purulent discharge (thick, opaque to white, or sanguineous secretion)		
1	The local infection involving only the skin and the subcutaneous tissue (without involvement of deeper tissues and without systemic signs as described below) Exclude other causes of an inflammatory response of the skin (e.g., trauma, gout, acute Charcot neuro-osteoarthropathy, fracture, thrombosis, venous stasis)		
2	Local infection (as described above) with erythema >2 cm or involving structures deeper than the skin and subcutaneous tissues (e.g., abscess, osteomyelitis, septic arthritis, fasciitis) No systemic inflammatory response signs (as described below)		
3	Local infection (as described above) with the signs of SIRS, as manifested by two or more of the following: • Temperature >38 °C or <36 °C • Heart rate >90 beats/min • Respiratory rate >20 breaths/min or $PaCO_2$ <32 mmHg • White blood cell count >12,000 or <4000 cu/mm or 10% immature (band) forms		

ABI ankle-brachial index, *$PaCO_2$* partial pressure of carbon dioxide, *SIRS* systemic inflammatory response syndrome, *$TcPO_2$* transcutaneous partial oxygen pressure, *TMA* transmetatarsal amputation, *TP* toe pressure

Table 17.8 Diabetic ulcer severity score (DUSS)

Variables	Score 0	Score 1
Palpable pedal pulses	Presence	Absence
Probing to the bone	No	Yes
Ulcer site	Toes	Foot
Ulcer number	Single	Multiple

17.5 Radiologic Evaluation

After physical examination radiologic assessment can be done. Depending on the physical examination findings, various diagnostic imaging techniques may be ordered:

- X-ray: Direct radiograph is the first and most common radiologic imaging approach used in diabetic foot patients. It is useful for determining the presence of osteomyelitis, Charcot neuroarthropathy (CN), and foreign bodies.

  The radiographs may be clean in the early stages of osteomyelitis, but after 2–3 weeks, alterations become visible [26] (Figs. 17.8 and 17.9).

  Possible findings of osteomyelitis on X-rays are:

  - Periosteal reaction (acute osteomyelitis)
  - Increased bone sclerosis (acute-subacute osteomyelitis)
  - Destruction, demineralization, and erosion of the bone (chronic osteomyelitis).
  - Presence of sequestrum (devitalized dense bone) and/or involucrum (layer of new bone) (chronic osteomyelitis).
  - Abnormal subcutaneous tissue or gas density under the skin.

  Similar findings can also be observed in Charcot arthropathy. In that case, magnetic resonance imaging or scintigraphy can be used for differentiation.
- Magnetic resonance imaging (MRI): If osteomyelitis is suspected, an MRI of the foot should be obtained. Aside from osteomyelitis, the morphology of soft tissues and joints, as well as deep infections and abscesses, may all be observed (Figs. 17.10 and 17.11).
- Ultrasonography: It can be used for diagnosis of foreign bodies, soft tissue infections, and abscesses. Limited information on the extent of infection can be gathered based on the performers' experience. It is generally preferred in patients who have contraindications for MRI.
- Computed tomography (CT): It can be used for evaluation of bone anatomy. CT can detect the presence of sequestrum and involucrum as indicators of osteomyelitis. In Charcot arthropathy, the structural integrity of the bones, fractures, and dislocations could be examined. It can be used in conjunction with IV contrast to investigate osteomyelitis [19].
- Nuclear medicine: Leukocyte-labeled scintigraphy and positron emission tomography are very sensitive but expensive imaging techniques. They can be used to diagnose osteomyelitis. They can be preferred particularly in patients who had contraindications for MRI [2, 26].

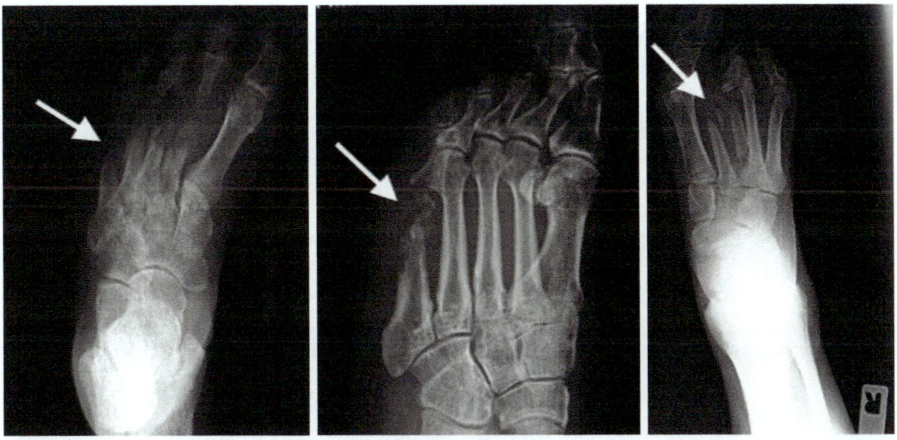

Fig. 17.8 Osteomyelitis in three patients. Note the destruction and osteolysis

Fig. 17.9 Ulcer on the fourth toe. The foot is swollen. There is osteolysis on x-ray which may indicate osteomyelitis

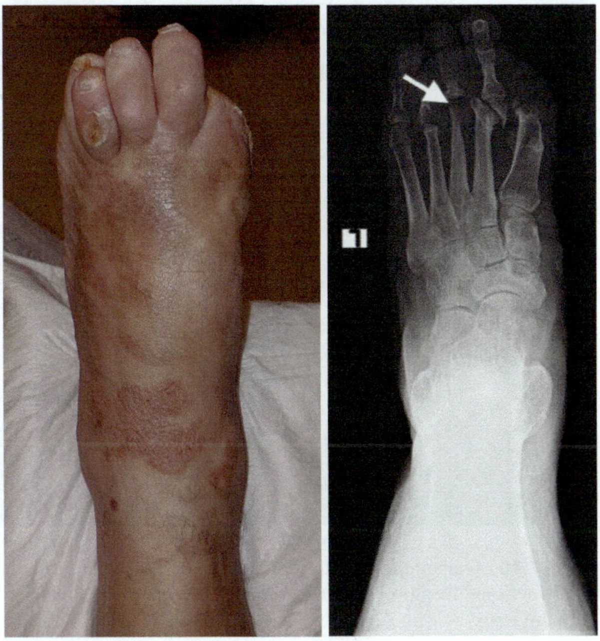

- Doppler ultrasonography: If the pedal pulses cannot be felt or there are signs of PAD, Doppler USG should be ordered. The arteries should be pulsatile with two or three sounds (biphasic or triphasic). It can be audible but not pulsatile (single sound/monophasic). Inaudible or monophasic arteries indicate poor circulation. Detailed assessment should be done in that case [17]. Color-coded duplex sonography, digital subtraction angiography, MRI angiography, and CT angiography can be preferred depending on the patient's condition (Fig. 17.12).

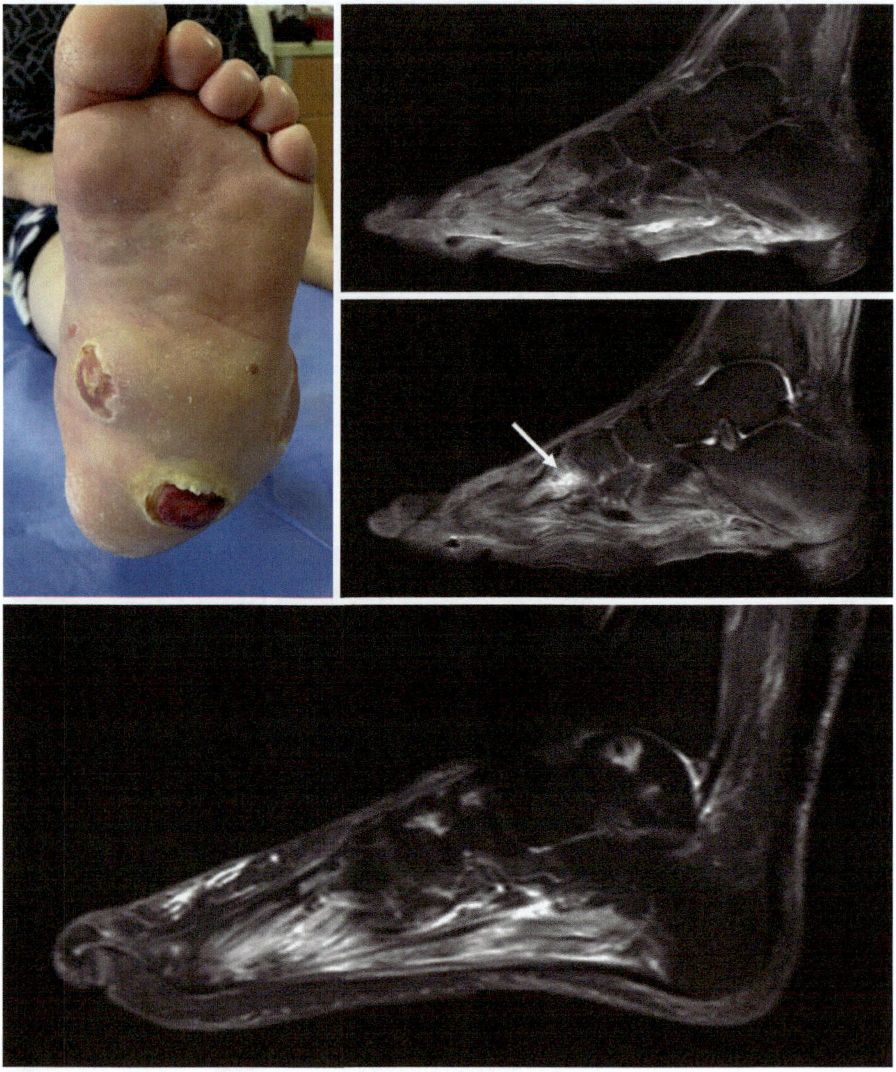

Fig. 17.10 Diabetic foot with ulcer on the heel. Bone marrow and soft tissue edema are seen on MRI sections. Depth of the ulcer is extending to the calcaneus; therefore it is more likely to be osteomyelitis

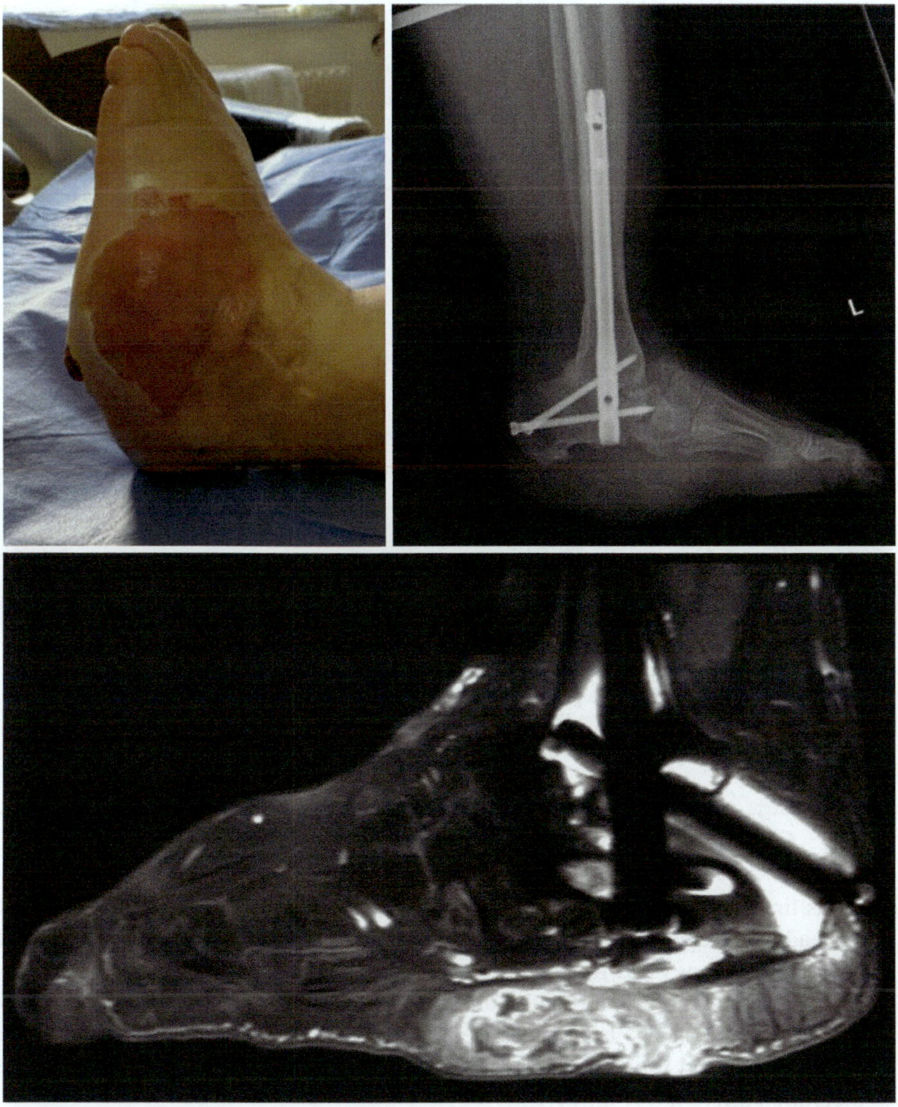

Fig. 17.11 Osteomyelitis and abscess in a diabetic patient

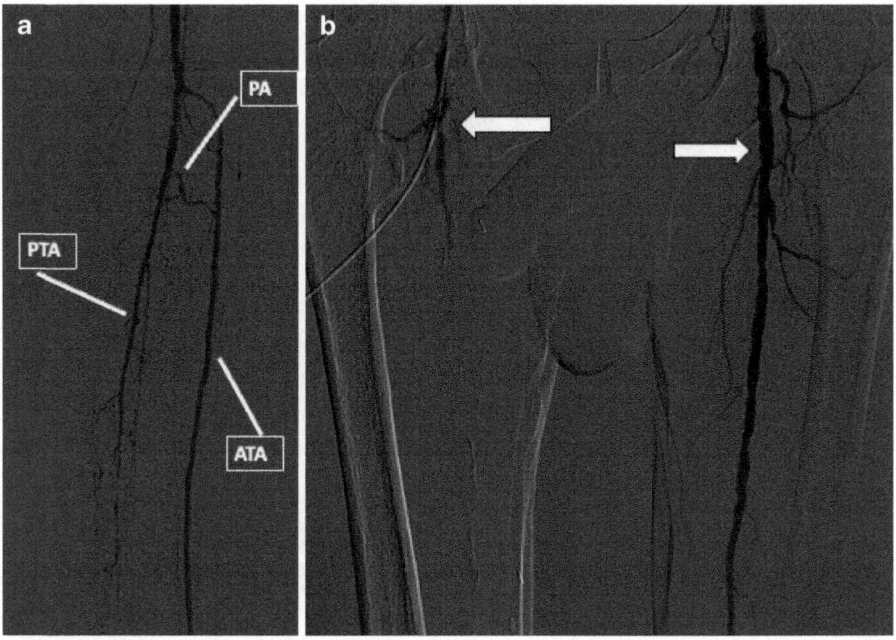

Fig. 17.12 (**a**) Vascular assessment of a diabetic patient. Note the cessation of blood flow in the proximal peroneal artery. Tibialis posterior artery becomes almost invisible distally. *PTA* posterior tibial artery, *PA* peroneal artery, *ATA* anterior tibial artery. (**b**) Angiography of a patient with occluded right femoral artery. Left femoral artery is narrowed in multiple points

17.6 Charcot Neuroarthropathy

CN was first described by Jean-Martin Charcot in a patient with syphilis. Diabetes is the leading cause of CN nowadays [27]. It is presented as dislocation of joints and collapse of the bones and foot. There is no need for apparent trauma. Local osteoporosis develops due to neuropathy and disturbed microvascular circulation (arteriovenous shunting). The foot is perfused but insensitive. Spontaneous or traumatic fractures occur, and progressive deformity develops. The foot becomes vulnerable to ulceration afterward (Fig. 17.13).

Brodsky classified CN anatomically. The affected part of the foot describes the type as midfoot type 1, hindfoot (subtalar, talonavicular, and calcaneonavicular joints) type 2, ankle joint type 3A, and tuber calcanei type 3B. Type 1 is the most common one (60%) followed by type 2 (30–35%) [28].

Sanders and Frykberg also developed an anatomical classification system. In type 1 metatarsophalangeal and interphalangeal joints, type 2 tarsometatarsal joints, type 3 tarsal bones and joints, type 4 subtalar and ankle joints, and type 5 calcaneus are affected. Type 2 (40%) and type 3 (30%) are the most common ones [29].

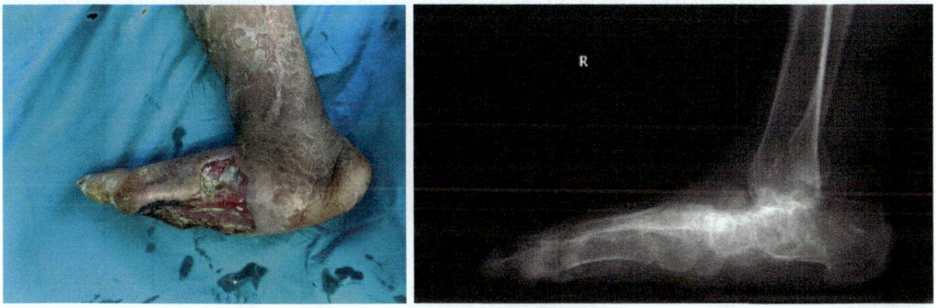

Fig. 17.13 Rocker-bottom foot deformity and ulceration in a patient with Charcot neuroarthropathy

Table 17.9 Modified Eichenholtz classification. Clinical and radiological findings of Charcot foot

Stage		Clinical findings	Radiology
0 (acute)	Prodromal (inflammatory)	Swelling, erythema, warmth	Normal
1 (acute)	Developmental (fragmentation)	Swelling, erythema, warmth	Osteopenia, fragmentation, fractures, subluxations/ dislocations
2 (chronic)	Coalescence	Improved clinic (decreased edema, erythema, and warmth)	Coalescence/fusion of fragments, callus formation, sclerosis of bone ends
3 (chronic)	Reconstruction	No warmth, swelling, or erythema Fixed-foot deformity	Consolidation and remodeling of fractures, decreased sclerosis of bone ends

Eichenholtz classified CN clinically and radiologically [30]. He staged the condition as inflammation (stage 1), repair (stage 2), and remodeling (stage 3) [30]. Later it was revised by addition of Stage 0 which describes radiologically undetectable changes [31] (Table 17.9).

The patients generally attend to physicians in the prodromal period. The foot is swollen, red, and warm. Depending on the severity of the neuropathy, the patient may experience pain. The condition can be confused with acute infection. In that scenario, an elevation test may be beneficial. Elevating the lower extremity for 2 min. reduces erythema, indicating reactive hyperemia in the CN. If there is no infection, leucocyte count, CRP, and ESR values are normal. Radiographs can be normal in the early stages. Osteolysis, fractures, dislocations, and new bone formation can be observed during follow-ups (Fig. 17.14). MRI and leucocyte-labeled (Indium-111) bone scintigraphy could help for early diagnosis and differentiation from osteomyelitis [26] (Fig. 17.15).

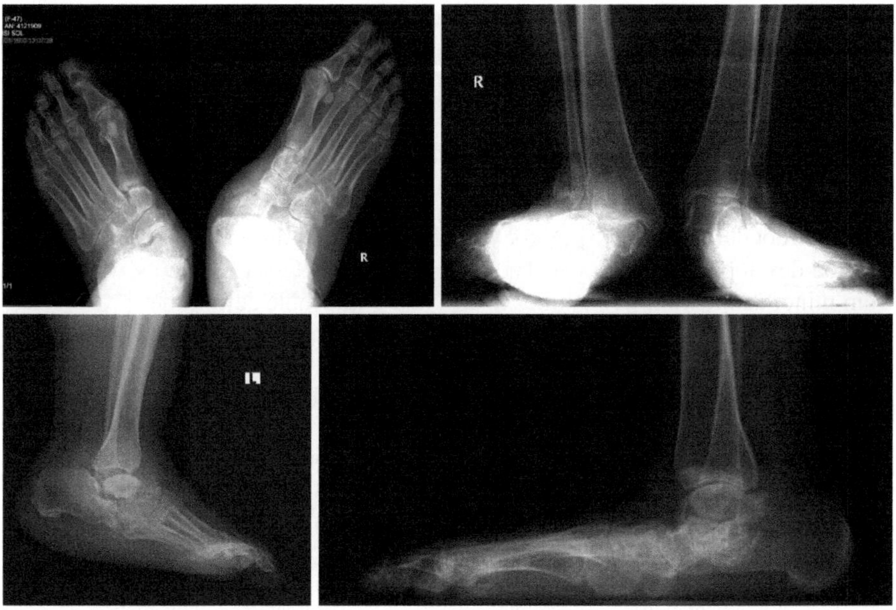

Fig. 17.14 Destruction of bones and joints in Charcot neuroarthropathy. The patient had neuropathy and tried to walk despite the ulceration

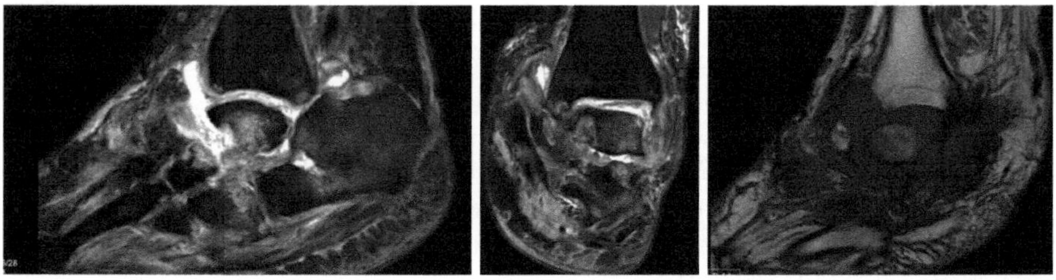

Fig. 17.15 MRI sections of a patient with Charcot neuroarthropathy. Destruction of the bones and bone marrow edema are obvious and can also be seen in osteomyelitis

17.7 Determination of the Risk Level

The IWGDF and ADA published guidelines for the management of diabetic foot. They developed classification systems to stratify the risk category. The aim is to help physicians by providing an algorithm for patient follow-up and referral. The ADA and IWGDF guidelines were revised in 2018 and 2023, respectively [3, 16, 23] (Tables 17.10 and 17.11).

Table 17.10 IWGDF risk stratification system

Category	Ulcer risk	Characteristics	Frequency
0	Very low	No LOPS and no PAD	Once a year
1	Low	LOPS or PAD	Once every 6–12 months
2	Moderate	LOPS + PAD or LOPS + foot deformity or PAD + foot deformity	Once every 3–6 months
3	High	LOPS or PAD and one or more of the following: (1) history of a foot ulcer, (2) a lower-extremity amputation (minor or major), and (3) end-stage renal disease	Once every 1–3 months

LOPS loos of protective sensation, *PAD* peripheral arterial disease

Table 17.11 Modified ADA diabetic foot risk classification [16]

Priority	Indications	Timeline	Suggested follow-up
Urgent (active pathology)	• Open wound or ulcerative area, with or without signs of infection • New neuropathic pain or pain at rest • Signs of active Charcot deformity (red, hot, swollen midfoot or ankle) • Vascular compromise (sudden absent DP/PT pulses or gangrene)	Immediate referral/ consultation	As determined by specialist
High (ADA risk category 3: the diabetic foot in remission)	• Presence of diabetes with a previous history of ulcer or lower-extremity amputation • Chronic venous insufficiency (skin color change or temperature difference)	Immediate or "next available" outpatient referral	Every 1–2 months
Moderate (ADA risk category 2)	• PAD ± LOPS • DP/PT pulses diminished • Presence of swelling or edema	Referral within 1–3 weeks (if not already receiving regular care)	Every 2–3 months
Low (ADA risk category 1)	• LOPS ± long-standing, non-changing deformity • Patient requires prescriptive or accommodative footwear	Referral within 1 month	Every 4–6 months
Very low (ADA risk category 0)	• No LOPS or PAD • Patient seeks education on topics such as routine foot care, athletic training, appropriate footwear, or injury prevention	Referral within 1–3 months	At least annually for all people with diabetes

LOPS loos of protective sensation, *PAD* peripheral arterial disease, *DP* dorsalis pedis, *TP* tibialis posterior

Key Points

- Foot ulcer is defined as a wound extending to the dermis. It is one of the most important diabetic complications, affecting 15–25% of the patients.
- Calluses and pressure ulcers are caused by bony prominences and loss of protective sensation, especially after wearing ill-fitting footwear.
- Peripheral neuropathy is the leading cause of diabetic foot ulcers. The first step of neurologic examination is to determine whether there is loss of protective sensation.
- The ulcer may be covered by a callus or necrotic tissue. Signs of infection should be checked and graded according to Infectious Diseases Society of America and International Working Group on the Diabetic Foot Classifications of Diabetic Foot Infection (ISDA/IWGDF).
- Direct radiograph is the first and most common radiologic imaging approach used in diabetic foot patients. It is useful for determining the presence of osteomyelitis, Charcot neuroarthropathy (CN), and foreign bodies.

References

1. NCD-RisC. Worldwide trends in diabetes since 1980: a pooled analysis of 751 population-based studies with 4.4 million participants. Lancet. 2016;387(10027):1513–30. https://doi.org/10.1016/S0140-6736(16)00618-8.
2. Kravitz SR, McGuire J, Shanahan SD. Physical assessment of the diabetic foot. Adv Skin Wound Care. 2003;16(2):68–75.
3. Schaper NC, van Netten JJ, Apelqvist J, Bus SA, Fitridge R, Game F, Monteiro-Soares M, Senneville E, IWGDF Editorial Board. Practical guidelines on the prevention and management of diabetes-related foot disease (IWGDF 2023 update). Diabetes Metab Res Rev. 2023;27:e3657. https://doi.org/10.1002/dmrr.3657.
4. Walicka M, Raczyńska M, Marcinkowska K, Lisicka I, Czaicki A, Wierzba W, Franek E. Amputations of lower limb in subjects with diabetes mellitus: reasons and 30-day mortality. J Diabetes Res. 2021;2021:8866126. https://doi.org/10.1155/2021/8866126.
5. Norvell DC, Czerniecki JM. Risks and risk factors for ipsilateral re- amputation in the first year following first major unilateral dysvascular amputation. Eur J Vasc Endovasc Surg. 2020;60(4):614–21.
6. Ciufo DJ, Thirukumaran CP, Marchese R, Oh I. Risk factors for reoperation, readmission and early complications after below knee amputation. Injury. 2019;50(2):462–6.
7. Dillingham TR, Pezzin LE, Shore AD. Reamputation, mortality and health care costs among persons with dysvascular lower limb amputations. Arch Phys Med Rehabil. 2005;86(3):480–6.
8. Gurney JK, Stanley J, Rumball-Smith J, York S, Sarfati D. Postoperative death after lower-limb amputation in a national prevalent cohort of patients with diabetes. Diabetes Care. 2018;41(6):1204–11.
9. Aulivola B, Hile CN, Hamdan AD, Sheahan MG, Veraldi JR, Skillman JJ, Campbell DR, Scovell SD, LoGerfo FW, Pomposelli FB. Major lower extremity amputation: outcome of a modern series. Arch Surg. 2004;139(4):395–9; discussion 399. https://doi.org/10.1001/archsurg.139.4.395.
10. Rogers LC, Lavery LA, Armstrong DG. The right to bear legs. J Am Podiatr Med Assoc. 2008;98:166–8.
11. Kerr M, Barron E, Chadwick P, Evans T, Kong WM, Rayman G, Sutton-Smith M, Todd G, Young B, Jeffcoate WJ. The cost of diabetic foot ulcers and amputations to the National Health Service in England. Diabet Med. 2019;36(8):995–1002. https://doi.org/10.1111/dme.13973.

12. Boulton AJ, Armstrong DG, Albert SF, Frykberg RG, Hellman R, Kirkman MS, Lavery LA, Lemaster JW, Mills JL, Mueller MJ, Sheehan P, Wukich DK, American Diabetes Association; American Association of Clinical Endocrinologists. Comprehensive foot examination and risk assessment: a report of the task force of the foot care interest group of the American Diabetes Association, with endorsement by the American Association of Clinical Endocrinologists. Diabetes Care. 2008;31(8):1679–85. https://doi.org/10.2337/dc08-9021.
13. Boulton AJM. Diabetic neuropathy and foot complications. Handb Clin Neurol. 2014;126:96–107.
14. Schaper NC, van Netten JJ, Apelqvist J, Bus SA, Hinchliffe RJ, Lipsky BA, IWGDF Editorial Board. Practical guidelines on the prevention and management of diabetic foot disease (IWGDF 2019 update). Diabetes Metab Res Rev. 2020;36(1):e3266. https://doi.org/10.1002/dmrr.3266.
15. Peters EJ, Armstrong DG, Lavery LA. Risk factors for recurrent diabetic foot ulcers: site matters. Diabetes Care. 2007;30:2077–9.
16. Boulton AJ, Armstrong DG, Kirsner RS, Attinger CE, Lavery LA, Lipsky BA, Mills JL, Steinberg JS. Diagnosis and management of diabetic foot complications. Arlington: American Diabetes Association; 2018.
17. Johnson R, Osbourne A, Rispoli J, Verdin C. The diabetic foot assessment. Orthop Nurs. 2018;37(1):13–21. https://doi.org/10.1097/NOR.0000000000000414.
18. Arsanjani Shirazi A, Nasiri M, Yazdanpanah L. Dermatological and musculoskeletal assessment of diabetic foot: a narrative review. Diabetes Metab Syndr. 2016;10(2):158–64. https://doi.org/10.1016/j.dsx.2016.03.004.
19. Reardon R, Simring D, Kim B, Mortensen J, Williams D, Leslie A. The diabetic foot ulcer. Aust J Gen Pract. 2020;49(5):250–5. https://doi.org/10.31128/AJGP-11-19-5161.
20. Armstrong DG, Lavery LA. Diabetic foot ulcers: prevention, diagnosis and classification. Am Fam Physician. 1998;57(6):1325–32.
21. Boulton AJ, Malik RA, Arezzo JC, Sosenko JM. Diabetic somatic neuropathies. Diabetes Care. 2004;27:1458–86.
22. Mishra SC, Chhatbar KC, Kashikar A, Mehndiratta A. Diabetic foot. BMJ. 2017;359:j5064. https://doi.org/10.1136/bmj.j5064.
23. Monteiro-Soares M, Hamilton EJ, Russell DA, Srisawasdi G, Boyko EJ, Mills JL, Jeffcoate W, Game F. Guidelines on the classification of foot ulcers in people with diabetes (IWGDF 2023 update). Diabetes Metab Res Rev. 2023;14:e3648. https://doi.org/10.1002/dmrr.3648.
24. Beckert S, Witte M, Wicke C, Königsrainer A, Coerper S. A new wound-based severity score for diabetic foot ulcers: a prospective analysis of 1,000 patients. Diabetes Care. 2006;29(5):988–92. https://doi.org/10.2337/diacare.295988.
25. Jeon BJ, Choi HJ, Kang JS, Tak MS, Park ES. Comparison of five systems of classification of diabetic foot ulcers and predictive factors for amputation. Int Wound J. 2017;14(3):537–45. https://doi.org/10.1111/iwj.12642.
26. Bowering CK. Diabetic foot ulcers. Pathophysiology, assessment, and therapy. Can Fam Physician. 2001;47:1007–16.
27. Jeffcoate WJ, Game F, Cavanagh PR. The role of proinflammatory cytokines in the cause of neuropathic osteoarthropathy (acute Charcot foot) in diabetes. Lancet. 2005;366(9502):2058–61.
28. Brodsky JW. The diabetic foot. In: Coughlin MJ, Mann RA, editors. Surgery of the foot and ankle. Eichenholtz classification. 7th ed. St. Louis: Mosby; 1999. p. 895–969.
29. Sanders LJ, Frykberg RG. Diabetic neuropathic osteoarthropathy: the Charcot foot. In: Levin ME, O'Neal LW, Bowker JH, editors. The high risk foot in diabetes mellitus. New York: Churchill Livingstone; 1991. p. 297–338.
30. Eichenholtz S. Charcot joints. Springfield: Charles C. Thomas; 1996. p. 3–8.
31. Trieb K. The Charcot foot: pathophysiology, diagnosis and classification. Bone Jt J. 2016;98(9):1155–9. https://doi.org/10.1302/0301-620X.98B9.37038.

Entrapment Neuropathies 18

Mahmut Kürşat Özşahin, Muhammed Yusuf Afacan,
Mert Can Çavuş, and Ahmet Ertaş

18.1 Introduction

Entrapment neuropathies of the foot and ankle are vulgarly clinical conditions occurring due to compression of the nerves around the foot and ankle for various reasons. The well-defined causes of nerve compression are trauma, repetitive stress in athletes, space-occupying masses like tumors and cysts, spur formations, accessory anatomic structures, fibrous thickenings, hypertrophied muscles, etc. [1].

In general, these entrapment neuropathies are rare. Nevertheless, some of them are well-known in the literature. Before elaborating on these entrapment neuropathies, it is necessary to have a brief knowledge of the nerve anatomy of the foot and ankle and the general nerve courses.

18.2 Nerve Anatomy

The sciatic nerve is the longest and the largest nerve in the human body, and it is one of the branches of the sacral plexus. The sciatic nerve gives off the tibial nerve and the common fibular nerve as its two terminal branches at the apex of the popliteal fossa. Common fibular nerve gives off two cutaneous branches proximal to the fibular head, the sural cutaneous nerve, and the lateral sural cutaneous nerve. The medial sural cutaneous branch of the tibial nerve and the sural communicating branch of

M. K. Özşahin · M. Y. Afacan (✉) · M. C. Çavuş
Cerrahpasa Faculty of Medicine, Department of Orthopaedics and Traumatology, Istanbul University-Cerrahpasa, Istanbul, Turkey
e-mail: muhammedyusuf.afacan@iuc.edu.tr

A. Ertaş
Cerrahpasa Faculty of Medicine, Department of Anatomy, Istanbul University-Cerrahpasa, Istanbul, Turkey

S. Sharma et al. (eds.), *Clinical and Radiological Examination of the Foot and Ankle*, https://doi.org/10.1007/978-981-97-4202-8_18

the common fibular nerve unite to form the sural nerve. The sural nerve may also be formed as the continuation of the lateral sural cutaneous nerve. The lateral calcaneal nerves are the branches of the sural nerve. The tibial nerve descends through the popliteal fossa running in the deep posterior compartment of the leg. After exiting the posterior leg compartment at the ankle [2], it passes through the tarsal tunnel where it gives off the medial calcaneal nerve. After leaving the tarsal tunnel, the tibial nerve enters the sole of the foot. Beneath the flexor retinaculum, the tibial nerve terminates by dividing into the medial and lateral plantar nerves. While the medial plantar nerve runs through the medial plantar tunnel in the sole of the foot, the lateral plantar nerve runs through the lateral plantar tunnel. The first branch of the lateral plantar nerve is the inferior calcaneal nerve, also known as Baxter's nerve. Confluence branches of the lateral and medial plantar nerves unite to form the interdigital nerve. After winding around the neck of the fibula, the common fibular nerve pierces the long peroneal muscle, and then it gives off its two terminal branches as the superficial and the deep peroneal nerves. The superficial peroneal nerve descends in the lateral leg compartment. After piercing the deep fascia at the distal third of the leg, it travels subcutaneously. After descending on the anterior surface of the interosseous membrane, the deep peroneal nerve crosses the distal end of the tibia. Under the extensor retinaculum, it passes through the anterior tarsal tunnel before entering the dorsum of the foot. There it divides into medial and lateral branches. The medial branch of the deep peroneal nerve goes to the first dorsal web [3].

18.3 Posterior Tarsal Tunnel

The posterior tarsal tunnel is a fibro-osseous space that is bordered superiorly by the medial malleolus, anteriorly by the tibia, posteriorly by the posterior process of the talus, laterally by the calcaneus, and inferiorly by the abductor hallucis muscle. The flexor retinaculum (laciniate ligament) creates an enclosed space by making the roof of the tunnel and by laying over the tibial nerve. The contents of the tarsal tunnel are the posterior tibial tendon, flexor digitorum longus tendon, posterior tibial artery and vein, tibial nerve, and flexor hallucis longus from anterior to posterior [4].

The tibial nerve is usually compressed in the posterior tarsal tunnel, and its two distal branches medial and lateral plantar nerves are occasionally compressed in their own sheathes distally to the tarsal tunnel. The compression of the tibial nerve in the posterior tarsal tunnel is named posterior tarsal tunnel syndrome [5].

18.4 Anterior Tarsal Tunnel

The anterior tarsal tunnel is bordered by the inferior extensor retinaculum and the fascia overlying the talus and the navicular. The contents of the anterior tarsal tunnel include the deep peroneal nerve and its branches, extensor hallucis longus, extensor digitorum longus, and the dorsalis pedis artery. The deep peroneal nerve can be compressed within the tunnel. The compression of the deep peroneal nerve in the anterior tarsal tunnel is named anterior tarsal tunnel syndrome [6].

18.5 Distal Tarsal Tunnel

Between the fascia of the abductor hallucis longus and the medial border of the quadratus plantae muscle lie the distal tarsal tunnel and the inferior calcaneal nerve (lateral plantar nerve's first branch) that can be compressed there. The compression of the inferior calcaneal nerve in the distal tarsal tunnel is named distal tarsal tunnel syndrome, in other words Baxter's nerve entrapment [5].

18.6 Knot of Henry

The knot of Henry is located between the abductor hallucis muscle and the anatomical crossover between the flexor digitorum longus and the flexor hallucis longus tendons along the plantar surface [7]. The medial plantar nerve which is one of the distal branches of the tibial nerve can be compressed there. The compression of the medial plantar nerve in the knot of Henry is named medial plantar entrapment neuropathy, in other words Jogger's foot [5].

For other anatomical landmarks, the match of the entrapment neuropathy, the entrapped nerve, and the site of entrapment are given in Table 18.1. To show the nerves' anatomy and courses, dissection on cadavers was performed (Figs. 18.1, 18.2, 18.3, 18.4, and 18.5).

Table 18.1 The match of the entrapment neuropathy, the entrapped nerve, and the site of entrapment

Entrapment neuropathy	Entrapped nerve	Site of entrapment
Tarsal tunnel syndrome	Tibial nerve	Tarsal tunnel, flexor retinaculum
Distal tarsal tunnel syndrome (Baxter's nerve entrapment)	Inferior calcaneal nerve (lateral plantar nerve's first branch)	Between the fascia of the abductor hallucis longus and the medial border of the quadratus plantae muscle
Medial plantar entrapment neuropathy (Jogger's foot)	Medial plantar nerve	The knot of Henry located between the abductor hallucis muscle and the anatomical crossover between the flexor digitorum longus and the flexor hallucis longus tendons along the plantar surface
Lateral plantar nerve entrapment neuropathy	Lateral plantar nerve	Lateral plantar tunnel

(continued)

Table 18.1 (continued)

Entrapment neuropathy	Entrapped nerve	Site of entrapment
Medial calcaneal nerve entrapment	Medial calcaneal nerve	The tight fascia at the origin of the abductor hallucis muscle and the calcaneus
Anterior tarsal tunnel syndrome	Deep peroneal nerve	Anterior tarsal tunnel, extensor retinaculum
Superficial peroneal neuropathy	Superficial peroneal nerve	Deep crural fascia between long and short peroneal muscles
Digital neuropathy (Morton's neuroma)	Interdigital nerve	Plantar to the transverse intermetatarsal ligament, between the heads of the third and fourth metatarsals
Sural entrapment neuropathy	Sural nerve	Musculotendinous junction of the gastrocnemius muscle and the Achilles tendon: superficial sural aponeurosis
Lateral sural cutaneous nerve entrapment neuropathy	Lateral sural cutaneous nerve	Tendon of the biceps femoris muscle (while piercing), peri-popliteal cystic bursitis

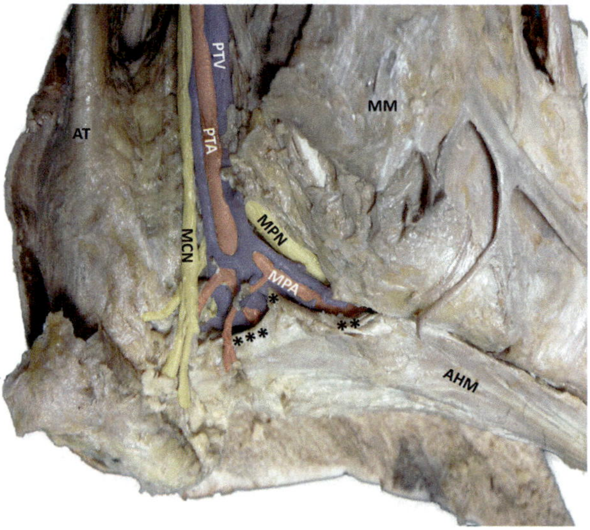

Fig. 18.1 Neurovascular structures within the tarsal tunnel. After dissecting the retinaculum musculorum flexorum in the left foot, the entrance of the neurovascular structures in the tarsal tunnel into the medial plantar tunnel and lateral plantar tunnel, which is formed by a septum made by the fascia under the musculus abductor hallucis, is seen [8]. *AT* achilles tendon, *MM* medial malleolus, *AHM* abductor hallucis muscle, *PTA* posterior tibial artery, *PTV* posterior tibial vein, *MCN* medial calcaneal nerve, *MPN* medial plantar nerve, *MPA* medial plantar artery. (*) represents the septum created by the fascia of the abductor hallucis muscle, (**) represents the medial plantar tunnel, and (***) represents the lateral plantar tunnel

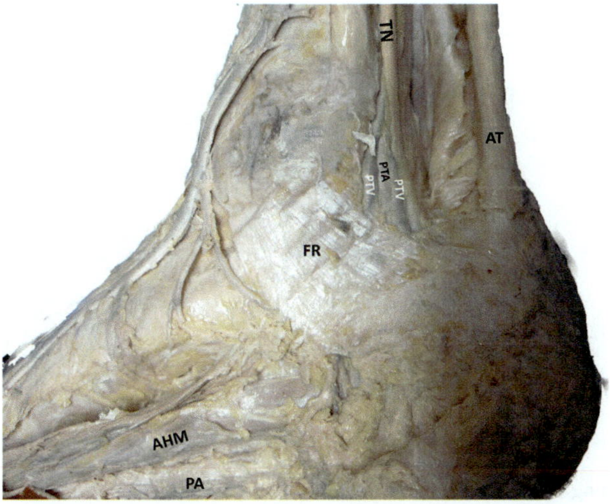

Fig. 18.2 Retinaculum musculorum flexorum. The fibers of the retinaculum musculorum flexorum running between the medial malleolus and the calcaneus are seen in the right foot. The part of the crural fascia associated with the retinaculum musculorum flexorum was dissected, and the tibial nerve, arteria tibialis posterior, and posterior tibial veins, which entered the tarsal tunnel, were exposed [8]. *AT* achilles tendon, *AHM* abductor hallucis muscle, *FR* flexor retinaculum, *PA* plantar aponeurosis, *PTA* posterior tibial artery, *PTV* posterior tibial vein, *TN* tibial nerve

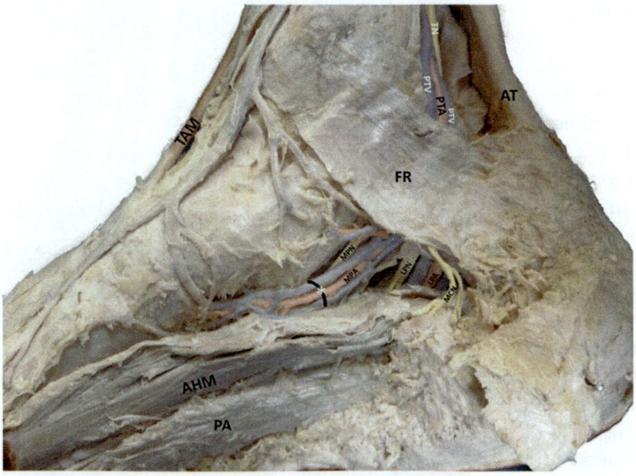

Fig. 18.3 Course of neurovascular structures proximal and distal to the tarsal tunnel. The figure shows the septum formed by the fascia of the musculus abductor hallucis extending toward the retinaculum musculorum flexorum and forming medial and lateral plantar tunnels. The fibers of the musculus abductor hallucis starting from the retinaculum musculorum flexorum are dissected, and the medial plantar nerve, medial plantar artery, and its associated veins are seen running through the medial plantar tunnel [8]. *AT* achilles tendon, *AHM* abductor hallucis muscle, *FR* flexor retinaculum, *TAM* tibialis anterior muscle, *PA* plantar aponeurosis, *PTA* posterior tibial artery, *PTV* posterior tibial vein, *TN* tibial nerve, *MCN* medial calcaneal nerve, *MPN* medial plantar nerve, *MPA* medial plantar artery, *LPN* lateral plantar nerve, *LPA* lateral plantar artery. (*) represents the medial plantar tunnel

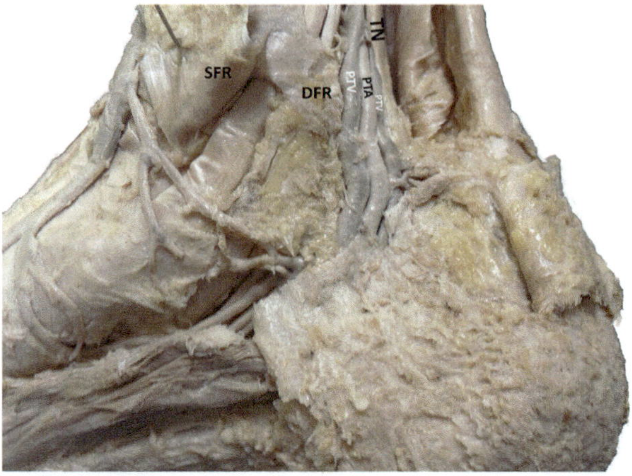

Fig. 18.4 Opening of the retinaculum musculorum flexorum. With an incision passing through the middle of the retinaculum musculorum flexorum, the superficial leaves are opened, and the pre-bifurcation course of the neurovascular structures is observed. The adipose tissue is seen between the superficial leaf and the deep leaf of the retinaculum musculorum flexorum [8]. *SFR* superficial leaf of the flexor retinaculum, *DFR* deep leaf of the flexor retinaculum, *PTA* posterior tibial artery, *PTV* posterior tibial vein, *TN* tibial nerve

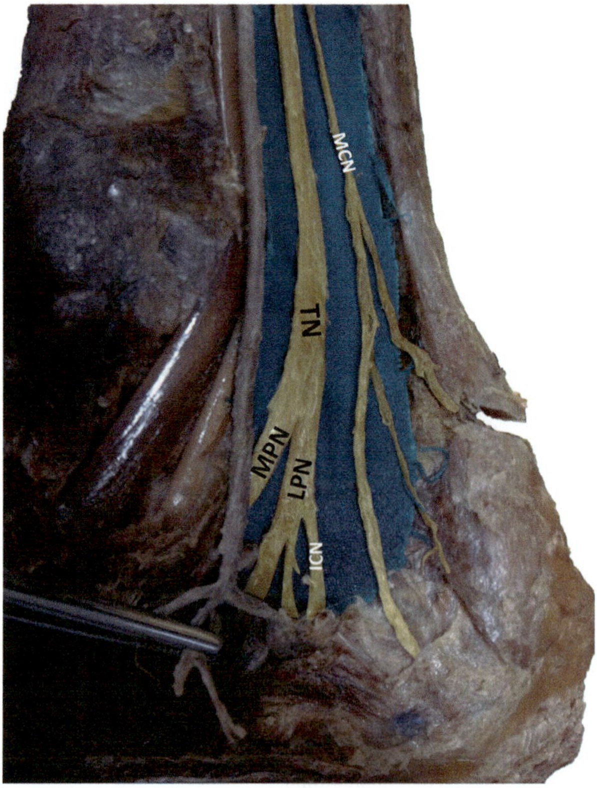

Fig. 18.5 Right foot containing tibial nerve, medial calcaneal nerve, medial plantar nerve, lateral plantar nerve, and inferior calcaneal nerve. The medial calcaneal nerve leaves the tibial nerve proximal to the tarsal tunnel entrance and gives off branches. The tibial nerve divides into two terminal branches: the medial plantar nerve and the lateral plantar nerve in the tarsal tunnel. The lateral plantar nerve gives off the inferior calcaneal nerve within the tarsal tunnel [8]. *TN* tibial nerve, *MCN* medial calcaneal nerve, *MPN* medial plantar nerve, *LPN* lateral plantar nerve, *ICN* inferior calcaneal nerve

18.7 Biomechanics

As the nerves pass through a closed osteofascial area during their course, they may not be able to tolerate the pressure increases due to the rigid tissues around them. In addition, nerves pass through various anatomical tunnels and narrow spaces during their course. Nerve entrapments occur frequently in these regions.

Generally, predisposing causes of nerve entrapment are diabetes mellitus; anatomical malformations of the foot and ankle that affect the load distribution and bony alignment such as hyperpronation, rheumatological, or degenerative arthropathies (osteophyte formations); use of inappropriate shoes; incorrect and compelling exercises; repetitive traumas; space-occupying formations (ganglion cyst, tumors, etc.); muscle hypertrophies; and fascia defects [1].

Although the pathophysiology is still unknown, local nerve ischemia, axonal degeneration, demyelination, and fibrosis are usually accused. The complaints of patients with nerve entrapment often increase at night. As muscle activities decrease during sleep at night, the effect of calf muscles on venous return decreases. Therefore, venous congestion during sleep may increase the symptoms of entrapment neuropathies at night [9]. While taking the patient's anamnesis, pain that increases at night and with exercise during the day should bring entrapment neuropathies to mind in the diagnosis.

18.8 Clinical Examination

18.8.1 General Examination

The examination of the foot and ankle to diagnose the entrapment neuropathies starts with the inspection after a successful anamnesis and history of the patient. The examiner should notice whether there is any obvious swelling, edema, redness, erythema, petechia, purpura, ecchymosis, venous regurgitation, or deformity. After a proper inspection, the palpation will be performed. The examiner should gently palpate the ankle and the foot region to remark whether there is any obvious palpable pathology like temperature increase, pathological bony prominence, masses, venous reflux, tenderness, or pathologic movement. Then, the examiner should note the range of motion for each and every joint. Afterward, muscles, tendons, ligaments, and the traces of arteries with their pulses, veins, and nerves should be palpated, and the muscle strengths should be evaluated to determine any pathology. The general examination ends with the alignment and gait analysis. After excluding other foot and ankle pathologies like cellulitis, septic arthritis, fractures, dislocations, etc., the examiner should begin with the specific examination and provocative tests in suspicion of entrapment neuropathies. Palpation of the anatomical prominent landmarks, tunnels, narrow osteofascial areas, and the course of each suspected nerve is essential in entrapment neuropathy examination. Dermatomes and myotomes of each suspected nerve shall also be examined for any sensory or motor neuropathy. The palpation and the percussion on the course of the nerves with the

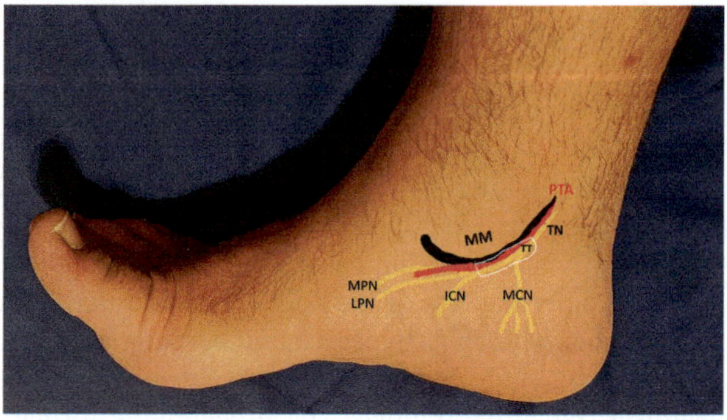

Fig. 18.6 Projection of the anatomical structures on the skin of the patient's right foot from the medial view. *MM* medial malleolus, *TT* tarsal tunnel, *PTA* posterior tibial artery, *TN* tibial nerve, *MCN* medial calcaneal nerve, *MPN* medial plantar nerve, *LPN* lateral plantar nerve, *ICN* inferior calcaneal nerve

resulting tingling sensation occurring in dermatomes will also supply hints for the side of the nerve compression. Projection of the anatomical structures on the skin is given in Fig. 18.6. Injecting local anesthetics to the determined entrapped side will also be helpful in the clinical diagnosis if the symptoms reveal after local anesthetics administration. The compression tests are also valuable to determine which nerve is entrapped.

We elaborated on each possible entrapped nerve and their specific examination methods below.

18.8.2 Tibial Nerve Entrapment Examination

The tibial nerve is commonly named the posterior tibial nerve on its course near the ankle. The tibial nerve is entrapped mostly by the flexor retinaculum (laciniate ligament) in the tarsal tunnel also known as the tibiotalocalcaneal tunnel, calcaneal tunnel, or Richet's tunnel. The other potential entrapment site is the tendinous arch between the heads of the soleus muscle and called the soleal sling syndrome [10]. This pathology is rare than the tarsal tunnel syndrome, but it can cause tibial nerve entrapment symptoms alone or cause a double crush effect on the tibial nerve with concomitant tarsal tunnel syndrome. In tarsal tunnel syndrome, pain, numbness, tenderness, and paresthesia can be seen in the medial ankle, heel, and sole of the foot. Paresis of the minimal muscles of the foot can be seen in cases where motor fibers are affected. Complaints may spread toward the medial calf, and in cases with soleal sling syndrome, the sense of the back of the knee and proximal calf and the foot flexor muscles of the posterior calf may also be affected. The tarsal tunnel syndrome is more common in patients with overpronation foot structure, and complaints may increase in passive overpronation, dorsiflexion of the ankle, and hind-food eversion. Palpation for palpable masses (Baker's cyst, tumors, etc.) and tenderness on the tibial nerve course and percussion in terms of Tinel's positivity can be helpful in physical examination. The triple compression stress test (Fig. 18.7a) is a specific provocative test for tarsal tunnel syndrome. To perform this test, dorsiflex the ankle maximally, evert the foot, apply digital pressure over the tibial nerve,

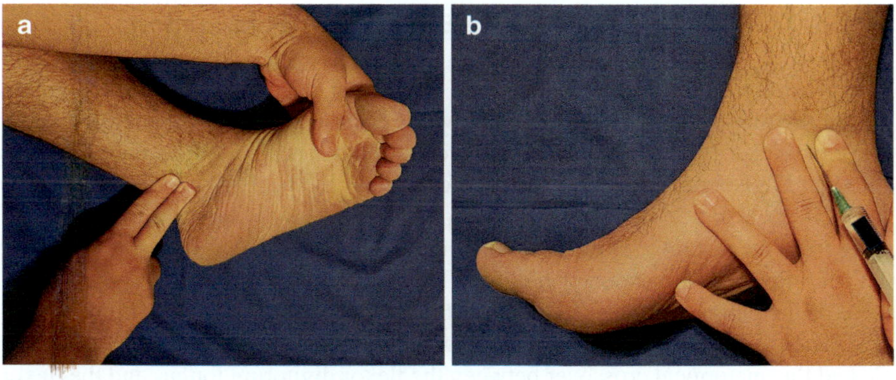

Fig. 18.7 Tibial nerve entrapment examination test and tibial nerve injection point on the skin. (**a**) The triple compression test with pressing on the tarsal tunnel while plantarflexing and inverting the ankle joint. (**b**) The injection point to enter the tarsal tunnel with a needle

and check for paresthesia and pain symptoms. The triple compression stress test has been shown to have a sensitivity of 85.9% and specificity of 100% for tarsal tunnel syndrome diagnosis [4, 11]. Injecting local anesthetics to the tarsal tunnel will also be helpful in the clinical diagnosis of the tarsal tunnel syndrome if the symptoms reveal after local anesthetics administration to the tarsal tunnel (Fig. 18.7b).

18.8.3 Medial Calcaneal Nerve Entrapment Examination

Medial calcaneal nerve entrapment can often be overlooked by considering diagnoses such as plantar fasciitis, heel fat pad pathologies, and calcaneal stress fractures [12]. The medial calcaneal nerve entrapment may be accompanied by entrapment of the tibial nerve and its other branches. Medial calcaneal nerve entrapment is the second most frequently affected nerve after the tibial nerve in heel pain of the neural origin. Due to its anatomical variations, the medial calcaneal nerve can travel through the tarsal tunnel or on the flexor retinaculum; in cases where it passes through the tarsal tunnel, it can be entrapped in this tunnel; or during its course, it can be entrapped between the tight fascia at the origin of the abductor hallucis muscle and the calcaneus. Patients with medial calcaneal nerve entrapment may complain of a burning sensation, sharp pain, tingling, and tenderness along the medial and inferior heel. On physical examination, the plantar fascia is palpated from the middle of the plantar fascia to the medial calcaneal tubercle, and the presence of plantar fasciitis is evaluated. Likewise, the Achilles tendon is palpated toward the attachment point posterior to the calcaneus. Force is applied to the calcaneus to assess the presence of a calcaneal stress fracture. Paresthesia may be experienced with palpation over the abductor hallucis muscle. Tinel's positivity can be seen with percussion, especially on the abductor hallucis muscle, along the medial calcaneal nerve course [4].

18.8.4 The Medial Plantar Nerve Entrapment Examination

After the tibialis posterior nerve gives off the medial calcaneal nerve, it divides into two as the medial and lateral plantar nerves. Of these branches, the medial plantar nerve is thicker [4]. Entrapment of the medial plantar nerve alone is rare; it can be entrapped in isolation or together with the medial calcaneal nerve or lateral plantar nerve. It can also become compressed as part of tarsal tunnel syndrome. While traveling from the distal tarsal tunnel to the medial plantar tunnel, it may be trapped by the deep fascia of the abductor hallucis. This condition is analogous to the entrapment of the palmar cutaneous branch of the median nerve in its tunnel in the wrist [13]. The medial plantar nerve is most trapped between the abductor hallucis muscle and the anatomical crossover between the flexor digitorum longus and the flexor hallucis longus tendons along the plantar surface, known as the knot of Henry or chiasma tendineum plantare. The specific name of this condition is called Jogger's foot [7]. Another location where the medial plantar nerve can theoretically be trapped is the medial plantar septum, but there is no reported case yet [1]. The medial plantar nerve innervates the flexor digitorum brevis, abductor hallucis, flexor hallucis brevis, and the first lumbrical muscle and carries the sense of the medial two-thirds of the plantar surface. In patients with medial plantar nerve entrapment, burning pain; paresthesia; numbness along the medial arch of the foot, extending to the first and second toes; and weakness in toe flexion can be observed [5]. Increased complaints with the use of medial arch support and increased pronation of the foot may be a warning in the diagnosis. Complaints may be provoked by pressing on the tarsal tunnel or the medial plantar tunnel, or a space-occupying mass may be palpable. Observe the pathologies in the ankle and foot anatomy by asking the patients to stand on their feet. Palpate the entire tibial nerve and its terminal branches while pronating the ankle with plantar flexion. This position especially narrows the medial and lateral plantar tunnels and may provoke the symptoms. Tinel's test may be positive on the course of the nerve, usually on the plantar aspect of the navicular tuberosity. Provocative tests mentioned in tarsal tunnel entrapment syndrome may be positive [4].

18.8.5 The Lateral Plantar Nerve Entrapment Examination

Isolated lateral plantar entrapment can be seen in the clinic, as well as combinations that can be seen with entrapments of the inferior calcaneal nerve, which is a branch of the lateral plantar nerve, or with the entrapment of the medial plantar nerve. The lateral plantar nerve carries the sense of the lateral aspect of the sole of the foot, and its dysfunction may cause burning pain, paresthesia, numbness, and balance defects. It also innervates all the intrinsic muscles from the sole except for the abductor hallucis, flexor digitorum brevis, flexor hallucis brevis, and the first lumbrical muscle. The lateral plantar nerve can be trapped due to anatomical formations in the lateral plantar tunnel and cause clinical symptoms which is functionally an analogous to

Guyon's canal in the wrist [13]. Due to its close proximity to the tendon of the abductor hallucis muscle, the physical examination may reveal tenderness and paresthesia in the lateral plantar nerve with palpation of its tendon. Tinel's test positivity can be seen on the tarsal tunnel and on the lateral plantar tunnel. To provoke symptoms, it may be helpful to apply pressure to the abductor hallucis tendon in the plantar flexion and inversion position or to apply pressure to the tarsal tunnel during ankle dorsiflexion and foot eversion [4].

18.8.6 The Inferior Calcaneal Nerve Entrapment Examination

Heel pain is common in clinical practice. Plantar fasciitis, plantar fascia rupture, calcaneal stress fracture, and heel fat pad atrophy are considered among the common causes of heel pain, but inferior calcaneal nerve entrapment should not be overlooked, which is thought to be the cause of 20% of chronic heel pain [14]. The presence of plantar fasciitis does not exclude the inferior calcaneal nerve entrapment but may be the cause. The inferior calcaneal nerve often branches off as the first branch of the lateral plantar nerve and may also be called Baxter's nerve or deep calcaneal nerve. The inferior calcaneal nerve entrapment may be accompanied by entrapment of the tibial nerve and its branches. The first of the two places where the inferior calcaneal nerve is frequently trapped is between the fascia of the abductor hallucis longus and the medial border of the quadratus plantae muscle, and the special name of this condition is called distal tarsal tunnel syndrome. Hypertrophy of the abductor hallucis longus muscle or increased foot pronation may cause entrapment of the inferior calcaneal nerve in this region. The inferior calcaneal nerve can be trapped more distally, between the flexor digitorum brevis and quadratus plantae muscles at the level of the medial calcaneal tubercle with or without the presence of a heel spur [15]. Patients with ICN entrapment have burning medial heel pain that sometimes radiates to the arch of the foot or to the ankle. On physical examination, the plantar fascia is palpated from the middle of the plantar fascia to the medial calcaneal tubercle, and the presence of plantar fasciitis is evaluated. Likewise, the Achilles tendon is palpated toward the attachment point posterior to the calcaneus. Force is applied to the calcaneus to assess the presence of a calcaneal stress fracture. To assess the motor fibers of the inferior calcaneal nerve, the little finger is asked to abduct, and muscle strength is compared to the contralateral foot. Nerve tracing can be percussed, but it is often not helpful in the diagnosis of inferior calcaneal nerve entrapment. The increase in complaints during dorsiflexion of the ankle is more likely to suggest plantar fasciitis, while the increase in complaints during plantar flexion and inversion suggests entrapment. The symptoms of plantar fasciitis are often more severe with the first steps in the morning and when resuming activity after rest, and night pain is not expected. The presence of sensory changes such as paresthesia, due to the lack of cutaneous innervation of the inferior calcaneal nerve, excludes the diagnosis of inferior calcaneal nerve entrapment.

18.8.7 Interdigital Nerve Entrapment (Morton's Neuroma) Examination

The branches of the medial and lateral plantar nerves form the interdigital nerves. The tissue around the interdigital nerve thickens and entraps the nerve called Morton's neuroma despite not being an actual tumor but interdigital neuralgia resulting from perineural fibrosis [5]. The most affected interdigital nerve is the third interdigital nerve, but it can also be seen in other interdigital nerves [16]. Palpate superficially for palpable masses and neuromas for every intermetatarsal web. Numbness, pain, and burning sensation radiate to the fingers when deepening palpation, a positive finding. Squeezing the metatarsal heads may cause clinical complaints, and a clicking sound may be heard. The presence of bursal click is a provocative test called Mulder's click (Fig. 18.8a) described for Morton's neuroma [17]. In some cases, Tinel's sign may be positive. Pain and tingling provoked by forcing the toes adjacent to the neuroma to dorsiflexion are supportive entrapment findings (Fig. 18.8b, c).

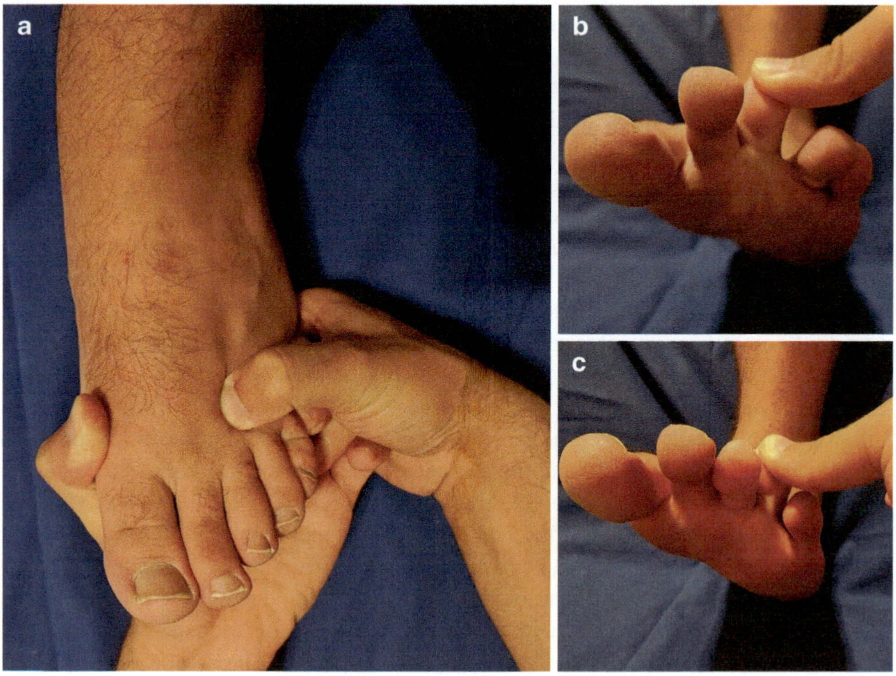

Fig. 18.8 Morton's neuroma examination. (**a**) Squeezing the metatarsal heads while pressing the web area between the third and the fourth metatarsal heads to provoke the bursal click called Mulder's click. (**b**) Forcing the third toe adjacent to the neuroma to dorsiflexion to provoke pain and tingling. (**c**) Forcing the fourth toe adjacent to the neuroma to dorsiflexion to provoke pain and tingling

18.8.8 Deep Fibular (Peroneal) Nerve Entrapment Examination

Observe the lateral compartment muscles for any atrophy or hypertrophy, especially tibialis anterior, extensor digitorum longus, peroneus tertius, and extensor hallucis longus. Inspect the patient's walking for the weakness of ankle dorsiflexion and eversion. Palpate the course of the nerve while passively performing plantar flexion of the ankle and dorsiflexion of the toes (similar to the position of wearing high-heeled shoes). This position stretches the deep fibular nerve over the talonavicular joint maximally [18]. The presence of anatomical structures such as osteophytes or bony growths of the foot is important in the examination. Percussion on the nerve, particularly over the extensor retinaculum, may result in Tinel's sign or sensational changes in the first web area. Compression of the deep fibular nerve under the extensor retinaculum represents the best-known entrapment of this nerve, anterior tarsal tunnel syndrome. The deep fibular nerve divides further into the medial and lateral branches usually 1 cm proximal to the ankle. The lateral branch of the deep fibular nerve innervates intrinsic toe extensors, extensor digitorum brevis and extensor hallucis brevis muscles, and the medial branch of the deep fibular nerve is the sensory nerve of the first web area of the foot. In case of entrapment of the medial branch, pain and decreased sensation may occur in the first web area of the foot, or in case of entrapment of the lateral branch, the muscles will be affected, and pain will radiate to the lateral tarsometatarsal joints. Deep peroneal nerve examination will be performed with squeezing the metatarsal heads while pressing the web area between the first and the second metatarsals (Fig. 18.9a).

18.8.9 Superficial Fibular (Peroneal) Nerve Entrapment Examination

The superficial fibular nerve entrapment occurs mostly due to two anatomic reasons: the presence of a long peroneal tunnel and a facial defect that cause herniation of lateral compartment muscles. The examination of a possible superficial fibular nerve entrapment should contain the examination of the previous nerve, the common fibular nerve. After checking the common fibular nerve for any pathologies, observe the lateral compartment muscles if there is atrophy or hypertrophy, and ask the patient to walk. Then check the patient's leg for any swelling, redness, or visible abnormalities. Evaluate the sensory pathologies of the lateral lower leg and dorsum of the foot except for the first web area (deep peroneal nerve's sensory area). Palpate the course of the superficial fibular nerve for any soft tissue tenderness or tumors while holding the foot in passively plantar flexion and inversion. If there is herniation of the lateral compartment muscles, it can be made more exaggerated by resisted dorsiflexion of the ankle. During palpation of the superficial fibular nerve, the distal part of the nerve can be visible by flexing the fourth toe passively (Fig. 18.9b), named fourth toe sign [19]. Percussion of the course of the superficial fibular nerve area, especially the nerve's exit through the fascia, can be helpful. A shooting pain (Tinel's sign), paresthesia, or tingling is a positive sign.

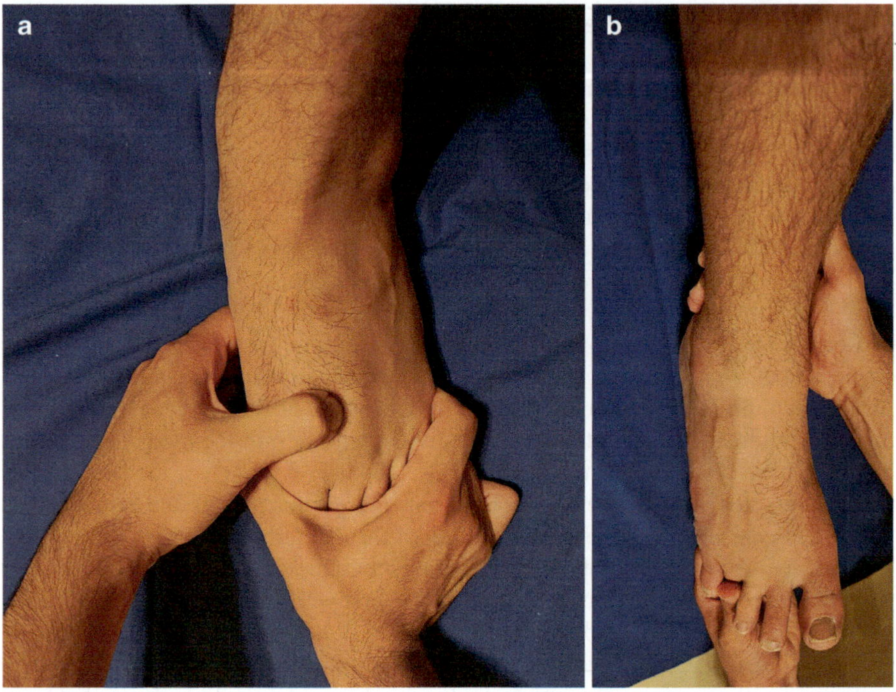

Fig. 18.9 Superficial and deep peroneal nerve examination. (**a**) Deep peroneal nerve examination with squeezing the metatarsal heads while pressing the web area between the first and the second metatarsals. (**b**) Hyperflexion of the fourth toe passively relieves the superficial peroneal nerve trace

18.8.10 Sural Nerve Entrapment Examination

The sural nerve also known as the short saphenous nerve is a pure sensory nerve, but in 4.5% of the population, it may contain motor fibers [20]. The sural nerve travels along the distal leg and ankle superficially, making it susceptible to traumas and iatrogenic injuries during surgery. The sural nerve may entrap in different locations on its course and may cause pain and paresthesia in the calf, lateral ankle, and foot. The sural nerve is typically trapped at the gastrocnemius musculotendinous junction and the Achilles tendon region, also known as the superficial sural aponeurosis at the ankle level and at the base of the fifth metatarsal. Due to its close proximity to the Achilles tendon, sural nerve entrapment may occur during Achilles tendon injuries and repair. On physical examination, tenderness with palpation is often observed in the entrapment area. Palpation between the gastrocnemius heads, lateral to the Achilles tendon, and lateral surface of the foot may be helpful. Passive plantar flexion and inversion may exacerbate the symptoms. Tinel's test is usually negative in sural neuropathies and is not useful in diagnosis.

18.8.11 Lateral Sural Cutaneous Nerve Entrapment Examination

The sural nerve complex shows many variations, and therefore the lateral sural cutaneous nerve has synonyms like the lateral cutaneous nerve of the calf and lateral sural nerve. The lateral sural cutaneous nerve entrapment is a rare condition because of its short and protected course away from rigid structures like bones and retinaculum. There are three case reports so far. The first one was due to biceps femoris tendon [21], the second one was due to peri-popliteal cystic bursitis [22], and the third one was due to a fibrotic band [23]. It can affect an area from the proximal lower extremity to the ankle. Sensory symptoms without loss of motor function may be cautionary. Increasing complaints during palpation and percussion along the nerve tracing may guide the physical examination.

18.9 Radiological Examination

The radiological examination should be started with a plain X-ray imaging, and the patient's bone structure, bone alignment, accessory bones, and osseous pathology (fractures, spurs, osteophytes, etc.) should be evaluated. In cases where X-ray is insufficient, computed tomography can be used as an advanced examination.

Magnetic resonance imaging (MRI) is used to evaluate soft tissue pathologies, stress fractures, space-occupying masses, bone marrow edema, and effusion. In addition, MRI can be used for the differential diagnosis of nerve pathologies that may originate from the proximal, such as central nervous system pathologies and radiculopathy. MRI findings of Morton's neuroma (Figs. 18.10, 18.11, and 18.12), ganglion cyst in the tarsal tunnel (Fig. 18.13), and the inflammation in the structures of the tarsal tunnel (Fig. 18.14) are given in Figs. 18.10, 18.11, 18.12, 18.13, and 18.14.

With the ultrasound evaluation, the relationship with the surrounding tissues and other anatomical formations along the nerve's tracing can be evaluated. Differential diagnosis and or treatment can be applied with local anesthetics applications to be performed around the nerve under ultrasound guidance. Moreover, some special signs can be encountered during ultrasound examination such as the Gingko leaf sign that may be specific for interdigital nerve entrapment (Morton's neuroma). Via the ultrasound exam of the intermetatarsal space in the coronal plane, a hypoechoic mass like a ginkgo leaf not directly contacting the bone can be seen in Morton's neuroma [24].

Although electromyography and nerve conduction velocity methods are not radiological imaging modalities, these electrophysiological tests are also used for diagnosis.

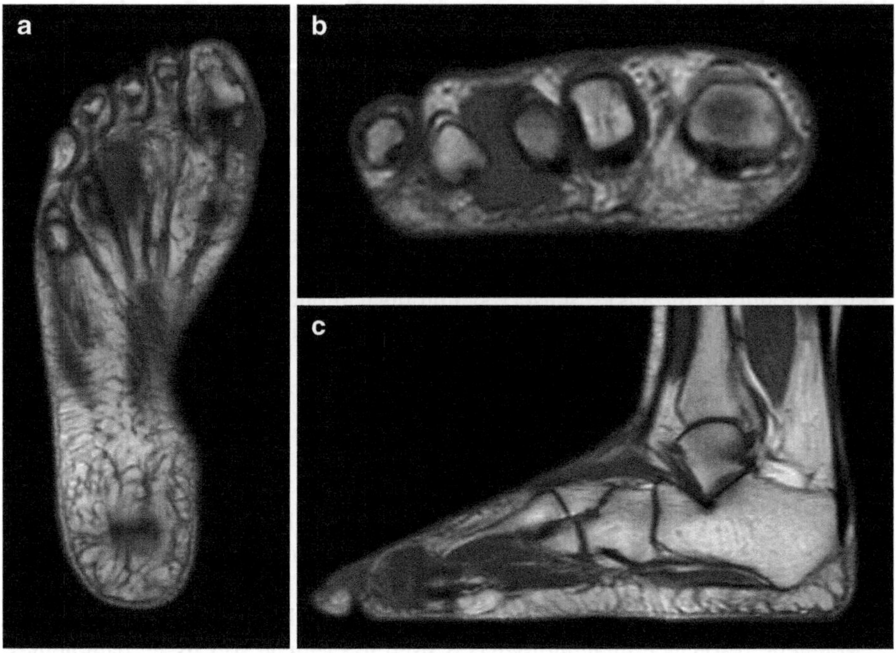

Fig. 18.10 Morton's neuroma located between the third and the fourth metatarsal heads. (**a**) T1-weighted axial section of the magnetic resonance imaging of the patient's mass located between the third and the fourth metatarsal heads. (**b**) T1-weighted coronal section of the magnetic resonance imaging of the patient's mass located between the third and the fourth metatarsal heads. (**c**) T1-weighted sagittal section of the magnetic resonance imaging of the patient's mass located between the third and the fourth metatarsal heads

Key Points

- The nerves around the foot and ankle can be compressed due to edema, inflammation of the surrounding tissues, bony prominences, space-occupying lesions, trauma, metabolic diseases, etc.
- Posterior tarsal tunnel for the tibial nerve, distal tarsal tunnel for the inferior calcaneal nerve, medial plantar tunnel and the knot of Henry for the medial plantar nerve, lateral plantar tunnel for the lateral plantar nerve, abductor hallucis muscle's fascia and the calcaneus for the medial calcaneal nerve, anterior tarsal tunnel for the deep peroneal nerve, deep crural fascia between the peroneal muscles for the superficial peroneal nerve, intermetatarsal area of the third and the fourth metatarsals for the interdigital nerve, superficial sural aponeurosis for the sural nerve, and the biceps femoris muscle's tendon for the lateral sural cutaneous nerve are the common sides for compression.
- Pressure over the nerve's course, injecting local anesthetics, and special provocative tests are helpful in clinical examination.
- Direct radiography, ultrasonographic evaluation of the nerve's course, and MRI are useful in radiological examination.

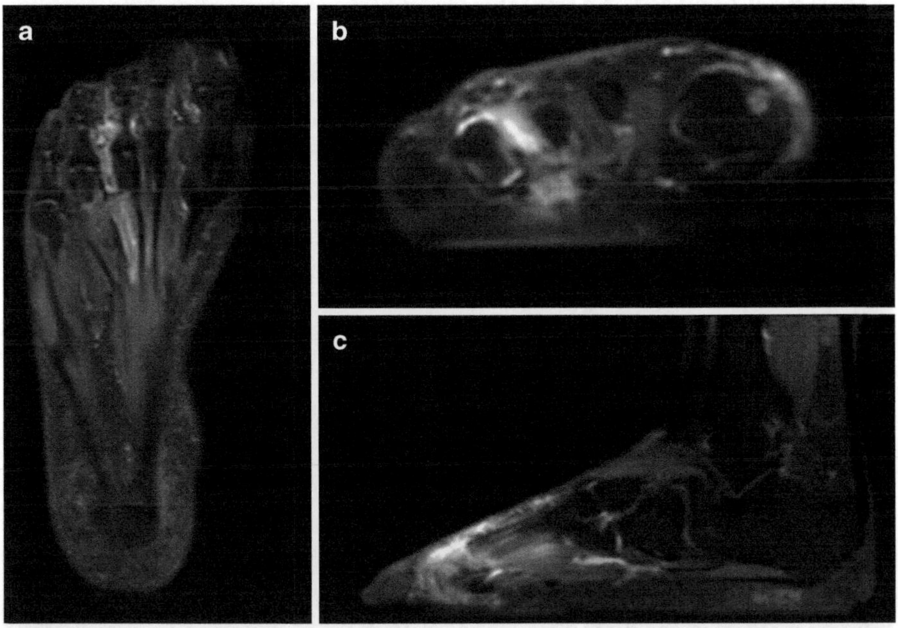

Fig. 18.11 Morton's neuroma located between the third and the fourth metatarsal heads. (**a**) T2-weighted axial section of the magnetic resonance imaging of the patient's mass located between the third and the fourth metatarsal heads. (**b**) T2-weighted coronal section of the magnetic resonance imaging of the patient's mass located between the third and the fourth metatarsal heads. (**c**) T2-weighted sagittal section of the magnetic resonance imaging of the patient's mass located between the third and the fourth metatarsal heads

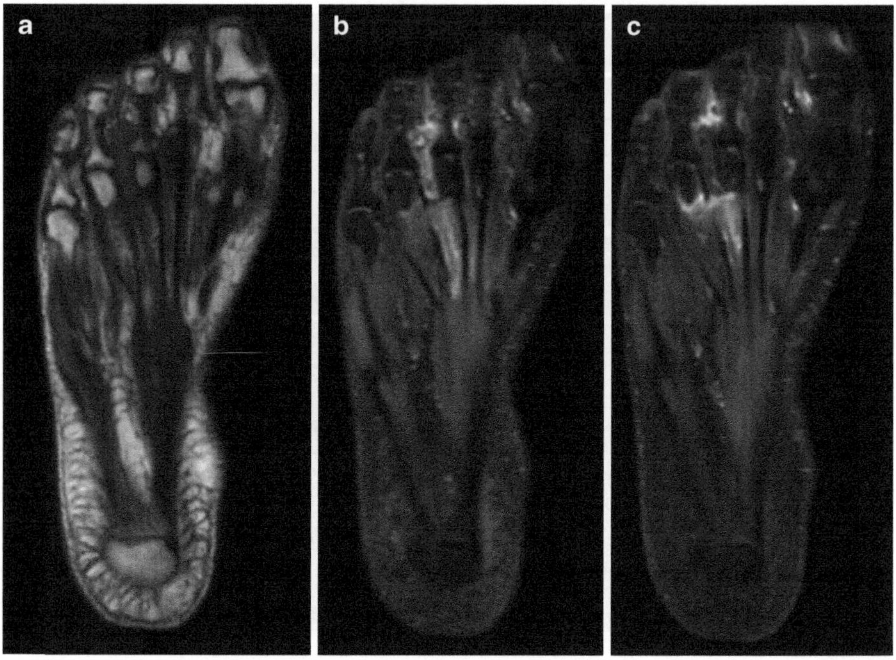

Fig. 18.12 Morton's neuroma located between the third and the fourth metatarsal heads. (**a**) T1-weighted axial section of the magnetic resonance imaging of the patient's mass located between the third and the fourth metatarsal heads. (**b**) T2-weighted axial section of the magnetic resonance imaging of the patient's mass located between the third and the fourth metatarsal heads. (**c**) T2-weighted axial section of the magnetic resonance imaging of the patient's mass located between the third and the fourth metatarsal heads after contrast infusion

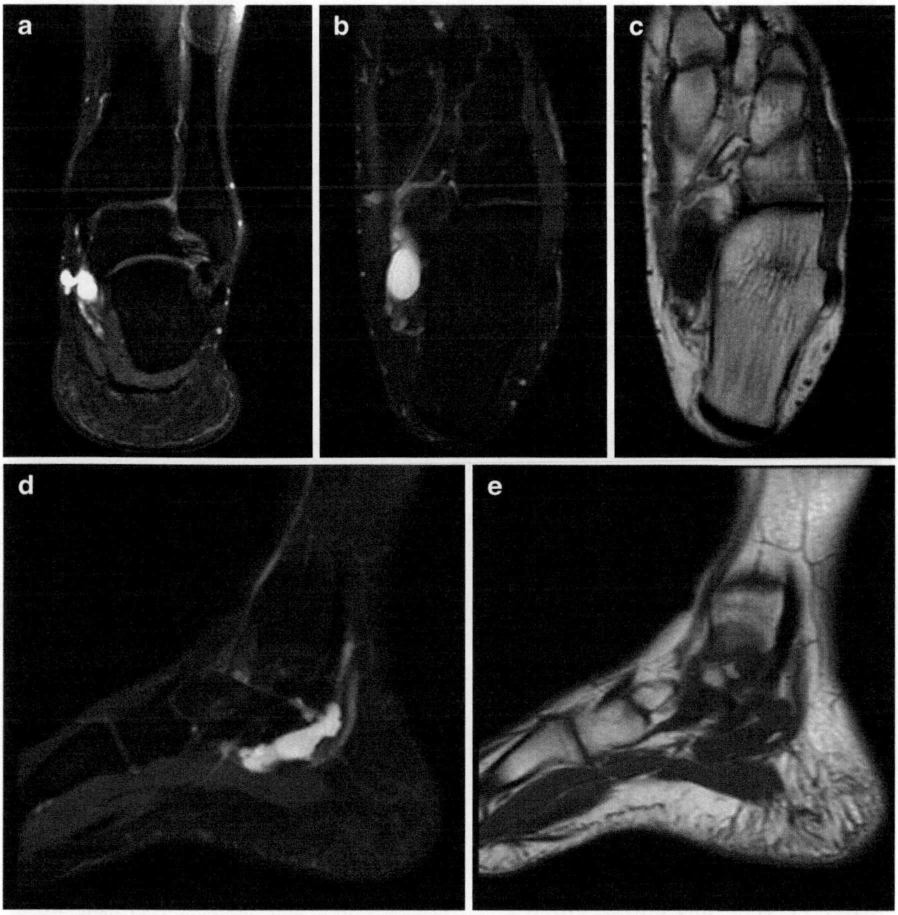

Fig. 18.13 Two giant lobulated ganglion cysts located in the tarsal tunnel causing tibial nerve entrapment. (**a**) T2-weighted coronal section of the magnetic resonance imaging of the patient's mass located within the tarsal tunnel. (**b**) T2-weighted axial section of the magnetic resonance imaging of the patient's mass located within the tarsal tunnel. (**c**) T1-weighted axial section of the magnetic resonance imaging of the patient's mass located within the tarsal tunnel. (**d**) T2-weighted sagittal section of the magnetic resonance imaging of the patient's mass located within the tarsal tunnel. (**e**) T1-weighted sagittal section of the magnetic resonance imaging of the patient's mass located within the tarsal tunnel

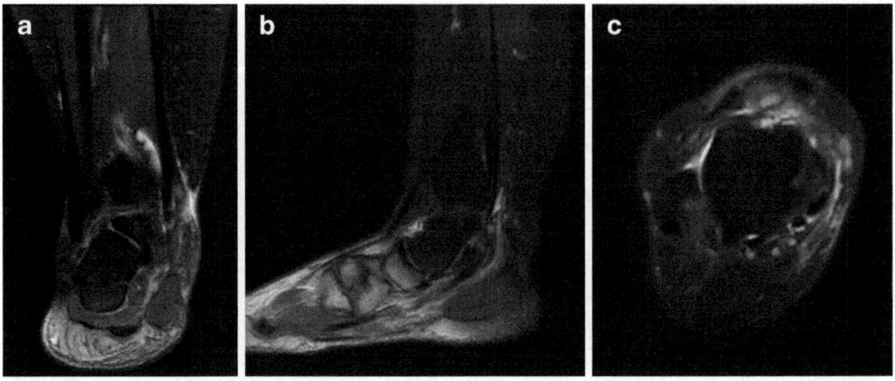

Fig. 18.14 Ankle magnetic resonance imaging of a patient suffering from the tibial nerve entrapment symptoms demonstrates inflammation of the structures within the tarsal tunnel. (**a**) T2-weighted coronal section of the magnetic resonance imaging demonstrating inflammation of the structures within the tarsal tunnel. (**b**) T2-weighted sagittal section of the magnetic resonance imaging demonstrating inflammation of the structures within the tarsal tunnel. (**c**) T2-weighted axial section of the magnetic resonance imaging demonstrating inflammation of the structures within the tarsal tunnel

References

1. Singh G, Kumar VP. Neuroanatomical basis for the tarsal tunnel syndrome. Foot Ankle Int. 2012;33(6):513–8. https://doi.org/10.3113/FAI.2012.0513.
2. Moore KL, et al. Moore clinically oriented anatomy. 8th ed. Alphen aan den Rijn: Wolters Kluwer; 2018.
3. Standring S. Gray's anatomy 41st edition: the anatomical basis of clinical practice. In: Susan S, editor. Gray's anatomy. 41st ed. London: Elsevier; 2016. p. 776–862.
4. Trescot AM. Peripheral nerve entrapments. Cham: Springer; 2016.
5. Ferkel E, Davis WH, Ellington JK. Entrapment neuropathies of the foot and ankle. Clin Sports Med. 2015;34(4):791–801. https://doi.org/10.1016/j.csm.2015.06.002.
6. Kopell HP, Thompson WAL. Peripheral entrapment neuropathies of the lower extremity. N Engl J Med. 1960;262(2):56–60. https://doi.org/10.1056/NEJM196001142620202.
7. De Maeseneer M, Madani H, Lenchik L, Kalume Brigido M, Shahabpour M, Marcelis S, de Mey J, Scafoglieri A. Normal anatomy and compression areas of nerves of the foot and ankle: US and MR imaging with anatomic correlation. RadioGraphics. 2015;35(5):1469–82. https://doi.org/10.1148/rg.2015150028.
8. Ertas A Topographic relationship of neurovascular structures in tarsal tunnel. Istanbul University-Cerrahpasa, Cerrahpasa Faculty of Medicine. 2014.
9. Rempel DM, Diao E. Entrapment neuropathies: pathophysiology and pathogenesis. J Electromyogr Kinesiol. 2004;14(1):71–5. https://doi.org/10.1016/j.jelekin.2003.09.009.
10. Mastaglia FL, Venerys J, Stokes BA, Vaughan R. Compression of the tibial nerve by the tendinous arch of origin of the soleus muscle. Clin Exp Neurol. 1981;18:81–5.
11. Abouelela AAKH, Zohiery AK. The triple compression stress test for diagnosis of tarsal tunnel syndrome. Foot. 2012;22(3):146–9. https://doi.org/10.1016/j.foot.2012.02.002.
12. Tu P, Bytomski JR. Diagnosis of heel pain. Am Fam Physician. 2011;84(8):909–16.

13. Barker AR, Rosson GD, Dellon AL. Pressure changes in the medial and lateral plantar and tarsal tunnels related to ankle position: a cadaver study. Foot Ankle Int. 2007;28(2):250–4. https://doi.org/10.3113/FAI.2007.0250.
14. Baxter DE, Pfeffer GB, Thigpen M. Chronic heel pain. Treatment rationale. Orthop Clin North Am. 1989;20(4):563–9.
15. Rondhuis JJ, Huson A. The first branch of the lateral plantar nerve and heel pain. Acta Morphol Neerl Scand. 1986;24(4):269–79.
16. Thomas JL, Blitch EL, Chaney DM, Dinucci KA, Eickmeier K, Rubin LG, Stapp MD, Vanore JV, Clinical Practice Guideline Forefoot Disorders Panel. Diagnosis and treatment of forefoot disorders. Section 3. Morton's intermetatarsal neuroma. J Foot Ankle Surg. 2009;48(2):251–6. https://doi.org/10.1053/j.jfas.2008.12.005.
17. Pastides P, El-Sallakh S, Charalambides C. Morton's neuroma: a clinical versus radiological diagnosis. Foot Ankle Surg. 2012;18(1):22–4. https://doi.org/10.1016/j.fas.2011.01.007.
18. Borges LF, Hallett M, Selkoe DJ, Welch K. The anterior tarsal tunnel syndrome. Report of two cases. J Neurosurg. 1981;54(1):89–92. https://doi.org/10.3171/jns.1981.54.1.0089.
19. Stephens MM, Kelly PM. Fourth toe flexion sign: a new clinical sign for identification of the superficial peroneal nerve. Foot Ankle Int. 2000;21(10):860–3. https://doi.org/10.1177/107110070002101012.
20. Amoiridis G, Schöls L, Ameridis N, Przuntek H. Motor fibers in the sural nerve of humans. Neurology. 1997;49(6):1725–8. https://doi.org/10.1212/wnl.49.6.1725.
21. Hackam DG, Zwimpfer TJ. Congenital entrapment of the lateral cutaneous nerve of the calf presenting as a personal sensory neuropathy. Can J Neurol Sci. 1998;25(2):168–70. https://doi.org/10.1017/s0317167100033813.
22. Ginanneschi F, Rossi A. Lateral cutaneous nerve of calf neuropathy due to peri-popliteal cystic bursitis. Muscle Nerve. 2006;34(4):503–4. https://doi.org/10.1002/mus.20595.
23. Khalil NM, Nicotra A, Kaplan C, O'Neill KS. Entrapment of the lateral cutaneous nerve of the calf. Case Rep Dermatol. 2013;2013:8847. https://doi.org/10.1136/bcr-2013-008847.
24. Park H-J, Kim SS, Rho M-H, Hong H-P, Lee S-Y. Sonographic appearances of Morton's neuroma: differences from other interdigital soft tissue masses. Ultrasound Med Biol. 2011;37(8):1204–9. https://doi.org/10.1016/j.ultrasmedbio.2011.05.008.

Ankle Impingement

19

Hayden Hartman, Peter Y. Joo, and Arianna L. Gianakos

19.1 Introduction

Ankle impingement is a condition of limited range of motion and pain when weight-bearing or when performing specific movements around the ankle joint [1]. This can be due to osseous obstruction from abnormal calcification (on the talus, tibia, or fibula) or formation of synovial scar tissue causing soft tissue impingement [2]. Impingement can occur both anteriorly and posteriorly within the ankle joint resulting in discomfort, reduced range of motion, and difficulty ambulating. This chapter will review the underlying etiologies for anterior and posterior ankle impingement, describe the utilization of various imaging modalities for diagnosis, and discuss treatment strategies for management.

19.1.1 Anterior

Anterior ankle impingement can be classified based on location: anterolateral, purely anterior, and anteromedial. Anterolateral impingement syndrome is characterized by soft tissue obstruction of the anterolateral gutter (ALG), a recess formed by the fibula, tibia, anteroinferior tibiofibular ligament (AITFL), calcaneofibular

H. Hartman
Department of Orthopaedic Surgery, Lincoln Memorial University, DeBusk College of Osteopathic Medicine, Knoxville, TN, USA
e-mail: hayden.hartman@lmunet.edu

P. Y. Joo · A. L. Gianakos (✉)
Yale Medicine, Orthopedics and Rehabilitation, Yale School of Medicine, New Haven, CT, USA
e-mail: Peter.joo@yale.edu; arianna.gianakos@yale.edu

S. Sharma et al. (eds.), *Clinical and Radiological Examination of the Foot and Ankle*, https://doi.org/10.1007/978-981-97-4202-8_19

ligament (CFL), the anterior talofibular ligament (ATFL), and joint capsule. An individual will typically remain asymptomatic when fluid fills the space, but obstruction with debris or a synovial reaction can cause limited dorsiflexion and pain [1–3]. The most common ankle injury is an ATFL sprain, which may on occasion result in anterolateral impingement (2% of cases) [1]. Typically, ankle sprains are successfully treated conservatively; however in 20–30% of patients, there is lingering pain and chronic instability that is often the origin of scar tissue formation leading to anterior impingement [4]. Other causes of impingement include small ganglia, anterolateral tibial plafond spurs/ossicles, avulsed osseus fragments, syndesmotic ligament complex injury, or osteochondral injury to the anterolateral talar dome [1].

Anterior impingement syndrome is typically found in soccer players and ballet dancers, both due to repetitive stress on the anterior structures of the ankle. These athletes present with limited dorsiflexion and anterior joint pain. Repetitive microtrauma to the talus and tibia results in the formation of osseous spurs causing impingement, synovial inflammation, irritation of the capsule, and the formation of fibrous bands [1]. Over time, soft tissue entrapment promotes hypertrophy and tissue fibrosis, which leads to the formation of a meniscoid lesion—a term for scar tissue that obstructs the ankle at the ATFL and CFL and is often preceded by an inversion ankle sprain [2, 3]. Inversion sprains can additionally initiate ligament laxity in the joint leading to impingement on the distal fascicle of the AITFL—a normal anatomic finding in most but which may cause symptomatic impingement in those with hyperlaxity [2, 5].

Anteromedial impingement syndrome is defined by bony spur formation and is anatomically located in the anteromedial recess formed by the anteromedial joint capsule, the talar dome, the body and neck, the medial malleolus, and the anterior tibiotalar ligament (ATTL) [1, 3]. After a prior plantar/inversion injury, there is damage to the ATTL resulting in thickening, osteophyte formation, synovitis, and recurrent microfractures leading to the formation of bony spurs along the talar neck, medial malleolus, or anteromedial tibial plafond [1]. These bony spurs then lead to impingement and pain when the joint is loaded.

19.1.2 Posterior

There are two types of posterior impingement syndrome: posteromedial and posterior. Anatomically, posteromedial impingement syndrome may result from fibrotic scar tissue entrapment within the posteromedial recess formed by the medial malleolus, posterior tibiotalar ligament (PTTL), talar dome, and posterior talus [1]. Typically, this is a sequela of plantarflexion, inversion, and internal rotation trauma resulting in compression and subsequent hypertrophy to the PTTL [1, 6]. The deltoid ligament is also at risk for entrapment with inversion ankle injury [2].

Posterior impingement syndromes are most often related to the posterior talus, the posterior inferior tibiofibular ligament (PITFL), the posterior talofibular ligament (PTFL), or the peroneal tendons. Most commonly this is due to a Stieda's process, an elongated lateral talar tubercle, resulting from repetitive plantarflexion trauma causing pain and inflammation [2]. The most common soft tissue etiologies include the PITFL, PTFL, and peroneal tendons, which may cause hypertrophy or tear, resulting in soft tissue entrapment, tendinopathy, possible inflammatory calcification, and limited plantarflexion [2]. Regarding the bony etiology of posterior impingement syndrome, the posterolateral talus secondary ossification center forms at 8–13 years of age, fusing 1 year afterward. In 7–14% of adults, non-fusion occurs, and an accessory bone remains, known as an os trigonum [1, 6]. Individuals who are predisposed to a symptomatic os trigonum impingement include those with an inverted pes cavus and an everted pes planus structure [2]. The ligamentous anatomy surrounding the posterior ankle can also be involved in pathology, including the PITFL, the transverse tibiofibular ligament, the posterior intermalleolar ligament, and the PTFL [2]. Injury mechanism is often either chronic repetitive microtrauma or an acute plantar hyperflexion injury leading to hypertrophy and compression of the soft tissues between the posterior talus and calcaneus [7]. The os trigonum can contribute to this space being even further narrowed, increasing compression and potential damage to local ligaments [1]. Acute fracture may occur in a plantarflexion event, resulting to the disruption of the synchondrosis [6]. Formation of osteophytes on the subtalar joint and posterior ankle can be due to a history of previous fractures and/or degenerative arthritis, whereas chronic soft tissue inflammation is often due to impinging calcifications [6].

19.2 Clinical Examination

19.2.1 Common Findings

Six characteristics to be considered at evaluation for ankle impingement include pain with forced dorsiflexion and eversion (anterior impingement) or plantarflexion and inversion (posterior impingement), ankle effusion, ankle joint tenderness, pain with activity, pain with a single-leg squat, and absence of mechanical instability [2] (Figs. 19.1, 19.2, and 19.3).

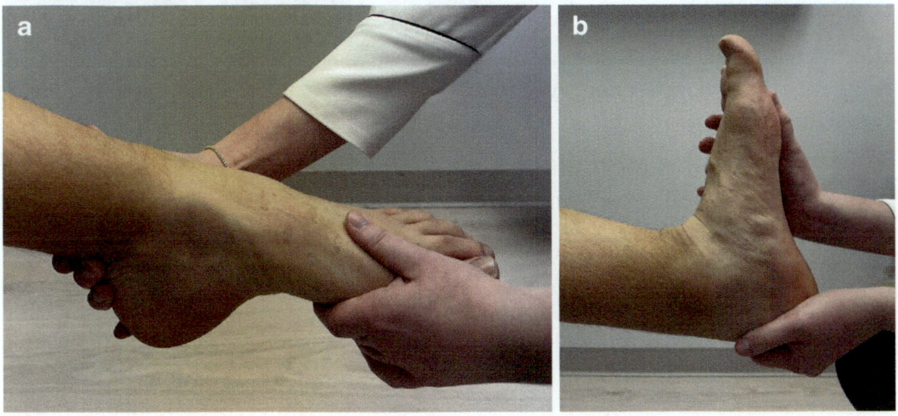

Fig. 19.1 (**a**) Demonstration of forced plantarflexion. (**b**) Demonstration of forced dorsiflexion

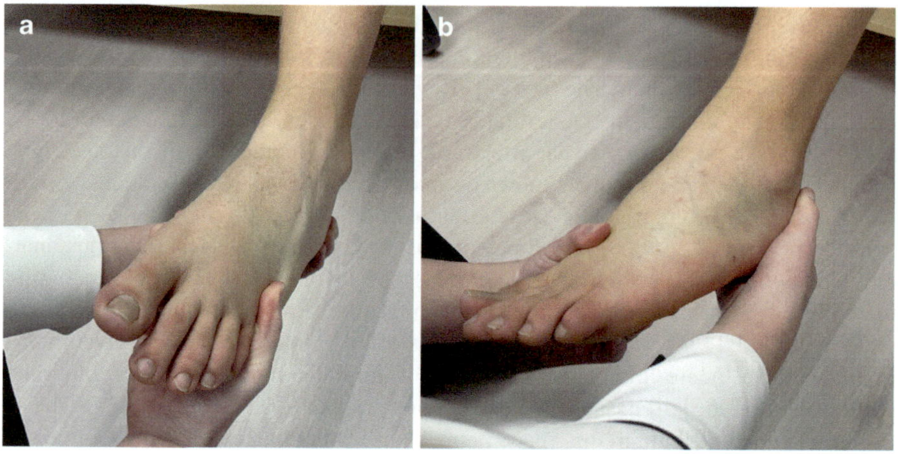

Fig. 19.2 (**a**) Demonstration of inversion. (**b**) Demonstration of eversion

Fig. 19.3 Demonstration of single-leg squat

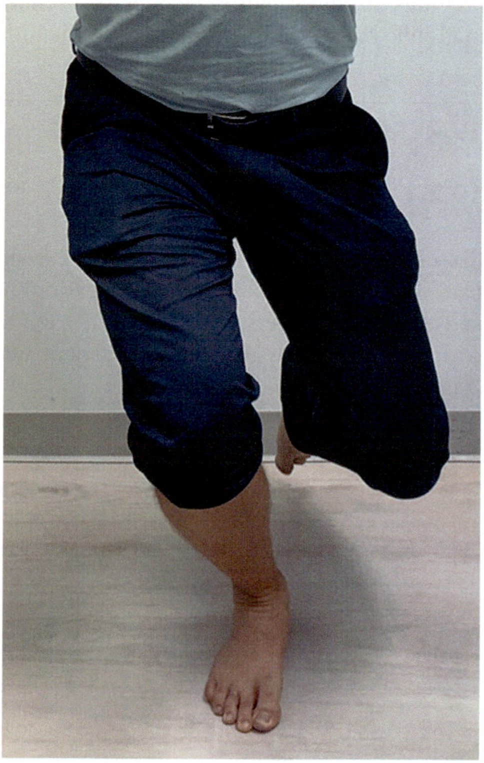

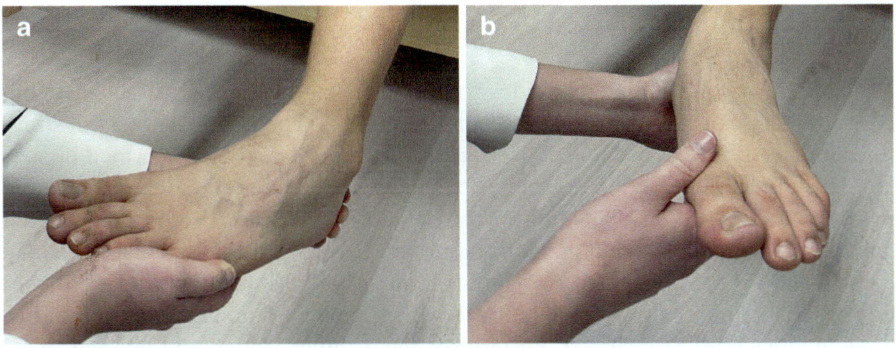

Fig. 19.4 Demonstration of internal (**a**) and external rotation (**b**)

19.2.2 Anterior

Anterior impingement syndrome is described as restricted dorsiflexion and pain during internal and external rotation (Fig. 19.4). Clinical examination findings are noted to have 95% sensitivity and 75% specificity in accurately detecting impingement [1]. If impingement of the AITFL is occurring, mechanical instability during activity is often noted and correlated to symptoms [2]. Anteromedial ankle impingement syndrome is classically found in dancers, soccer players, and cross-country runners presenting with pain with inversion and dorsiflexion [1].

19.2.3 Posterior

Posteromedial impingement syndrome presents with pain, tenderness, and fullness over the posteromedial aspect of the ankle in active and passive plantarflexion and inversion [1, 6]. Patients with posterior impingement syndrome experience swelling within the posterior ankle and chronic pain. Though often asymptomatic, those partaking in repetitive plantarflexion sports often experience pain related to the os trigonum [6]. Symptomology with an os trigonum steers diagnosis as mere presence on radiography without symptoms does not indicate diagnosis of posterior impingement syndrome [8]. This is often experienced in activities requiring significant plantarflexion or push-off activities, such as ballet, downhill running, football, soccer, and wearing high heels [1, 6]. Physical exam reproduces tenderness on palpation to the posterior ankle and pain with passive plantarflexion. Posteromedial pain with resisted plantarflexion of the first metatarsophalangeal joint often indicates flexor hallucis longus abnormality [6]. Posterolateral pain with resisted plantarflexion is more likely to indicate pathology associated with the os trigonum [8].

19.2.4 Additional Physical Exam

Visual comparison of bilateral ankles for any visible effusions/soft tissue swelling or asymmetry can clue the provider in on location of pathology and potential structures involved [9]. Passive and active range of motion should be performed bilaterally, including dorsiflexion, plantar flexion, external rotation, internal rotation, eversion, and inversion. Palpation of the ankle provides information on irritation and involvement of particular tendons or nerves. Laterally, the peroneal tendon may be palpated for pain, deformity, or subluxation, as well as the sural nerve distribution for irritation/sensitivity. Medially, the tarsal tunnel, tibial nerve, and posterior tibial tendons may be assessed. Posteriorly, the Achilles tendon and retrocalcaneal bursa may be assessed for fusiform enlargement or bursitis, respectively. Ankle ligamentous laxity may be assessed with the anterior drawer and talar tilt tests assessing for tibiotalar joint instability [9]. Other physical exam tests may be utilized to rule out pathologies not related to the ankle, such as the straight leg raise to exclude L5 or S1 radiculopathy, Silfverskiold test to differentiate gastrocnemius vs soleal or bony pathology, and palpation/squeeze of bony prominences to rule out fracture [10].

19.2.5 Imaging

19.2.5.1 X-ray

Primary imaging studies for both anterior and posterior ankle impingement include three-view (anteroposterior (AP), mortise, and lateral views) weight-bearing ankle radiographs, which are useful to view osteophytes or osseous bodies over the ALG or anterolateral tibial plafond, as well as the presence of an os trigonum or a Stieda's process. Initial radiographs in anterior impingement syndrome may show osteophytes of the anterior tibia and talar neck and are useful to evaluate the tibiotalar joint space [11] (Fig. 19.5). Radiography will serve the primary role in imaging for anteromedial impingement. Anteromedial bone spurs may be difficult to visualize in standard lateral radiographs, in which case an oblique 45° view with 30° of external rotation and foot in plantarflexion may provide better visualization of the anteromedial structures, including the tibial plafond, medial malleolus, and medial talar osteophytes [12]. The lateral view can show osseus impingement and loss of joint space, which is necessary to evaluate as secondary osteoarthritis is related to a poor postoperative prognosis. Due to its soft tissue nature, radiography does not provide a large diagnostic value for anterolateral impingement [11].

In posterior impingement syndrome, initial radiography may appear to be without abnormalities; however the lateral radiographs often demonstrate the presence of an os trigonum or a Stieda's process, as well as possible soft tissue swelling or effusions [1]. It is necessary to evaluate if the os trigonum is intact, assesses for acute or chronic fracture of the trigonal process, and screens for calcifications of the posterior ankle [6]. Acute trigonal fractures often display triangular, rough, jagged edges, in comparison to chronic trigonal fractures of which are well corticated [13].

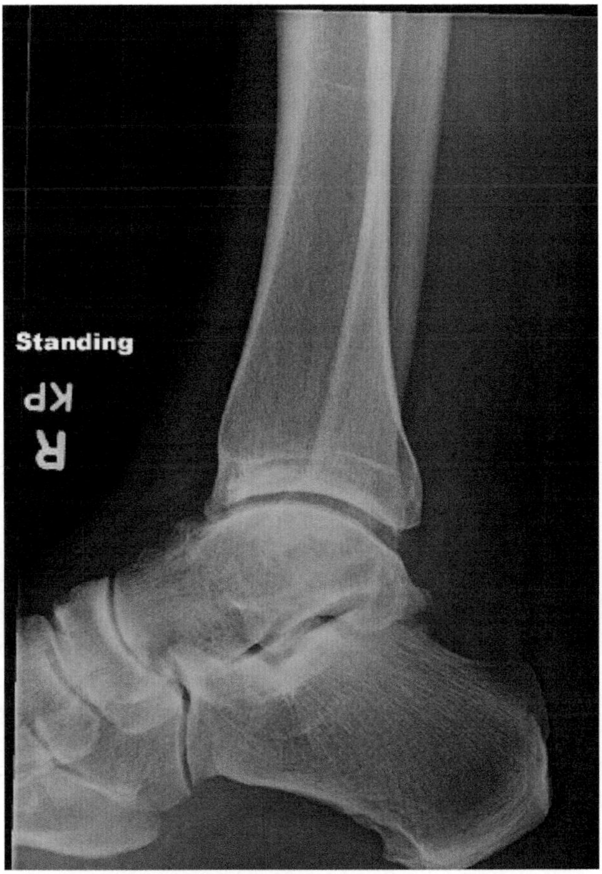

Fig. 19.5 Lateral X-ray demonstrating osteophytes over the anterior talar body and neck

Though less common, superior calcaneal tuberosity prominences may also be seen [14]. Radiography is often initially performed in posteromedial impingement, but given the ligamentous and/or soft tissue nature of this pathology, other imaging options are likely needed for diagnosis. In a small group, up to 12% of cases, avulsed bone fragments in the medial malleolus can be detected on radiography [6].

19.2.5.2 CT

Computed topography (CT) scans are used to assess and localize bony intra-articular fragments, osteochondral lesions, or osteophytes and are especially useful for surgical planning [1]. It can be effective for fracture diagnosis when looking to confirm the presence of a minimally or nondisplaced fractured, as CT can demonstrate fractures that may not be visible on conventional radiography [3, 6]. CT is superior to MRI in the case of osteochondral lesions, especially when assessing for lesion size and location, presence of loose bodies in the joint space, and when ruling out specific joint pathology, such as osteochondritis dissecans [3, 13]. CT is especially helpful in identifying and determining the size of the anteromedial and posterior bone spurs that may be difficult to visualize on standard radiographic views [12]. More so than radiography, the CT scan is particularly useful in guiding surgical decision-making and planning. Location of small osteophytes around the ankle joint as well as loose bodies can inform the surgeon on accessibility with ankle arthroscopy vs open debridement and may even help determine if the patient may be a candidate for a needle arthroscopy.

19.2.5.3 MRI

For anterolateral and posteromedial impingements, MRI remains the most effective imaging modality due to its soft tissue contrast and multiplanar capabilities. The MRI is appreciated for its efficacy in imaging the ALG and ability to evaluate ligamentous integrity, signs of tissue inflammation, and characterize the nature of lesions [1, 7]. Protocols for routine MRI should include a fluid-sensitive sequence and an anatomic sequence (usually T1) in three anatomic planes [14].

MRI is useful in anterior impingement to visualize synovial thickening, soft tissue pathology, and intra-articular fibrous bands as well as to localize osteophytes [1] (Fig. 19.6). A reliable sign for detecting anterolateral impingement is scar tissue obstruction of the anterolateral recess [11]. Synovitis is typically seen as well and should prompt examination of nearby ligaments, such as the ATFL and AITFL, specifically looking for thickening, disruption, or other post-traumatic changes [14]. Large imaging studies are lacking in anteromedial impingement, though the presence of synovial hyperplasia, post-traumatic synovitis, scaring of the ATTL, and capsular thickening all may be seen. Concomitant injuries are commonly found, and MRI is useful to assess for any other related pathologies.

In the case of posterior impingement syndrome, MRI is the modality of choice because it can elucidate bone marrow edema, synovitis, thickening of the posterior

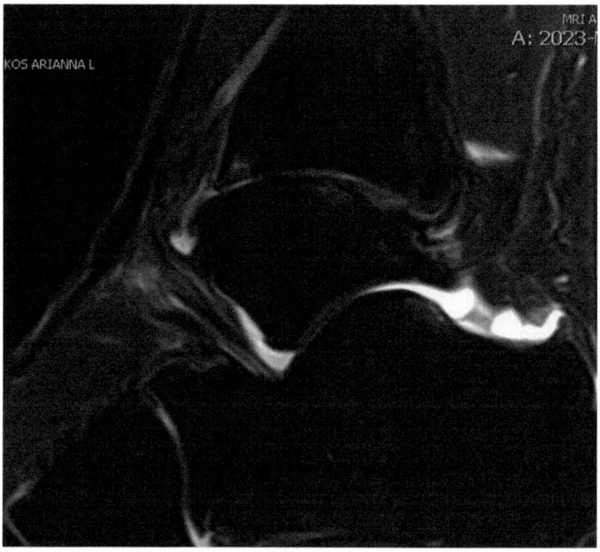

Fig. 19.6 Sagittal fluid-sensitive sequence MRI of the ankle demonstrating cystic changes of the anterior tibial plafond and synovial reaction around the anterior dome of the talus

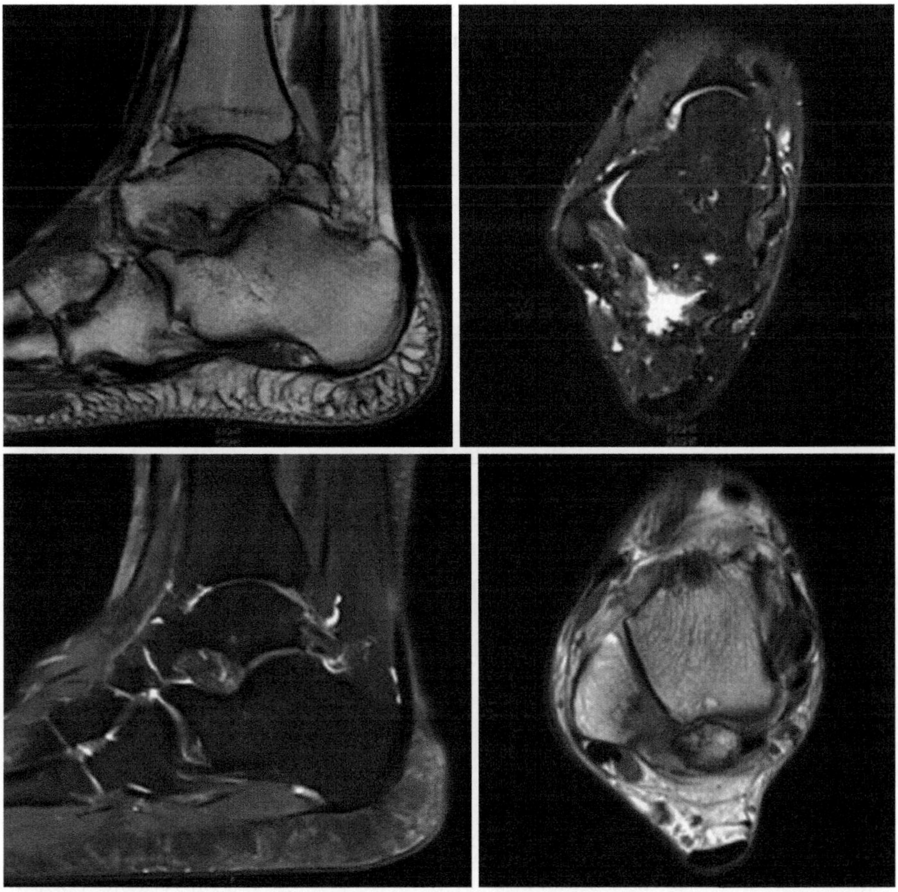

Fig. 19.7 Sagittal and axial T1 and T2 MRIs of the right ankle demonstrating an os trigonum with bony edema and increase in fluid signal around the os trigonum indicating posterior impingement

ligaments, soft tissue inflammation, and presence of ganglia in the posterior subtalar or tibiotalar areas [1, 6] (Fig. 19.7). Bone contusions of the trigonal process are common in posterior impingement patients, a condition that would not present on radiography despite edema, microtrabecular fracture, and bone marrow hemorrhage occurring [13]. Bone marrow edema, as well as fluid in the synchondrosis on MRI, is most frequently associated with symptomatic, unstable, or fractured os trigonum [3, 14]. Intravenous infusion of gadolinium contrast is often administered to visualize areas of focal synovitis around posterior ligaments [3]. In posteromedial impingement, MRI is frequently utilized to assess the fan-shaped deltoid ligament, attaching on the calcaneus, talus, and navicular. Pathology of the deltoid ligament and/or PTTL may be present as thickening, loss of the normal striated fibrillar patterns, synovitis, fluid collection, hypertrophy, and intermediate intensity on sequences with fluid intensity [1, 6]. MRI often also reveals additional pathology that may be occurring simultaneously, such as chondral injury, sinus tarsi syndrome, intra-articular bodies, marrow contusions, Achilles tendinopathy, retrocalcaneal bursitis, and flexor hallucis longus tenosynovitis, among others [6, 11].

19.2.5.4 Other Imaging

Ultrasonography is a helpful tool when delivering local injections of corticosteroids or anesthetic blocks [1, 13]. It has also been shown to detect soft tissue abnormality extruding from the ALG in the case of anterolateral impingement [11]. Appreciated for efficiency, portability, and low cost, ultrasound also offers the ability to view soft tissue structures during dynamic maneuvers providing additional diagnostic capability [14]. Hypoechoic thickening in the ALG is found when there is underlying ATFL abnormality, and depending on injury chronicity, Doppler imaging may be seen [14]. Patients with symptomatic posteromedial synovitis and no chondral defects or ligamentous instability are candidates for ultrasound-guided injections [11]. Hypoechoic posteromedial capsular thickening is expected to be found, often with the tibialis posterior tendon displaced and obscuring the PTTL. Bone scintigraphy can be an additional diagnostic resource when seeking to rule out os trigonum pathology, as it has been demonstrated that all patients with an asymptomatic os trigonum presented without scintigraphic abnormality, indicating that a normal bone scan is able to effectively rule out a pathologic os trigonum [13].

19.3 Treatment

19.3.1 Conservative

Ankle impingement can be initially managed with conservative intervention for 3–6 months [5]. This includes rest, ankle bracing or taping, shoe modification, physical therapy, nonsteroidal anti-inflammatory drugs, and local corticosteroid injection [1, 2]. If radiographs display evidence of an acute fracture, immobilization is indicated [6]. In posterior impingement syndrome, injections of corticosteroids and anesthetics have been reported to relieve symptoms in up to 84% of patients [6].

19.3.2 Surgical

If symptoms persist after 3–6 months of nonoperative treatment, surgical treatment may be considered [6]. The gold standard for anterolateral, posteromedial, and posterior impingement syndrome treatment is arthroscopic debridement of the bony prominences causing pain or the associated inflamed soft tissue [1] (Fig. 19.8). In posterior impingement, the surgical approach includes excision of the os trigonum, tenosynovectomy of the FHL if necessary, and posterior ankle arthroscopy to resect a calcaneal tuberosity [6, 15]. AITFL impingement can be arthroscopically treated by debriding the impinging ligament to allow for a full range of motion [2]. Patients

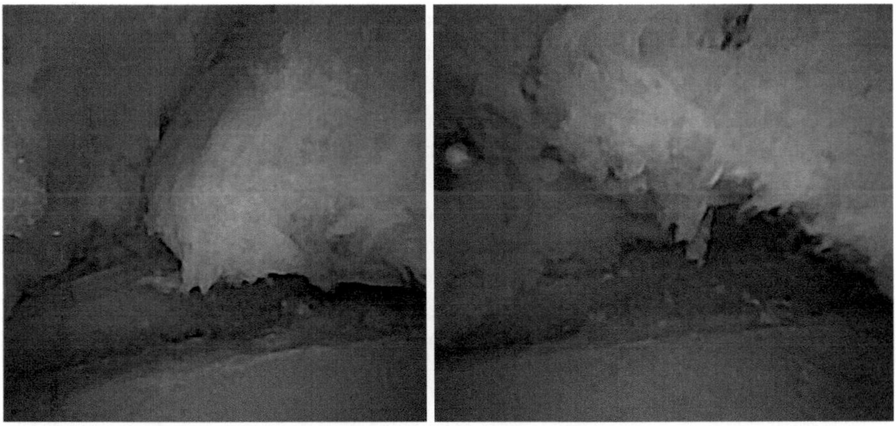

Fig. 19.8 Arthroscopic images demonstrating synovitis and anterior soft tissue impingement

should begin rehabilitation at 2 weeks postoperatively, continuing for 3 months, with early mobilization of the joint, weight-bearing as tolerated, and functional progressions [2].

19.4 Conclusion

Impingement syndrome, a condition that occurs both anteriorly and posteriorly, is characterized by osseus and/or soft tissue obstruction resulting in pain and decreased function. Anterior impingement is composed of three subtypes—anterior, anteromedial, and anterolateral—in which patients experience painful dorsiflexion and eversion. Patients with posterior impingement, subcategorized into posterior and posteromedial, have pain with plantarflexion and inversion. Though a clinical diagnosis can often be made, standard radiographs are the first line for diagnosis, and MRI is utilized for assessing soft tissue impingement and ligamentous integrity. CT scans are useful for specific osteophyte localization and preoperative planning, whereas ultrasound is primarily used when delivering therapeutic corticosteroid treatment. Initially, it is recommended to pursue conservative nonoperative treatment before progressing to surgical arthroscopic debridement.

Key Points

- Among the subtypes of anterior impingement syndrome, anterolateral typically presents with soft tissue obstruction, whereas anterior and anteromedial are characterized by the formation of osseus spurs.
- Posteromedial impingement may result from fibrotic scar tissue impingement, and posterior impingement can be related to bony impingement, often due to an os trigonum.

- Ankle effusion, pain with activity, pain with a one-legged squat, pain with forced dorsiflexion, and eversion for anterior or plantarflexion and inversion for posterior impingement are expected on physical examination.
- Assessment of bony intra-articular osteophytes is primarily completed through standard weight-bearing ankle radiographs, with additional CT imaging to further localize spurs.
- MRI is the most effective imaging modality to assess soft tissue pathology, ligamentous integrity, and bone marrow edema, which are most commonly seen with anterolateral, anterior, posteromedial, and posterior impingements.
- Conservative treatment with rest, taping, physical therapy, and local corticosteroid injection is the first line for ankle impingement syndrome.
- If symptoms persist after 3–6 months of nonoperative treatment, surgical debridement is recommended.

Competing Interests None.

References

1. Berman Z, Tafur M, Ahmed SS, Huang BK, Chang EY. Ankle impingement syndromes: an imaging review. Br J Radiol. 2017;90(1070):20160735. https://doi.org/10.1259/bjr.20160735.
2. Hess GW. Ankle impingement syndromes: a review of etiology and related implications. Foot Ankle Spec. 2011;4(5):290–7. https://doi.org/10.1177/1938640011412944.
3. Russo A, Zappia M, Reginelli A, Carfora M, D'Agosto GF, La Porta M, Genovese EA, Fonio P. Ankle impingement: a review of multimodality imaging approach. Musculoskelet Surg. 2013;97(Suppl 2):S161–8. https://doi.org/10.1007/s12306-013-0286-8.
4. Cerezal A, Ocampo R, Llopis E, Cerezal L. Ankle instability update. Semin Musculoskelet Radiol. 2023;27(3):231–44. https://doi.org/10.1055/s-0043-1767767.
5. Ribbans WJ, Ribbans HA, Cruickshank JA, Wood EV. The management of posterior ankle impingement syndrome in sport: a review. Foot Ankle Surg. 2015;21(1):1–10. https://doi.org/10.1016/j.fas.2014.08.006.
6. Giannini S, Buda R, Mosca M, Parma A, Di Caprio F. Posterior ankle impingement. Foot Ankle Int. 2013;34(3):459–65. https://doi.org/10.1177/1071100713477609.
7. Ishibashi MA, Doyle MD, Krcal CE Jr. Posterior ankle impingement syndrome. Clin Podiatr Med Surg. 2023;40(1):209–22. https://doi.org/10.1016/j.cpm.2022.07.014.
8. McAlister JE, Urooj U. Os trigonum syndrome. Clin Podiatr Med Surg. 2021;38(2):279–90. https://doi.org/10.1016/j.cpm.2020.12.011.
9. Alazzawi S, Sukeik M, King D, Vemulapalli K. Foot and ankle history and clinical examination: a guide to everyday practice. World J Orthop. 2017;8(1):21–9. https://doi.org/10.5312/wjo.v8.i1.21.
10. Barouk P, Barouk LS. Clinical diagnosis of gastrocnemius tightness. Foot Ankle Clin. 2014;19(4):659–67. https://doi.org/10.1016/j.fcl.2014.08.004.
11. Sellon E, Robinson P. MR imaging of impingement and entrapment syndromes of the foot and ankle. Magn Reson Imaging Clin N Am. 2017;25(1):145–58. https://doi.org/10.1016/j.mric.2016.08.004.
12. van Dijk CN, Wessel RN, Tol JL, Maas M. Oblique radiograph for the detection of bone spurs in anterior ankle impingement. Skeletal Radiol. 2002;31(4):214–21. https://doi.org/10.1007/s00256-002-0477-0.

13. Maquirriain J. Posterior ankle impingement syndrome. J Am Acad Orthop Surg. 2005;13(6):365–71. https://doi.org/10.5435/00124635-200510000-00001.
14. LiMarzi GM, Khan O, Shah Y, Yablon CM. Imaging manifestations of ankle impingement syndromes. Radiol Clin North Am. 2018;56(6):893–916. https://doi.org/10.1016/j.rcl.2018.06.005.
15. Theodoulou MH, Ravine M. Posterior ankle impingement. Clin Podiatr Med Surg. 2023;40(3):413–24. https://doi.org/10.1016/j.cpm.2022.12.002.

Cartilage Disorders of the Ankle 20

Mete Ozer, Matthias Peiffer, Arne Burssens,
and Bedri Karaismailoglu

20.1 Introduction

Ankle joint cartilage is provides shock absorption and a smooth surface for joint movement, which contributes to functions like standing and walking. It's possible to categorize ankle cartilage diseases under two headings: "osteochondral defects" and "osteoarthritis of the ankle."

Cartilage is composed of cells known as chondrocytes. Cartilage matrix contains collagen, hyaluronic acid, proteoglycans, and a certain amount of glycoproteins. Cartilage is avascular and is nourished by synovial fluid. Cartilage does not contain lymph vessels or nerves and has a slow metabolism [1]. Although the ankle joint is prone to frequent injuries, clinically significant ankle osteoarthritis (OA) is not as common as in other weight-bearing joints. This is due to the specific anatomical, biochemical, and biomolecular characteristics of ankle cartilage. Biologically, ankle cartilage is believed to have greater self-repair capacity than knee cartilage. Additionally, it possesses more hardness and less permeability due to higher water

M. Ozer
Department of Orthopaedics and Traumatology, Istanbul University-Cerrahpasa, Istanbul, Turkey

M. Peiffer · A. Burssens
Department of Orthopaedics and Traumatology, Ghent University Hospital, Ghent, Belgium

B. Karaismailoglu (✉)
Department of Orthopaedics and Traumatology, Istanbul University-Cerrahpasa, Istanbul, Istanbul, Turkey

FARIL - Foot and Ankle Research and Innovation Laboratory, Mass General Hospital - Harvard Medical School, Boston, USA

CAST - Cerrahpasa Research Simulation and Design Laboratory, Istanbul University-Cerrahpasa, Istanbul, Turkey
e-mail: bedrikio@iuc.edu.tr

S. Sharma et al. (eds.), *Clinical and Radiological Examination of the Foot and Ankle*, https://doi.org/10.1007/978-981-97-4202-8_20

and proteoglycan content. Moreover, the extracellular matrix is denser, enhancing load-bearing capacity and reducing sensitivity to mechanical damage. However, when joint congruence is disrupted and asymmetric stress and force distribution occur due to fractures, instability, or malalignment, degeneration can occur [2–7]. Malunion after ankle fractures affects the contact surface area in the tibiotalar region. As the contact surface area decreases, the load on the remaining cartilage will increase. It has been showed that a 1 mm lateral talus shift reduced the contact area by 42%, while a 2 mm lateral shift reduced the contact area by 56% [8–11].

20.2 Osteochondral Defects

Osteochondral defects (OCDs) of the ankle are characterized by abnormalities in the articular cartilage accompanied by subchondral bone pathology, often observed on the talar side of the tibiotalar joint [12]. The incidence of talar OCD has been reported to be 0.09%. OCD is most commonly seen in the second decade of life, but clinical symptoms can manifest later as well [13]. Symptomatic OCD is typically observed in adults. The reason for OCD being more common in adults compared to adolescents is that the disease often remains asymptomatic for a considerable period and eventually becomes evident in adulthood [12].

OCDs can cause deep ankle pain during weight-bearing. It may impact daily activities and sports participation. Symptoms might not always be evident on standard radiographs. Therefore, there can be a delay in diagnosis if the disease is not suspected. Prolonged deep ankle pain following supination trauma should raise suspicion of OCD during differential diagnosis.

20.2.1 Etiology

The etiology of OCD is not yet fully understood, and there are various etiological theories. It is believed to be traumatic, ischemic, or microtraumatic in nature. The most common contributing etiological factor, although not always reported by patients, is trauma. It has been suggested that repeated microtraumas could contribute to lesion development. Nontraumatic lesions are associated with ischemia and likely genetic factors. The ankle joint has a high congruency, and decreased joint congruency can increase the contact stress per unit area, leading to osteochondral problems. Similarly, alignment disorders also negatively affect joint congruency and contribute to chondral defect formation [10, 14].

The most prominent theory for nontraumatic lesions is the ischemic theory, supported by histological studies in animals. During endochondral ossification responsible for the growth of long bones, the epiphyseal cartilage is nourished through vessels derived from the perichondrium, passing through channels called cartilage canals. As cartilage tissue matures, these channels gradually close, and vessels in the ossification front continuing from epiphyseal cartilage become the main source of nourishment. The formation of anastomoses between vessels in the epiphyseal cartilage and vessels in the ongoing ossification front is crucial during this transition period. However, these anastomoses can fail, particularly due to local factors (such as chemical or mechanical support deficiencies). As a result, this leads to vascularization and nutritional deficiencies in a specific subchondral bone region, resulting in avascular necrosis [15, 16].

20.2.2 Clinical Examination

In acute cases, OCD may not be recognized due to swelling and pain caused by ligamentous injuries on the lateral side following ankle trauma. If symptoms of ligament injury persist for more than 4–6 weeks, OCD should be suspected. Locking and catching indicate a displaced fragment, and further evaluation should be made [10].

Chronic lesions often manifest as continuous or intermittent deep ankle pain during or after activity. In patients with OCD, detecting the pain described during weight-bearing, in the physical examination, is often challenging. Laterally located OCDs are mostly due to acute trauma, while medial lesions are usually chronic lesions.

Osteochondral lesions in the talus are generally localized on the anterolateral or posteromedial aspects of the talar dome. Lesions on the lateral side are often oval-shaped and result from an acute trauma. In contrast, medial lesions are usually deep and cup-shaped, resulting from torsional impaction and axial loading mechanisms [17–19].

Acute lesions may exhibit pain during palpation of the talar dome performed with the ankle taken into maximum plantar flexion (Fig. 20.1). Range of motion should be examined and compared to the contralateral side.

The patients should also be examined regarding any accompanying ankle instability which might be a predisposing factor and should be addressed during surgical planning.

The disease often remains asymptomatic for a long period, and even in adulthood, the condition can be incidentally discovered when an X-ray is taken for other reasons related to the ankle. When symptomatic, the patient presents with intermittent weight-bearing pain and mentions that the pain worsens during running.

Osteochondral fractures are often the result of a significant ankle injury, frequently occurring through inversion and internal rotation of the foot. Consequently, the lateral edge of the talus is the most common site for acute post-traumatic injuries.

Symptoms become more pronounced when an osteochondral fragment displaces, leading to intense pain, joint swelling, instability while walking, and even symptoms like locking. The immediate onset of symptoms after an ankle injury might suggest situations like osteochondral fractures or the displacement of an OCD fragment after trauma [12].

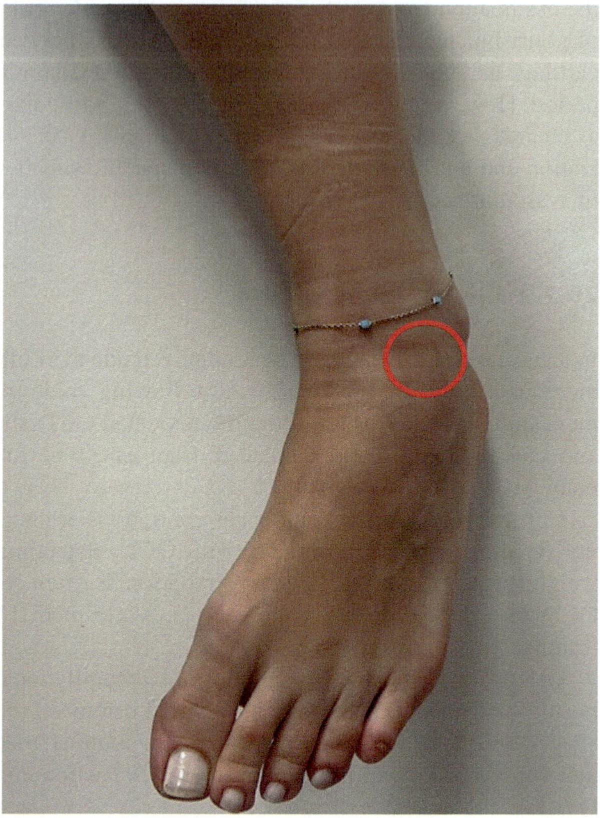

Fig. 20.1 The location to palpate the anterolateral surface of the talar dome

20.2.3 Radiological Examination

The initial imaging method to be used for diagnosis should be standard radiographs, although OCD may not always present findings on standard radiographs. Computed tomography (CT) and magnetic resonance imaging (MRI) are more helpful in definitive diagnosis.

20.2.3.1 Conventional Radiography

Conventional radiographs consist of anteroposterior (AP), AP mortise, and lateral views of the ankle taken with weight-bearing. The AP mortise view is the AP projection of the ankle taken with 15 degrees of internal rotation. This view is particularly useful for examining the superolateral corner of the talus, as it is free from fibular overlap in this position (Fig. 20.2).

Radiographs can reveal bone detachment, but the damage might be initially too small to be detectable. In the anteroposterior projection, a subchondral halo is a sign that identifies the osteochondral fragment (Fig. 20.3). Stress radiographs can be considered to evaluate instability but may not be possible to perform in acute situations.

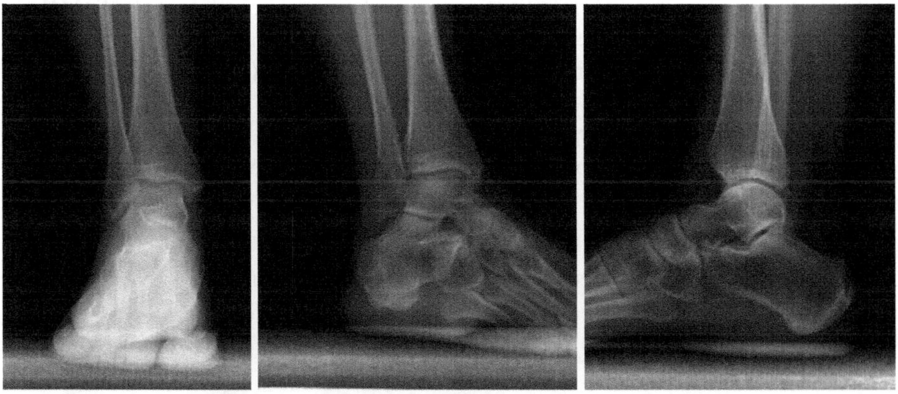

Fig. 20.2 Weight-bearing ankle anteroposterior (AP), AP mortise, and lateral images. Pay attention to the improved visibility of the talar dome in the AP mortise view, free from fibular overlap

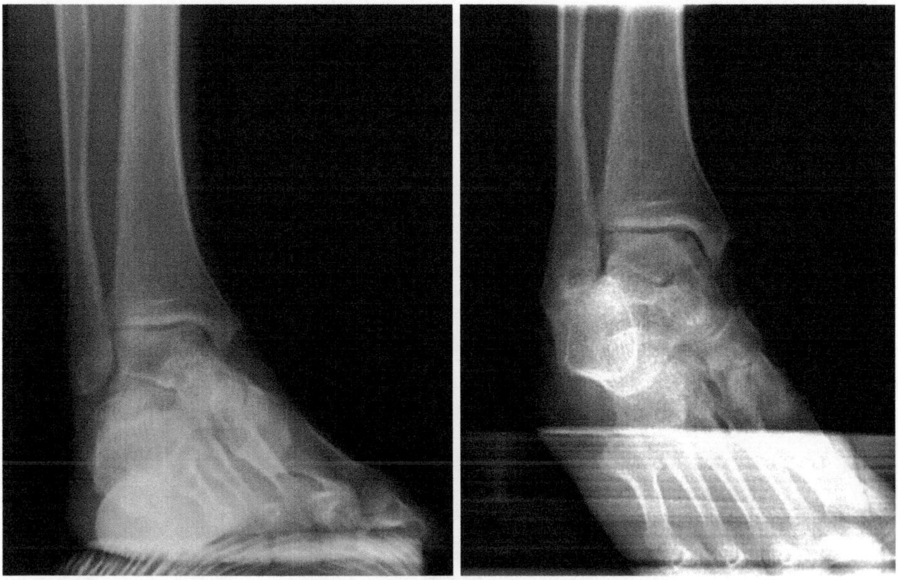

Fig. 20.3 AP mortise and "heel-rise" AP mortise views of the same ankle. Note the accentuation of the lesion on the posteromedial part of the talar dome in the "heel-rise" view

Routine radiological examination cannot detect 30–50% of OCDs. Most undiagnosed OCDs are reported to be in the posterior region. Therefore, the heel-rise view was developed to visualize the posterior part of the talar dome [20] (Fig. 20.3). This view is an AP mortise view taken with the heel raised by 4 cm. The diagnostic ratio of the heel-rise view is twice that of standard radiographs.

Over the years, various classifications have been proposed using CT, MR, or arthroscopy, but the Berndt and Harty classification [21] remains the most commonly used which is based on direct radiographic findings (Table 20.1).

20.2.3.2 Computed Tomography

CT is not an effective method for imaging cartilage tissue. However, the pain of OCD originates in the subchondral bone, and the integrity of the subchondral bone

Table 20.1 Berndt and Harty classification of talar osteochondral defects

Native	No pathology	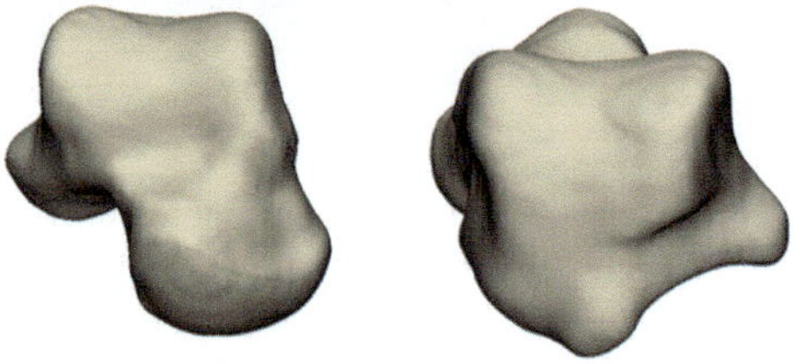
Stage 1	Small area of subchondral bone compression	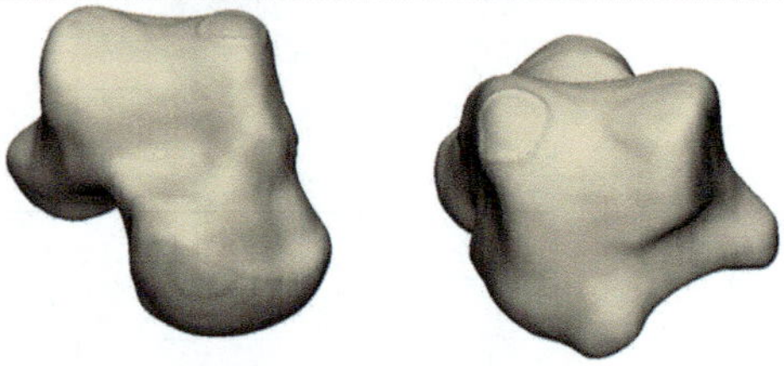
Stage 2	Partially detached osteochondral fragment	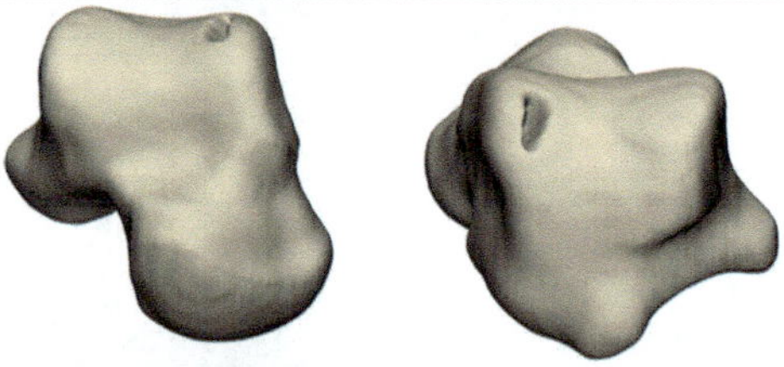
Stage 3	Complete fragment detachment but not displaced	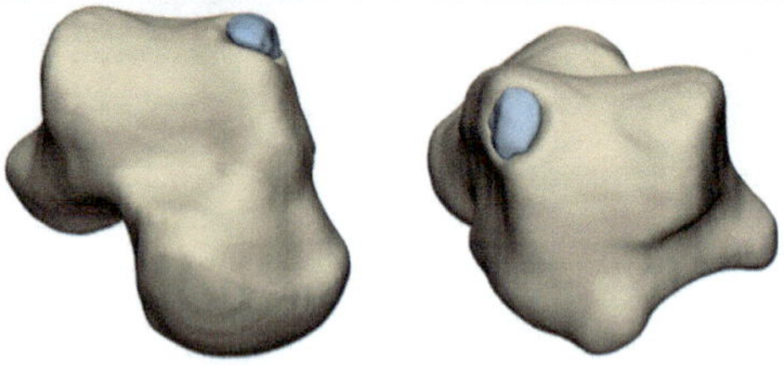
Stage 4	Displaced, free fragment	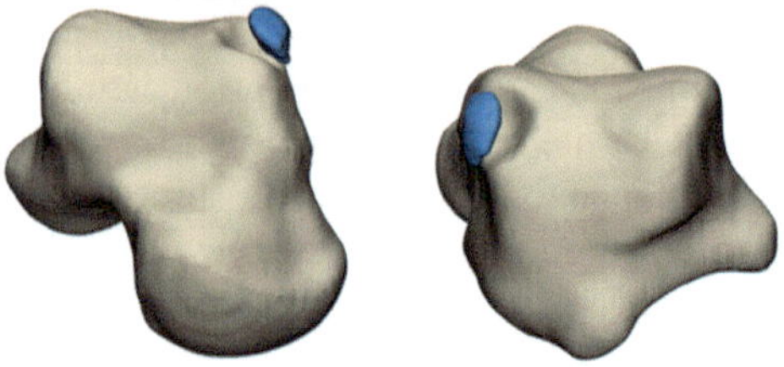

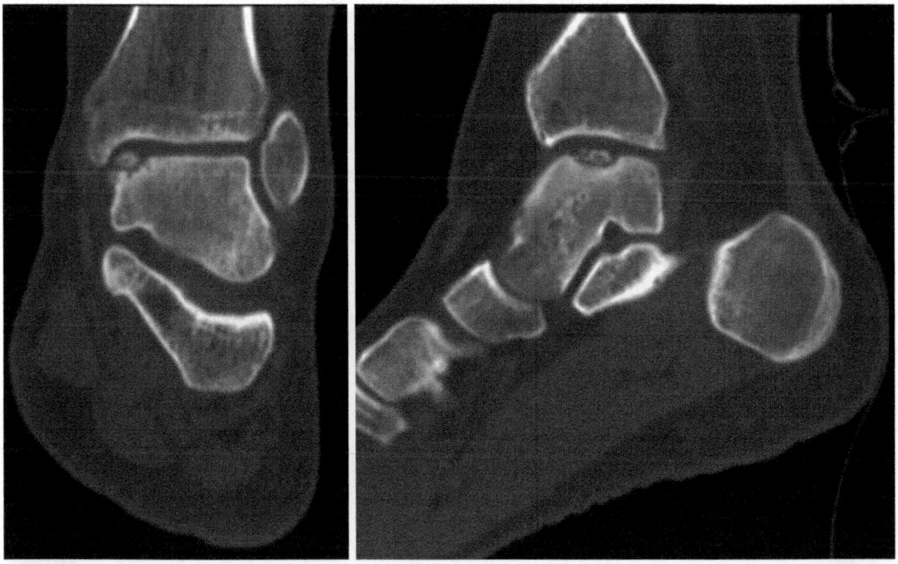

Fig. 20.4 Osteochondral defect on the superomedial part of the talus, detected by CT in a 17-year-old male patient

layer is crucial for the vitality of the joint cartilage [10, 22, 23]. CT can determine the size and location of bone defects or the displacement of a fragment (Fig. 20.4).

Especially in the presence of posteriorly located lesions or limited ankle joint range of motion, preoperative CT scanning in plantar flexion is recommended for evaluating the areas accessible by the anterior arthroscopic approach [23].

20.2.3.3 Magnetic Resonance Imaging

MRI can visualize the articular cartilage and subchondral bone. MRI has a high diagnostic rate for OCD (Figs. 20.5 and 20.6). In a study by Mintz et al., which examined the correlation between MRI findings and arthroscopic imaging, MRI was reported to have 100% specificity and 95% sensitivity in diagnosing OCDs. However, lesions can appear larger on MRI images due to bone edema, which can affect treatment decisions (Fig. 20.5) [24, 25]. Bone injuries are observed as hypointense on T1-weighted MRI and hyperintense on T2-weighted MRI.

Osteochondral lesions in the talus are generally localized on the anterolateral or posteromedial aspects of the talar dome. Lesions on the lateral side are often oval-shaped and result from an acute trauma (Fig. 20.7). In contrast, medial lesions are usually deep and cup-shaped, resulting from torsional impaction and axial loading mechanisms (Fig. 20.8) [17–19].

MRI is valuable for comprehending associated conditions such as subchondral cysts (Fig. 20.9) and for determining the presence of edema around the lesion (Fig. 20.10), helping assess its acuteness and its potential as a source of pain.

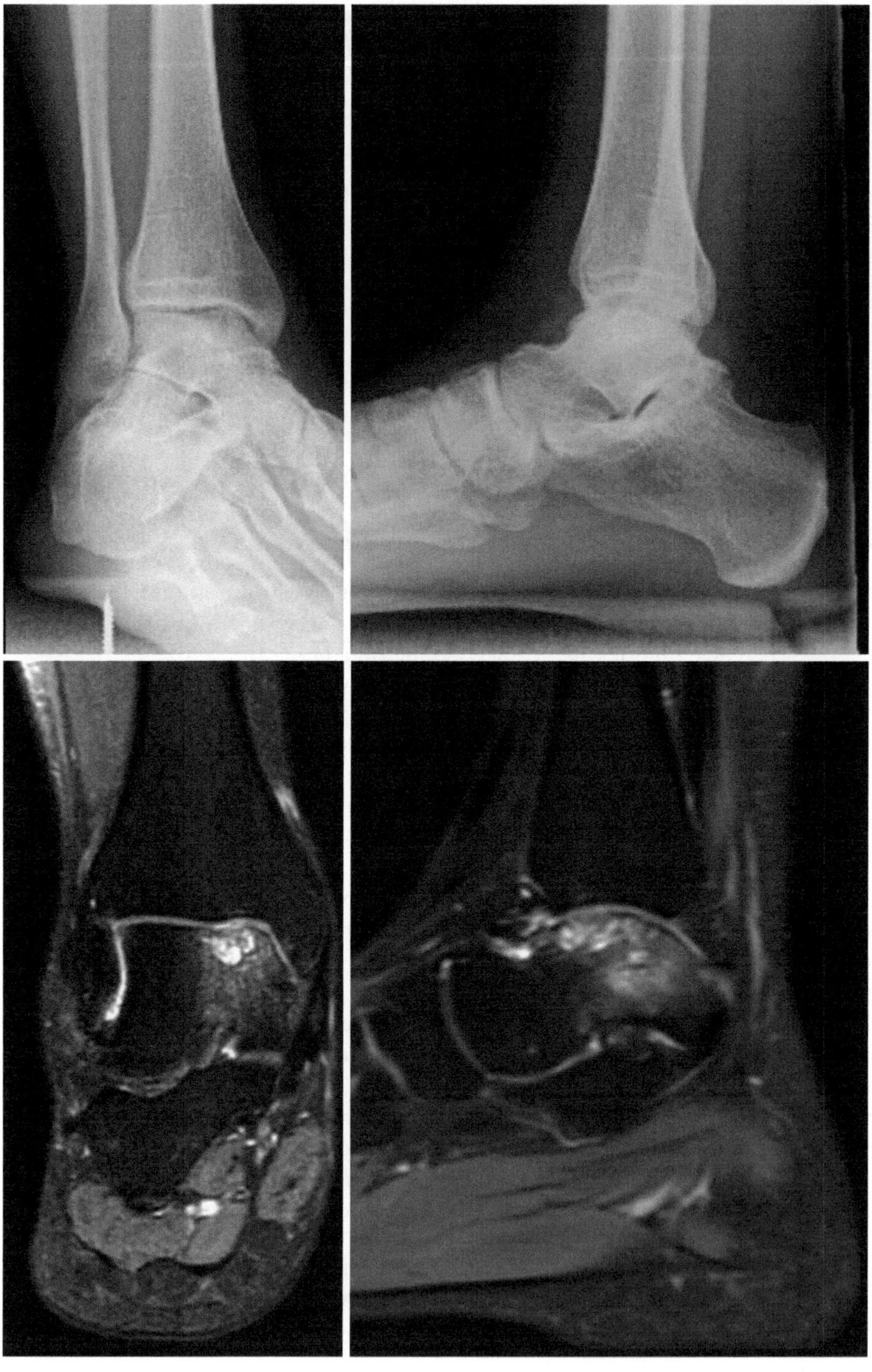

Fig. 20.5 OCD on the anteromedial part of the talus in a 23-year-old male hemophilia patient. More significant MRI findings were observed compared to conventional radiography

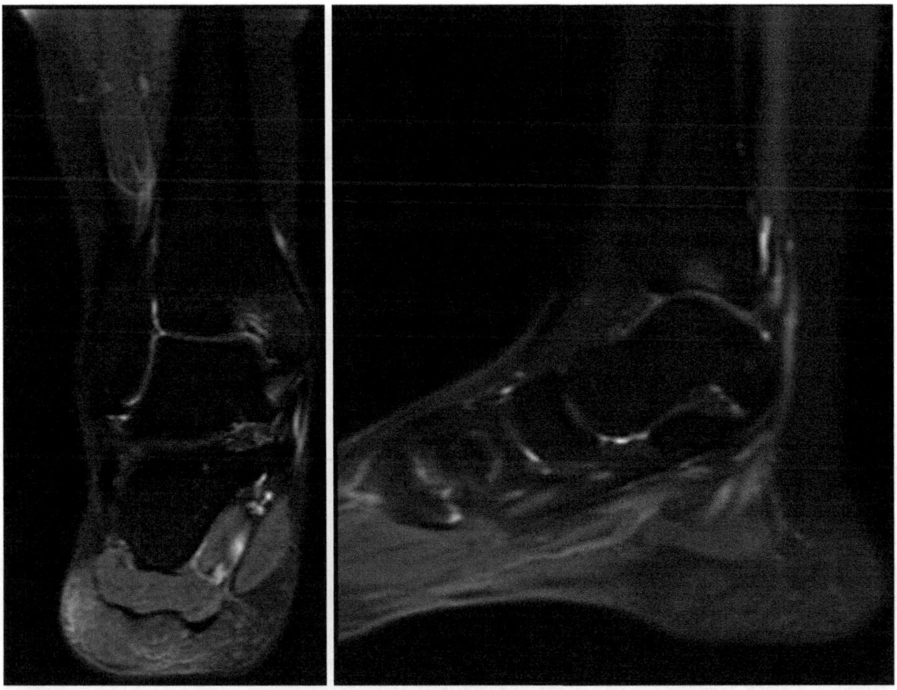

Fig. 20.6 OCD on the tibial side of the tibiotalar joint in a 22-year-old female patient. Although rare, OCDs can also be observed on the tibial side

MRI T2 mapping assessment is a biochemical imaging technique that complements morphological imaging in monitoring the biomechanical properties of cartilage repair tissue. This technique does not require the injection of contrast agents and is sensitive to collagen fiber network, collagen concentration, orientation, and integrity, as well as the water content in the repair tissue. Good collagen organization indicates maturation of the repair tissue and is associated with good clinical outcomes, while poor cartilage organization predicts weak clinical outcomes [26–28].

dGEMRIC is another qualitative MRI imaging technique that is considered a reliable tool for evaluating cartilage condition. This technique can provide a direct measurement of glycosaminoglycan (GAG) concentration in articular cartilage using T1 relaxation time [29]. dGEMRIC requires the injection of a negatively charged intravenous gadolinium-based contrast agent. Multiple studies have shown the potential of dGEMRIC in assessing the quality of cartilage repair tissue following cartilage repair techniques.

20.2.3.4 Alternative Imaging Methods

Arthrography Intra-articular contrast agent injection can be added during CT and MRI evaluations. Contrast agents can enhance the evaluation of osteochondral lesions. However, because arthrography is an invasive technique carrying a risk of complications and does not substantially affect treatment decisions, it is not routinely recommended [23].

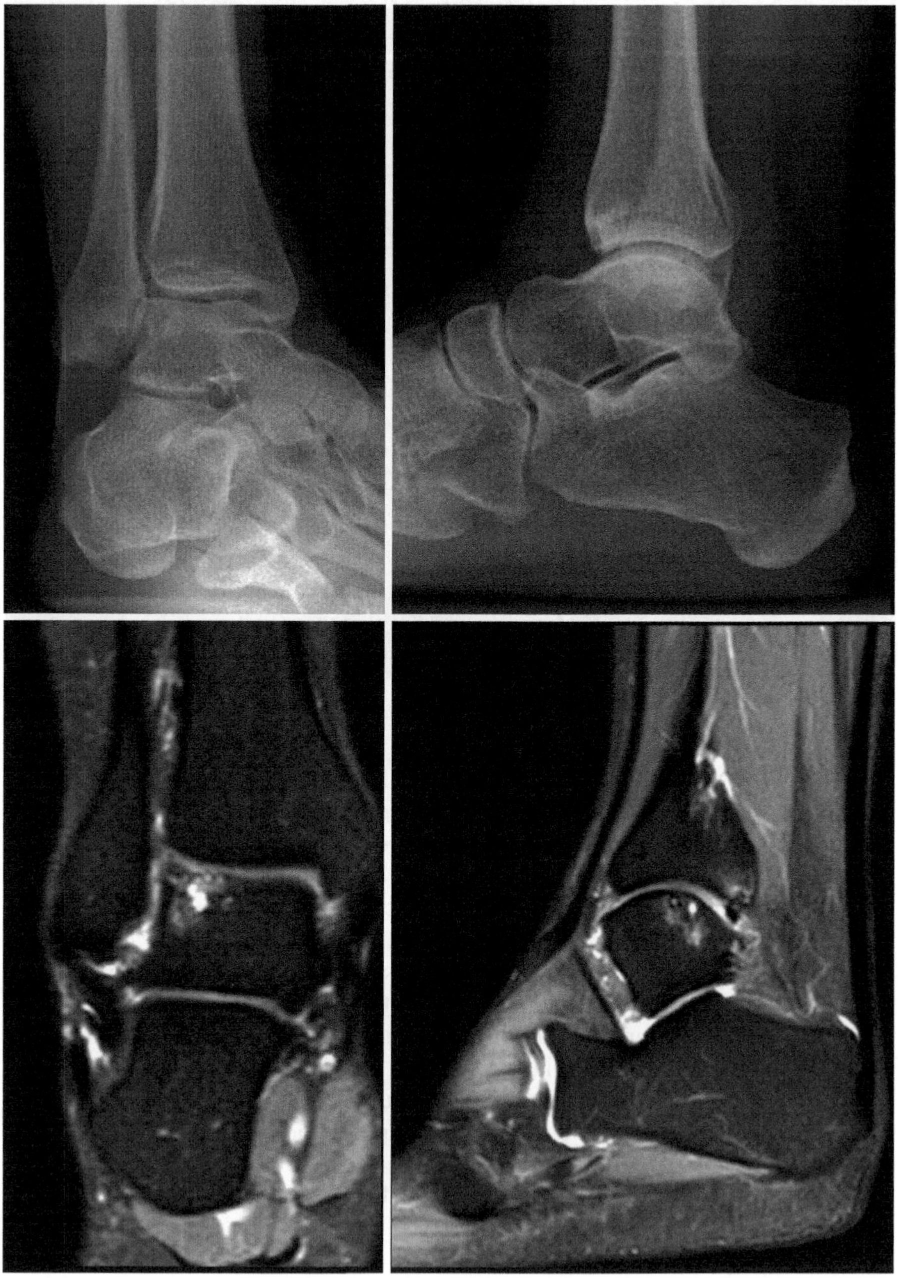

Fig. 20.7 OCD on the lateral side of the talar dome in a 23-year-old female patient

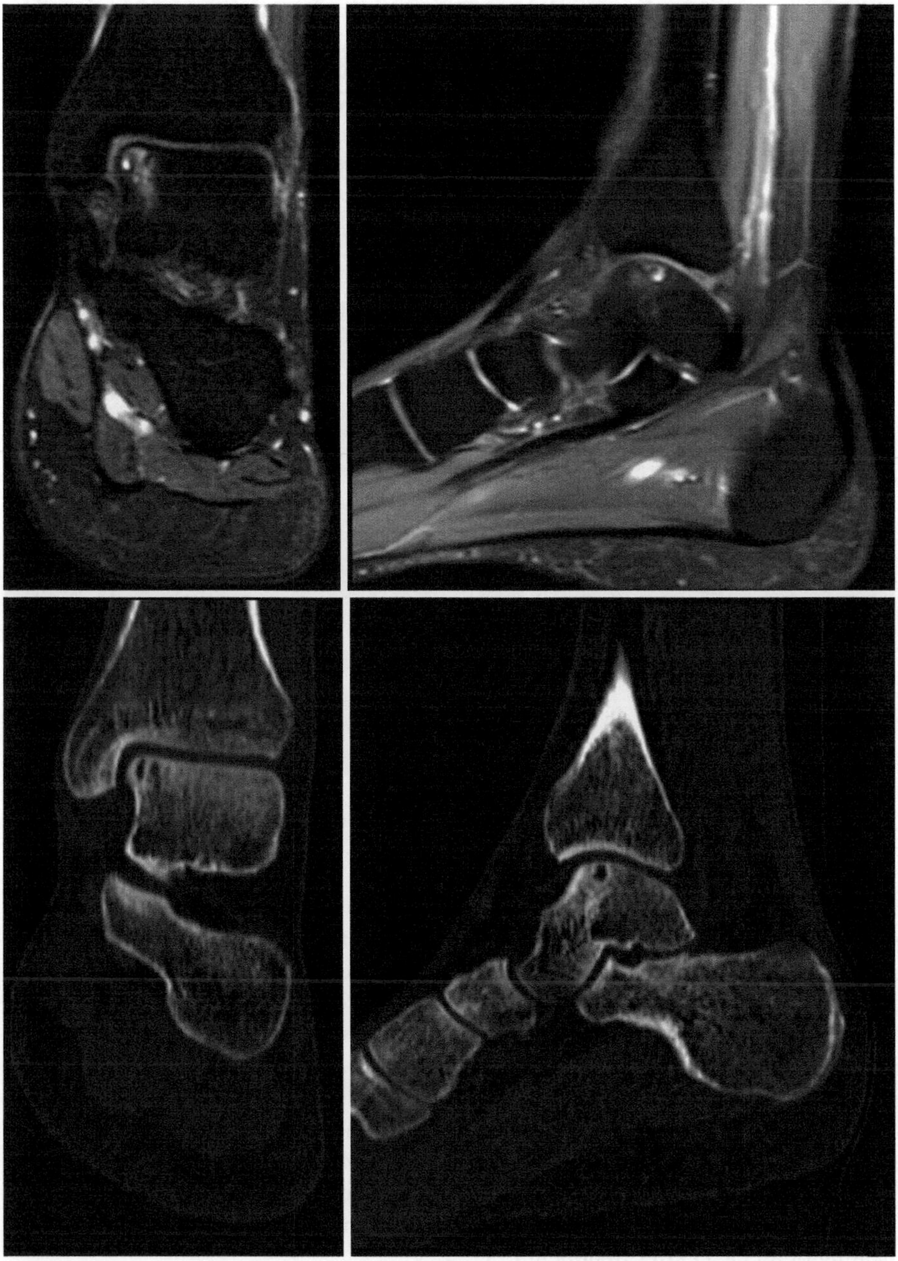

Fig. 20.8 OCD on the superomedial part of the talus with subchondral cyst formation in a 43-year-old male patient

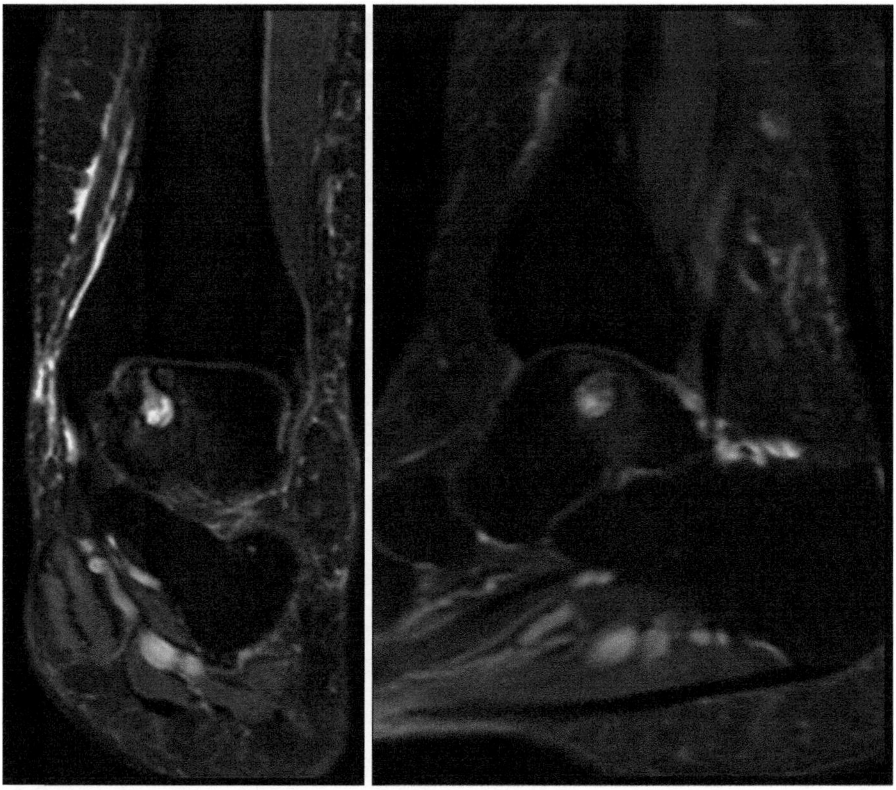

Fig. 20.9 Medial talar OCD with subcortical cyst in a 59-year-old female patient

SPECT SPECT-CT can differentiate between acute and chronic OCD, which is useful when a patient has chronic OCD and pain after a recent trauma. It can indicate the early onset of OCD by detecting activity in the presence of minimal MRI findings. In cases where MRI overestimates the size of the lesion due to bone marrow edema, SPECT-CT can be helpful to understand the true size of the lesion and make treatment decisions [24].

Ultrasound Ultrasound can detect typical OCD findings such as cortical irregularities and free fragments. It is inadequate for visualizing lesions located in the posterior region. However, while ultrasound is not yet generally applicable for detecting ankle OCDs, it has the potential to be used as a noninvasive monitoring tool for the healing of an OCD after treatment [23, 24].

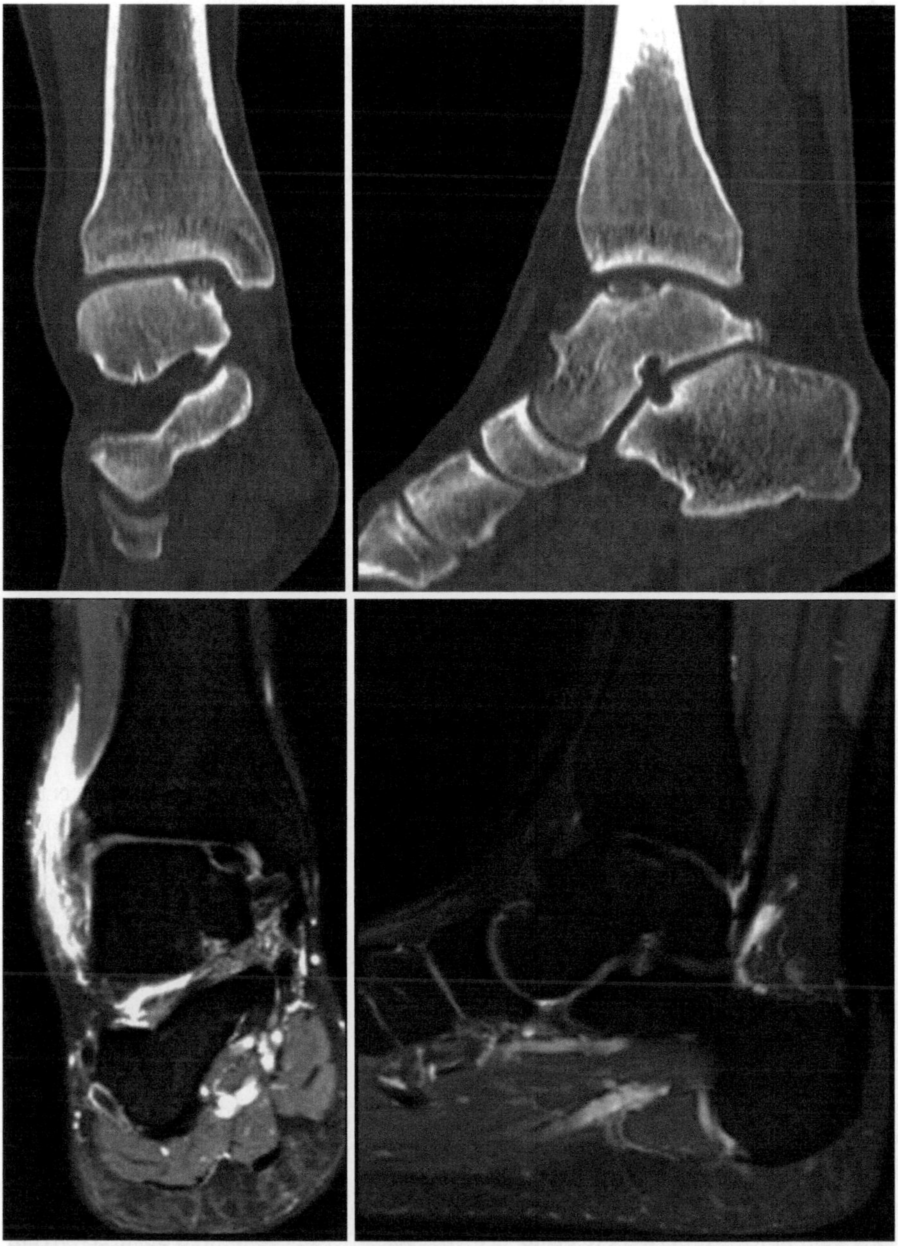

Fig. 20.10 Incidentally detected chronic OCD on the medial part of the talar dome while examining after an acute ankle distortion. The edema on the lateral side of the ankle indicates acute trauma, but there is no sign of inflammation around the OCD area which is medially located

20.3 Ankle Osteoarthritis

Osteoarthritis (OA) is a global public health issue affecting around 23% of the world's population [30]. Aside from causing pain, OA also contributes to disability. This condition not only leads to treatment costs but also imposes an economic burden due to reduced productivity and work impairment [31]. With the increasing life expectancy and rising obesity rates, the prevalence of OA is expected to rise [32].

Ankle osteoarthritis, on the other hand, is a chronic condition that affects approximately 1% of the world's population, with an estimated incidence of 30 per 100,000 (Fig. 20.11). It accounts for 2–4% of all OA cases [33, 34].

20.3.1 Etiology

Unlike hip and knee OA, the most common etiology of ankle OA is trauma (Fig. 20.12) [33, 35]. Traumatic etiology is present in nearly 80% of cases. Fractures around the ankle, particularly malleolar fractures, account for 62% of cases. Among other traumatic etiologies, chronic ligament instability increases stress on the ankle joint, leading to cartilage degeneration that can result in OA and is referred to as ligamentous ankle OA (Fig. 20.13) [35, 36].

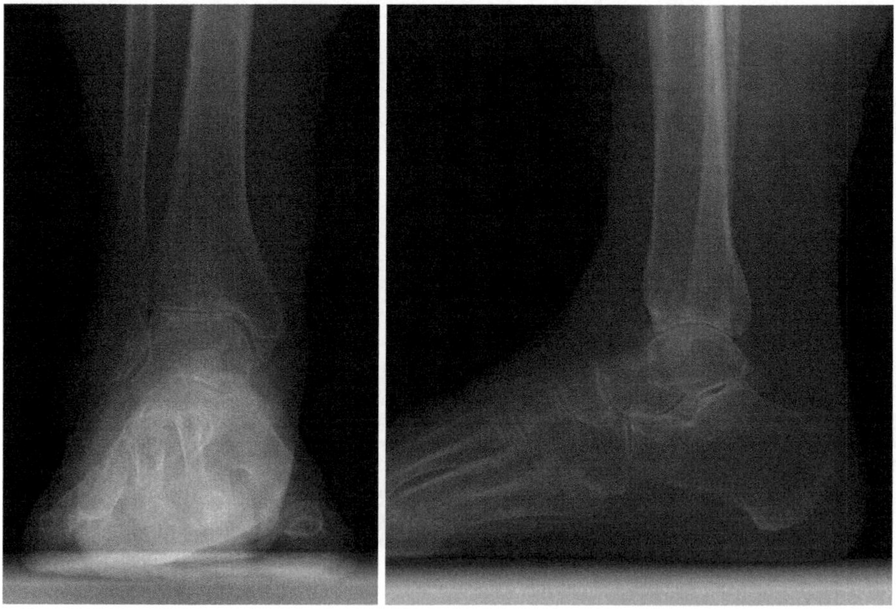

Fig. 20.11 Conventional radiographs showing osteoarthritis in the tibiotalar joint in a 73-year-old female patient

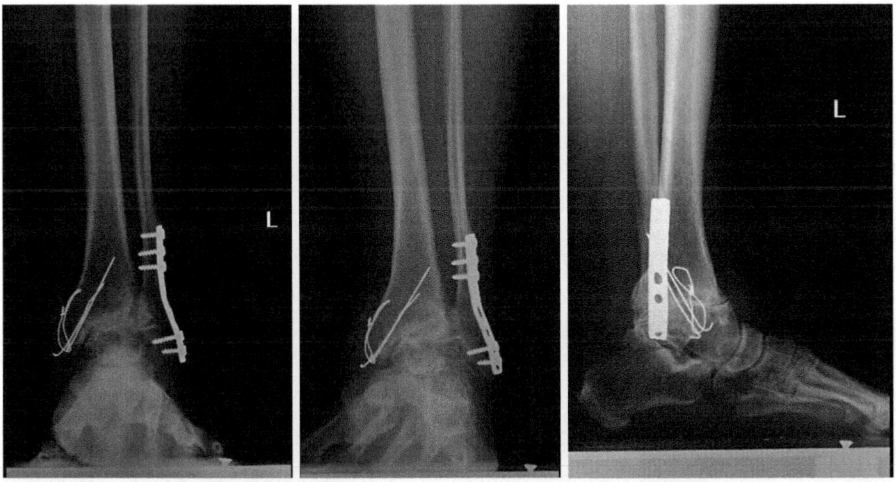

Fig. 20.12 Direct radiographs showing post-traumatic ankle osteoarthritis in a 46-year-old male patient who was operated due to ankle fracture

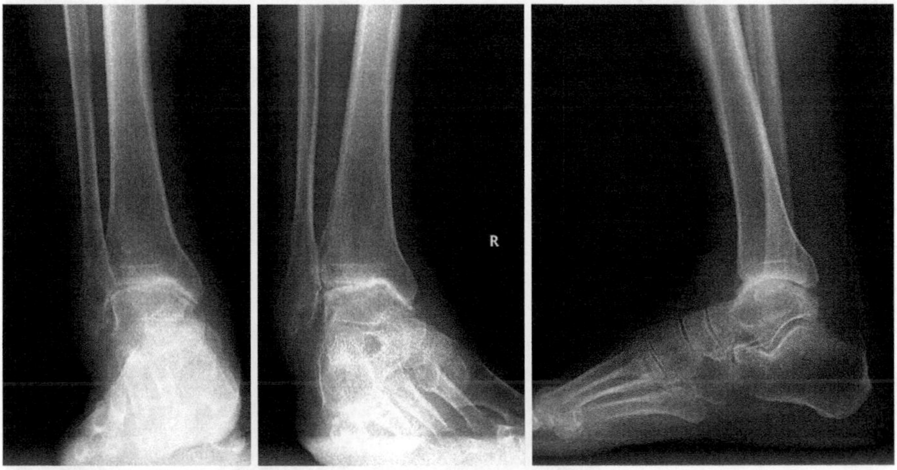

Fig. 20.13 Conventional weight-bearing radiographs of a 77-year-old male patient with chronic ankle instability leading to the development of ankle OA

Around 7–9% of cases are idiopathic OA, while 13% are attributed to other secondary causes (such as rheumatoid arthritis, hemochromatosis, hemophilia, deformity, or osteonecrosis) (Figs. 20.14 and 20.15).

Due to its primarily post-traumatic etiology, individuals with ankle osteoarthritis (OA) tend to be younger, typically between 18 and 44 years old, in contrast to other degenerative joint diseases. Furthermore, functional impairment progresses more

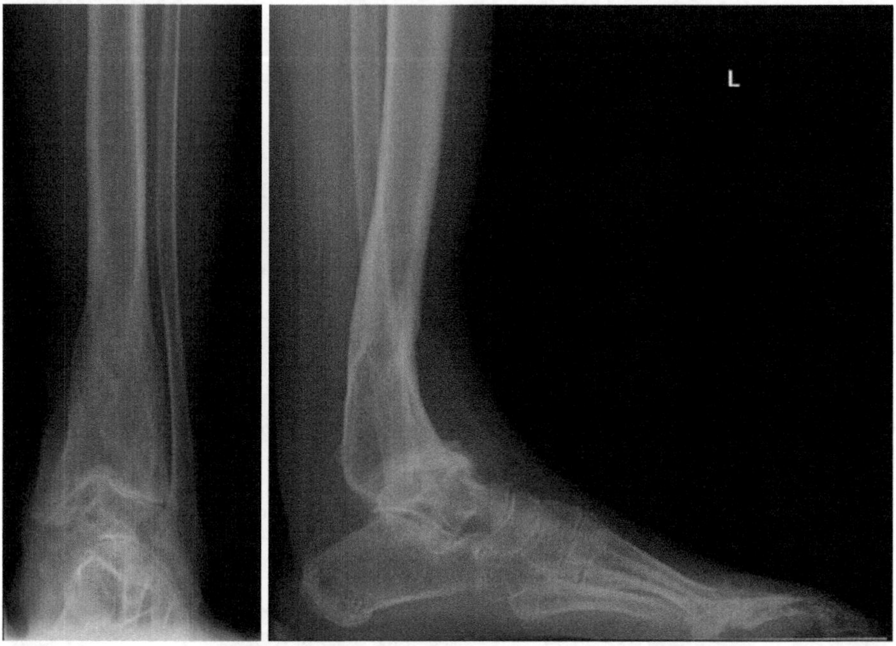

Fig. 20.14 Conventional radiographs displaying tibiotalar osteoarthritis in a 49-year-old female patient with a significant distal tibia deformity resulting from childhood trauma

Fig. 20.15 Advanced destruction in both ankle joints due to hemophilic arthropathy in a 63-year-old male patient

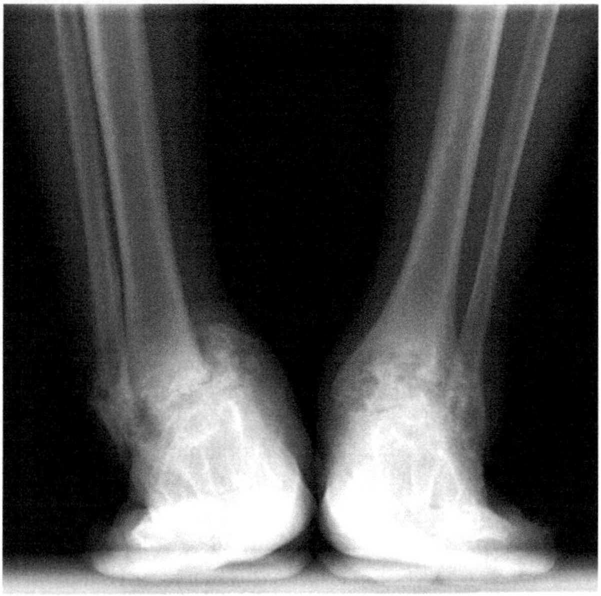

rapidly, with a transition to advanced stages of ankle OA typically occurring 10–20 years after the onset of the traumatic injury. However, clinical studies indicate that the initial degenerative changes following a traumatic injury develop within 12–18 months after the injury [36, 37].

The negative impact of ankle OA on health, physical function, and quality of life has been shown to be particularly associated with decreased employment in males, making it the only OA region significantly linked to employment reduction [38]. This is significant considering that half of OA patients are of working age.

The relationship between talus osteochondral lesions and the development of ankle OA is debated. Weigelt et al. reported that talus osteochondral lesions respond well to non-operative treatment, rarely cause chronic symptoms, and do not lead to ankle OA [39]. In contrast, Stufkens et al. suggested that anterolateral talar, posteromedial tibial, and medial malleolar osteochondral lesions increase the likelihood of ankle OA development [40].

20.3.2 Clinical Examination

In clinical diagnosis, it is essential to assess the mechanical characteristics of joint pain. Detecting pain during joint movement and identifying a decrease in joint range of motion compared to the contralateral side are crucial diagnostic factors. Notably, ankle dorsiflexion is frequently impacted. A normal ankle's overall range of motion in the sagittal plane typically falls between 65 and 75°, encompassing a range from 10 to 20° of dorsiflexion to 40–55° of plantar flexion (Fig. 20.16).

The pain is described as deep and at the level of ankle joint (Fig. 20.17). The presence or absence of effusion and the presence of deformities are important. Findings such as deformity, scar tissues from previous surgeries, muscle atrophy, and gait disturbances should also be assessed.

20.3.3 Radiological Examination

Anteroposterior and lateral radiographs obtained with weight-bearing are initial assessments (Fig. 20.11).

Takakura-Tanaka classification uses radiological findings to classify the stage of ankle OA (Table 20.2) [41]. It defines the disease in four stages and proposes limits for joint-preserving surgery based on the degree of joint degeneration. However, there's also an opposing view that Tanaka and other classification systems (Van Dijk, Kellgren–Lawrence) aren't always reliable decision-making tools for treatment [42].

Diagnosing early-stage conditions can be challenging using radiographs alone. 3D imaging methods such as CT and MRI (Fig. 20.18) are more useful for demonstrating damage to subchondral bone and cartilage tissue.

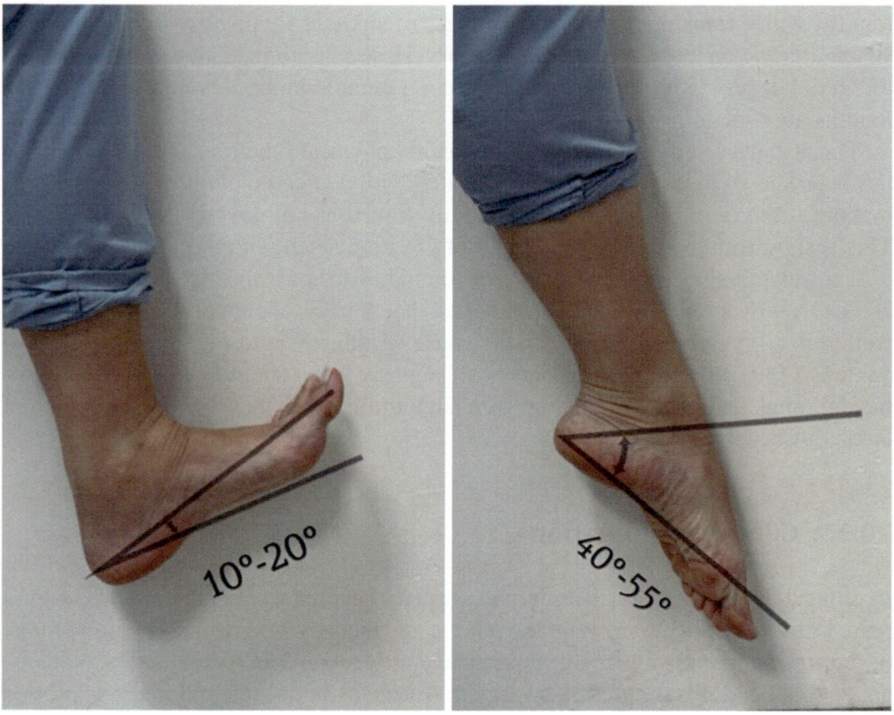

Fig. 20.16 Normal range of motion of the ankle joint

New MRI modalities like T2 mapping, though primarily useful for assessing postoperative repair tissue quality, can also be used for early ankle OA diagnosis [43].

In recent years, the use of single-photon emission CT (SPECT-CT) has gained popularity, adding anatomical details to the activity of OA or cartilage lesions [44]. Next-generation SPECT-CT scans have the ability to precisely locate degenerative lesions. SPECT-CT can detect increased subchondral metabolic bone activity, enabling the identification of early degenerative changes before clinical symptoms become apparent.

In more recent years, the development of weight-bearing CT (Fig. 20.19) scans has enabled a better 3D understanding of ankle OA and associated supra- and inframalleolar deformities in weight-bearing positions [45–49].

Key Points

- In acute cases, after an ankle injury, swelling and pain due to lateral ligamentous injury can obscure the diagnosis of OCD. If symptoms of ligament injury persist for more than 4–6 weeks, suspicion of chondral defect should arise.

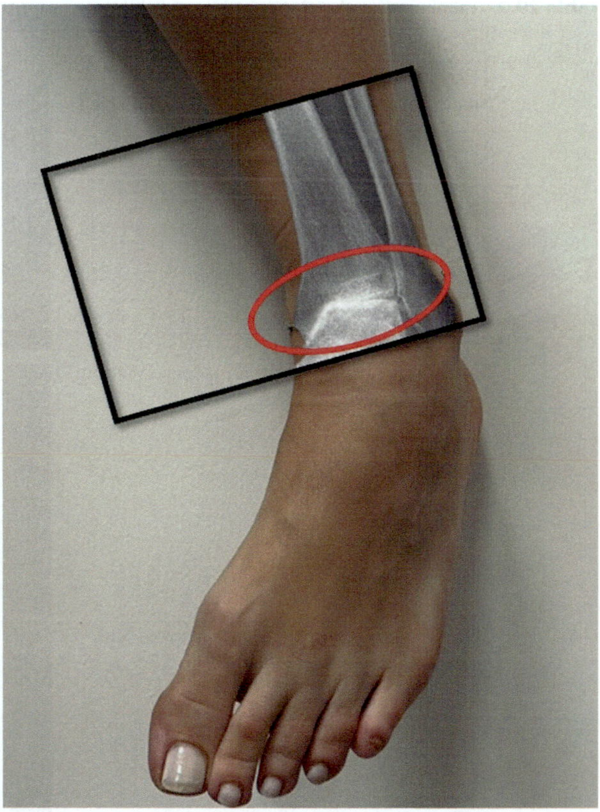

Fig. 20.17 The usual location of pain due to ankle osteoarthritis

- Locking and catching indicate a displaced fragment. In the presence of these findings, considering that early radiographic findings might be absent, evaluation of cartilage tissue using 3D imaging techniques like CT and MRI is recommended.
- Standard radiographs might be inadequate for visualizing lesions in the posterior talus. The heel-rise AP mortise view can provide better visualization of the talar dome.
- SPECT-CT, capable of detecting chondral defects early, can also distinguish between acute and chronic defects.
- The most common etiology of ankle osteoarthritis is trauma. Unlike other weight-bearing joints, it is observed at a younger age and can lead to workforce loss in the young population.

Table 20.2 Takakura-Tanaka classification of ankle osteoarthritis

Stage	Radiological finding	
1	Early sclerosis and formation of osteophytes, no narrowing of the joint space	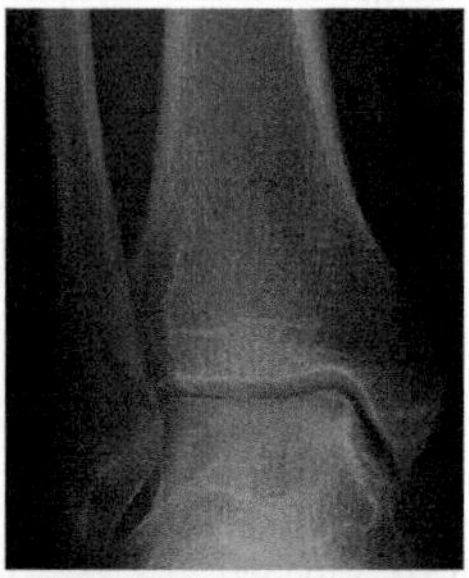
2	Narrowing of the medial joint space without subchondral contact	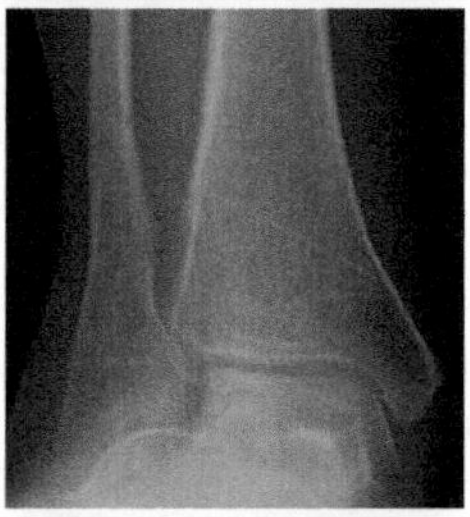
3A	Obliteration of the medial joint space with subchondral contact	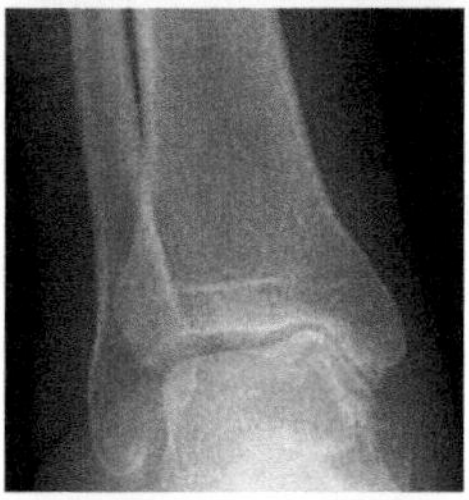
3B	Obliteration of the medial joint space and talus dome with subchondral contact	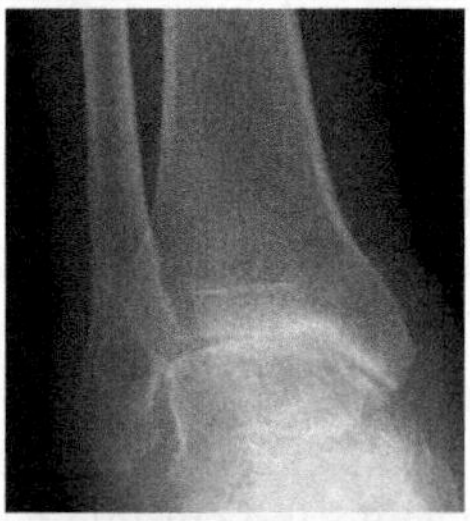
4	Complete obliteration of the joint space with full tibiotalar contact	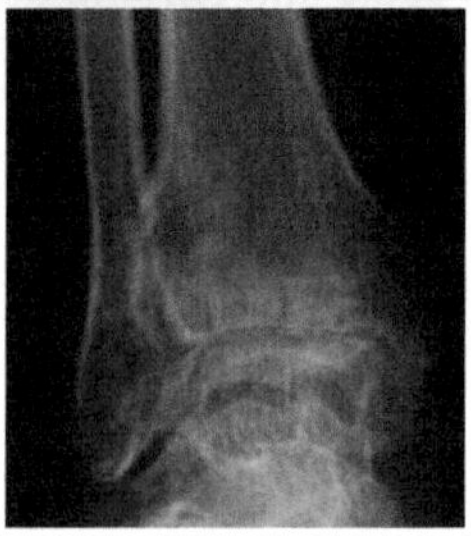

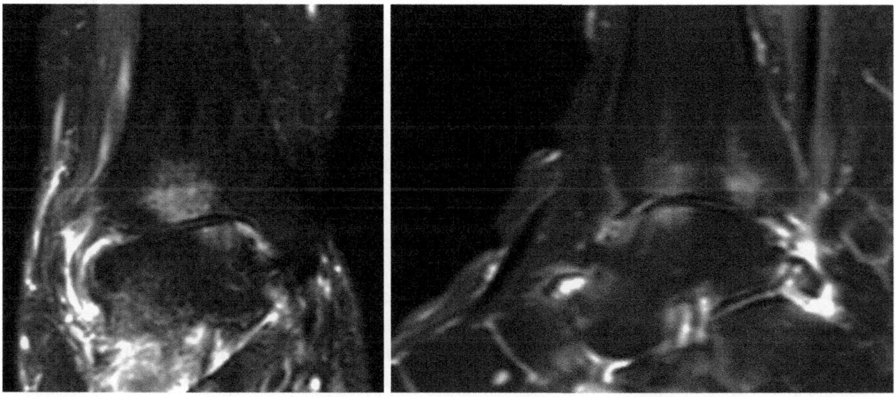

Fig. 20.18 MRI evaluation can reveal increased signal around cartilage tissue indicating cartilage pathology

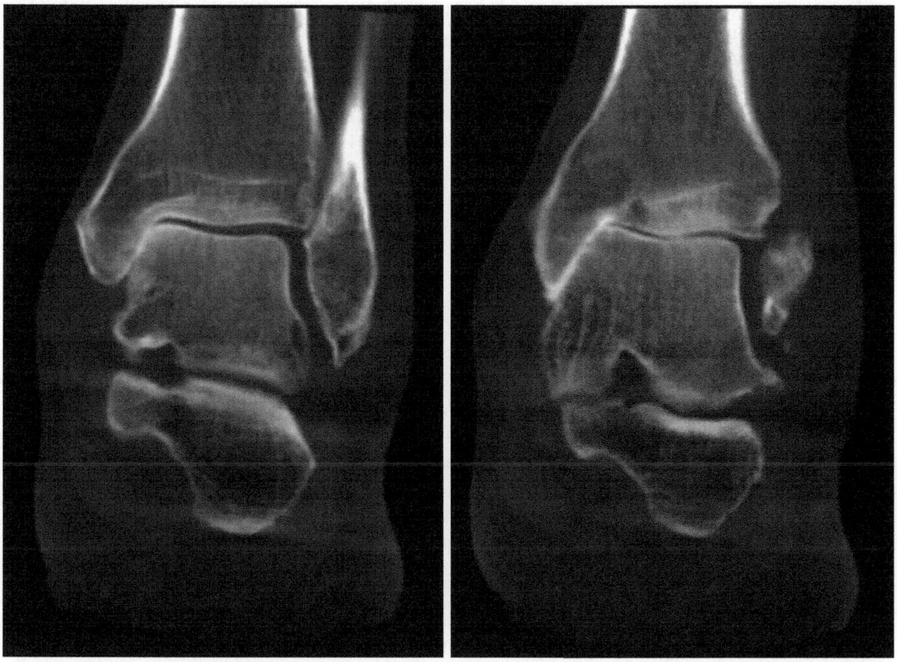

Fig. 20.19 A weight-bearing ankle CT scan reveals a narrowed tibiotalar joint space, offering a more comprehensive 3D understanding of the deformity

References

1. Mescher AL, Junqueira LCU. Junqueira's basic histology: text and atlas. New York: McGraw Hill Education; 2021.
2. Adams SB, Setton LA, Bell RD, et al. Inflammatory cytokines and matrix metalloproteinases in the synovial fluid after intra-articular ankle fracture. Foot Ankle Int. 2015;36:1264–71. https://doi.org/10.1177/1071100715611176.
3. Kraeutler MJ, Kaenkumchorn T, Pascual-Garrido C, et al. Peculiarities in ankle cartilage. Cartilage. 2017;8:12–8. https://doi.org/10.1177/1947603516642572.
4. Kuettner KE, Cole AA. Cartilage degeneration in different human joints. Osteoarthr Cartil. 2005;13:93–103. https://doi.org/10.1016/j.joca.2004.11.006.
5. Schmal H, Salzmann GM, Langenmair ER, et al. Biochemical characterization of early osteoarthritis in the ankle. Sci World J. 2014;2014:434802. https://doi.org/10.1155/2014/434802.
6. Tochigi Y, Rudert MJ, McKinley TO, et al. Correlation of dynamic cartilage contact stress aberrations with severity of instability in ankle incongruity. J Orthop Res. 2008;26:1186–93. https://doi.org/10.1002/JOR.20589.
7. Treppo S, Koepp H, Quan EC, et al. Comparison of biomechanical and biochemical properties of cartilage from human knee and ankle pairs. J Orthop Res. 2000;18:739–48. https://doi.org/10.1002/JOR.1100180510.
8. Procter P, Paul JP. Ankle joint biomechanics. J Biomech. 1982;15:627–34. https://doi.org/10.1016/0021-9290(82)90017-3.
9. Ramsey PHW. Changes in tibiotalar area of contact caused by lateral talar shift. J Bone Joint Surg. 1976;58:356–7.
10. van Dijk CN, Reilingh ML, Zengerink M, van Bergen CJA. Osteochondral defects in the ankle: why painful? Knee Surg Sports Traumatol Arthrosc. 2010;18:570–80. https://doi.org/10.1007/S00167-010-1064-X.
11. Whitesides TE. Orthopaedic basic science. Biology and biomechanics of the musculoskeletal system. J Bone Joint Surg. 2001;83:482. https://doi.org/10.2106/00004623-200103000-00040.
12. Zanon G, Di Vico G, Marullo M. Osteochondritis dissecans of the talus. Joints. 2014;2:115–23. https://doi.org/10.11138/JTS/2014.2.3.115.
13. Bruns J, Rosenbach B. Osteochondrosis dissecans of the talus. Comparison of results of surgical treatment in adolescents and adults. Arch Orthop Trauma Surg. 1992;112:23–7. https://doi.org/10.1007/BF00431039.
14. Lloyd J, Elsayed S, Hariharan K, Tanaka H. Revisiting the concept of talar shift in ankle fractures. Foot Ankle Int. 2006;27:793–6. https://doi.org/10.1177/107110070602701006.
15. McCoy AM, Toth F, Dolvik NI, et al. Articular osteochondrosis: a comparison of naturally-occurring human and animal disease. Osteoarthr Cartil. 2013;21:1638–47. https://doi.org/10.1016/J.JOCA.2013.08.011.
16. Olstad K, Ytrehus B, Ekman S, et al. Epiphyseal cartilage canal blood supply to the tarsus of foals and relationship to osteochondrosis. Equine Vet J. 2008;40:30–9. https://doi.org/10.2746/042516407X239836.
17. Flick AB, Gould N. Osteochondritis dissecans of the talus (transchondral fractures of the talus): review of the literature and new surgical approach for medial dome lesions. Foot Ankle. 1985;5:165–85. https://doi.org/10.1177/107110078500500403.
18. Verhagen RAW, Struijs PAA, Bossuyt PMM, Van Dijk CN. Systematic review of treatment strategies for osteochondral defects of the talar dome. Foot Ankle Clin. 2003;8:233–42. https://doi.org/10.1016/S1083-7515(02)00064-5.
19. Zengerink M, Struijs PAA, Tol JL, van Dijk CN. Treatment of osteochondral lesions of the talus: a systematic review. Knee Surg Sports Traumatol Arthrosc. 2010;18:238–46. https://doi.org/10.1007/S00167-009-0942-6.
20. Thompson JP, Loomer RL. Osteochondral lesions of the talus in a sports medicine clinic. A new radiographic technique and surgical approach. Am J Sports Med. 1984;12:460–3. https://doi.org/10.1177/036354658401200611.

21. Berndt MH. Transchondral fractures (osteochondritis dissecans) of the talus. J Bone Joint Surg Am. 1959;41:988–1020.
22. Nakasa T, Adachi N, Kato T, Ochi M. Appearance of subchondral bone in computed tomography is related to cartilage damage in osteochondral lesions of the talar dome. Foot Ankle Int. 2014;35:600–6. https://doi.org/10.1177/1071100714528493.
23. van Bergen CJ, Gerards RM, Opdam KT, et al. Diagnosing, planning and evaluating osteochondral ankle defects with imaging modalities. World J Orthop. 2015;6:944. https://doi.org/10.5312/WJO.V6.I11.944.
24. Leumann A, Valderrabano V, Plaass C, et al. A novel imaging method for osteochondral lesions of the talus–comparison of SPECT-CT with MRI. Am J Sports Med. 2011;39:1095–101. https://doi.org/10.1177/0363546510392709.
25. Mintz DN, Tashjian GS, Connell DA, et al. Osteochondral lesions of the talus: a new magnetic resonance grading system with arthroscopic correlation. Arthroscopy. 2003;19:353–9. https://doi.org/10.1053/jars.2003.50041.
26. Domayer SE, Apprich S, Stelzeneder D, et al. Cartilage repair of the ankle: first results of T2 mapping at 7.0 T after microfracture and matrix associated autologous cartilage transplantation. Osteoarthr Cartil. 2012;20:829–36. https://doi.org/10.1016/J.JOCA.2012.04.015.
27. Glaser C. New techniques for cartilage imaging: T2 relaxation time and diffusion-weighted MR imaging. Radiol Clin North Am. 2005;43:641–53. https://doi.org/10.1016/J.RCL.2005.02.007.
28. Trattnig S, Mamisch TC, Welsch GH, et al. Quantitative T2 mapping of matrix-associated autologous chondrocyte transplantation at 3 Tesla: an in vivo cross-sectional study. Investig Radiol. 2007;42:442–8. https://doi.org/10.1097/01.RLI.0000262088.67368.49.
29. Rutgers M, Bartels LW, Tsuchida AI, et al. dGEMRIC as a tool for measuring changes in cartilage quality following high tibial osteotomy: a feasibility study. Osteoarthr Cartil. 2012;20:1134–41. https://doi.org/10.1016/J.JOCA.2012.07.001.
30. Pereira D, Peleteiro B, Araújo J, et al. The effect of osteoarthritis definition on prevalence and incidence estimates: a systematic review. Osteoarthr Cartil. 2011;19:1270–85. https://doi.org/10.1016/J.JOCA.2011.08.009.
31. Murphy L, Helmick CG. The impact of osteoarthritis in the United States: a population-health perspective: a population-based review of the fourth most common cause of hospitalization in U.S. adults. Orthop Nurs. 2012;31:85–91. https://doi.org/10.1097/NOR.0B013E31824FCD42.
32. Cross M, Smith E, Hoy D, et al. The global burden of hip and knee osteoarthritis: estimates from the global burden of disease 2010 study. Ann Rheum Dis. 2014;73:1323–30. https://doi.org/10.1136/ANNRHEUMDIS-2013-204763.
33. Goldberg AJ, MacGregor A, Dawson J, et al. The demand incidence of symptomatic ankle osteoarthritis presenting to foot & ankle surgeons in the United Kingdom. Foot. 2012;22:163–6. https://doi.org/10.1016/J.FOOT.2012.02.005.
34. Valderrabano V, Horisberger M, Russell I, et al. Etiology of ankle osteoarthritis. Clin Orthop Relat Res. 2009;467:1800–6. https://doi.org/10.1007/S11999-008-0543-6.
35. Herrera-Pérez M, González-Martín D, Vallejo-Márquez M, et al. Ankle osteoarthritis aetiology. J Clin Med. 2021;10:194489. https://doi.org/10.3390/JCM10194489.
36. Herrera-Pérez M, Valderrabano V, Godoy-Santos AL, et al. Ankle osteoarthritis: comprehensive review and treatment algorithm proposal. EFORT Open Rev. 2022;7:448–59. https://doi.org/10.1530/EOR-21-0117.
37. Lindsjo U. Operative treatment of ankle fractures. Acta Orthop Scand Suppl. 2014;189:1–131. https://doi.org/10.3109/ORT.1981.52.SUPPL-189.01.
38. Sayre EC, Li LC, Kopec JA, et al. The effect of disease site (knee, hip, hand, foot, lower back or neck) on employment reduction due to osteoarthritis. PLoS One. 2010;5:470. https://doi.org/10.1371/JOURNAL.PONE.0010470.
39. Weigelt L, Laux CJ, Urbanschitz L, et al. Long-term prognosis after successful nonoperative treatment of osteochondral lesions of the talus: an observational 14-year follow-up study. Orthop J Sports Med. 2020;8:6.
40. Stufkens SA, Knupp M, Horisberger M, et al. Cartilage lesions and the development of osteoarthritis after internal fixation of ankle fractures: a prospective study. J Bone Joint Surg. 2010;92:279–86. https://doi.org/10.2106/JBJS.H.01635.

41. Tanaka Y, Takakura Y, Hayashi K, et al. Low tibial osteotomy for varus-type osteoarthritis of the ankle. J Bone Joint Surg. 2006;88:909–13. https://doi.org/10.1302/0301-620X.88B7.17325/LETTERTOEDITOR.
42. Claessen FM, Meijer DT, van den Bekerom MPJ, et al. Reliability of classification for post-traumatic ankle osteoarthritis. Knee Surg Sports Traumatol Arthrosc. 2016;24:1332–7. https://doi.org/10.1007/S00167-015-3871-6/TABLES/6.
43. Schreiner MM, Mlynarik V, Zbýň Š, et al. New technology in imaging cartilage of the ankle. Cartilage. 2017;8:31–41. https://doi.org/10.1177/1947603516632848.
44. Paul J, Barg A, Kretzschmar M, et al. Increased osseous (99m)Tc-DPD uptake in end-stage ankle osteoarthritis: correlation between SPECT-CT imaging and histologic findings. Foot Ankle Int. 2015;36:1438–47. https://doi.org/10.1177/1071100715596745.
45. Burssens A, Susdorf R, Peterhans U, Ruiz R, Barg A, Hintermann B. Supramalleolar osteotomy alters the subtalar joint alignment. Foot Ankle Int. 2022;43(9):1194–203.
46. Conti MS, Ellis SJ. Weight-bearing CT scans in foot and ankle surgery. J Am Acad Orthop Surg. 2020;28:e595–603. https://doi.org/10.5435/JAAOS-D-19-00700.
47. Li J, Fang M, Van Oevelen A, Peiffer M, Audenaert E, Burssens A, International Weightbearing CT Society. Diagnostic applications and benefits of weightbearing CT in the foot and ankle: a systematic review of clinical studies. Foot Ankle Surg. 2023;6:S1268.
48. Peiffer M, Belvedere C, Clockaerts S, Leenders T, Leardini A, Audenaert E, Victor J, Burssens A, WBCT ISG. Three-dimensional displacement after a medializing calcaneal osteotomy in relation to the osteotomy angle and hindfoot alignment. Foot Ankle Surg. 2020;26(1):78–84.
49. Raes L, Peiffer M, Leenders T, Kvarda P, Ahn J, Audenaert E, Burssens A. Medializing calcaneal osteotomy for progressive collapsing foot deformity alters the three-dimensional subtalar joint alignment. Foot Ankle Surg. 2023;29:S1268.

Emerging Technologies in the Diagnosis of Foot and Ankle Pathologies

21

Samir Ghandour and Soheil Ashkani-Esfahani

In recent years, the field of foot and ankle orthopedic surgery has come across a multitude of diagnostic technological advancements. These emerging technologies have revolutionized the way orthopedic surgeons and care providers approach the clinical and radiological examinations of the foot and ankle, paving the way for more accurate and efficient diagnostic methods. The importance of timely and accurate diagnosis cannot be overstated, as it directly impacts the effectiveness of treatment interventions. A delayed or inaccurate diagnosis can lead to inappropriate management strategies, ineffective treatments, complications, and increased healthcare costs. Therefore, it is crucial to explore emerging advanced technologies that can enhance diagnostic accuracy and efficiency, enabling clinicians to provide more targeted and timely interventions [1]. Some of the most significant advancements in imaging technologies and examination tools include but are not limited to the development of portable handheld ultrasound devices, weight-bearing computed tomography (WBCT) devices, minimally invasive arthroscopy devices, computer-assisted simulations and 3D modeling of pathologies, and artificial intelligence (AI) algorithms for prediction and early detection of diseases as well as improving the accuracy and speed of diagnosis. These technologies can be used at different levels of healthcare, including prevention, diagnosis, and post-treatment care and rehabilitation. This chapter aims to explore several of the latest developments in this realm,

S. Ghandour
Foot & Ankle Research and Innovation Laboratory (FARIL), Massachusetts General Hospital, Harvard Medical School, Boston, MA, USA

S. Ashkani-Esfahani (✉)
Foot & Ankle Research and Innovation Laboratory (FARIL), Massachusetts General Hospital, Harvard Medical School, Boston, MA, USA

Department of Orthopaedic Surgery, Massachusetts General Hospital, Harvard Medical School, Boston, MA, USA
e-mail: sashkaniesfahani@mgh.harvard.edu

S. Sharma et al. (eds.), *Clinical and Radiological Examination of the Foot and Ankle*, https://doi.org/10.1007/978-981-97-4202-8_21

shedding light on the innovative techniques and tools that have emerged to aid in the diagnosis of foot and ankle conditions.

21.1 Artificial Intelligence and Computer-Assisted Diagnosis

AI and computer-assisted diagnostic methods have the potential to revolutionize preventive measures, diagnosis, prediction of outcomes, and clinical decision-making for foot and ankle surgeons. Computer-assisted methods and AI were shown to be powerful tools for developing, organizing, analyzing, and reporting outcomes from large and granular databases and data registries, particularly in the field of foot and ankle surgery, with high accuracy and speed. However, developing these large and granular databases needs expert supervision and sometimes manual interference and annotation. Missing or misinterpreted data can harm the algorithms and lead to misleading outcomes [2, 3]. On the other hand, the more a database is created with high accuracy and fewer flaws, the more accurate, reliable, and reproducible the outcomes are.

Computer-assisted analysis improves the accuracy, validity, and reliability of a diagnostic method if designed appropriately. Improved interobserver agreement and accuracy of diagnosis of syndesmotic instability using WBCT are examples of this [4, 5]. In these studies, authors have used computer-based algorithms to interpret WBCT images and to detect instabilities of the ankle syndesmosis using 3D image analysis. Their approach not only improved the accuracy and interobserver agreement but also showed significant improvement in the speed of diagnosis. Adding AI to these computer-based approaches has also shown promise for improving the precision and quality of diagnosis in foot and ankle surgery. AI-based prediction algorithms for various complications of foot and ankle traumas, such as surgical site infections or venous thromboembolism—two of the most common and fatal complications—have been investigated by different groups of researchers. Prediction algorithms for surgical site infections after ankle fractures were shown to be helpful for clinicians in the process of decision-making and whether to administer more extensive antibiotics and additional care. These studies were also successful in determining risk factors for surgical site infections [6, 7]. For the prediction of venous thromboembolism, various studies have been done and guidelines proposed; however, the implementation of AI algorithms has added a new dimension to the prediction and decision-making process for foot and ankle providers. A group of scientists developed a prediction algorithm for venous thromboembolism after ankle fracture that showed up to 88% accuracy, 94% sensitivity, and 36% specificity [8]. The literature has shown that through the inclusion of patient-specific data such as demographics, socioeconomic status, past medical history, medications, patient-reported outcomes, and disease-specific clinical findings, the performance of prediction algorithms can be improved considerably. In order to achieve this goal, databases should be more comprehensive and include not only tabular and narrative data but also imaging and more complex inputs. Moreover, they should be updated routinely according to today's needs.

Advances in the realm of AI have also expanded to image interpretation and decision support for radiological methods [1]. Deep learning, which is a subset of machine learning, has been widely used to detect various pathologies in the foot and ankle. These pathologies include but are not limited to ankle fractures, Lisfranc joint instability, syndesmotic instability, and metatarsal fractures. Deep learning AI algorithms have shown an accuracy of over 90% in the detection of these pathologies on X-rays and WBCT [9–11]. In most of these studies, one of the concerns mentioned by the authors was the quality and quantity of the imaging data. They suggested that in order to improve the performance of these diagnostic algorithms, while a large and granular dataset is necessary, the images should have been reviewed and annotated by experts. There must be a consensus regarding the diagnosis of training datasets in order to set a ground truth that we can compare and improve the outcomes of the AI algorithms. Moreover, other than internal validation and testing of these algorithms in each institution, these technologies should be externally validated using datasets from other institutions at various other areas of the world. Thus, collaborative multicentric initiatives involving researchers, clinicians, and industry stakeholders are essential for harnessing the full potential of AI in the diagnosis and management of foot and ankle pathologies.

21.2 New Imaging Technologies

21.2.1 Weight-Bearing CT Scan

WBCT is an innovative imaging technique that merges the advantages of traditional CT scans with the added marvel of weight-bearing imaging (Fig. 21.1). WBCT emerged as an important diagnostic tool in the field of foot and ankle surgery, offering improved accuracy and sensitivity in detecting subtle injuries and joint instabilities. By capturing 3D images in this dynamic position, WBCT unmasks subtle

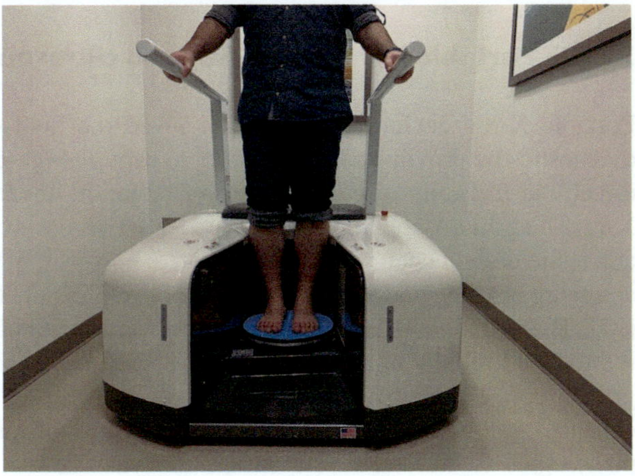

Fig. 21.1 Computed tomography acquisition of the foot and ankle under weight-bearing (physiological) conditions

injuries, joint instabilities, and changes in alignment that would otherwise remain hidden under non-weight-bearing conditions [1]. Three-dimensional WBCT was found to be better than gravity stress radiographs and gravity stress CT in detecting ankle instability after supination external rotation ankle fractures [12]. WBCT measurements, including 3D volumetric and 2D area measurements, have shown higher accuracy and reliability compared to conventional distance measurements in the detection of 3D pathologies such as syndesmosis instability and Lisfranc joint instability [13–15]. Automation of these techniques using computer-assisted algorithms, as previously mentioned, has led to even further superior outcomes compared to other imaging modalities, particularly for subtle injuries that might not be detectable on conventional radiographic and CT images. In deformities such as Charcot-Marie-Tooth (CMT), cavovarus foot, and acquired flatfoot deformity, WBCT has provided a more accurate assessment of foot alignment and deformity compared to 2D manual measurements [16]. It has also been used to characterize hindfoot alignment and differentiate between the different types of foot deformities, especially those of the foot arch [17, 18]. Viewing the joints in a 3D manner under physiologic load is helpful not only for diagnostic purposes but also for pre-operative planning and post-operative monitoring [19, 20]. However, a significant drawback of WBCT is the cost associated with the device and with implementing these advanced imaging techniques, particularly in limited resource settings [21–23]. Studies have shown that the initial use of WBCT instead of non-weight-bearing radiographs to diagnose foot and ankle conditions was cost-effective when looking at metrics such as quality-adjusted life-years, incremental cost-effectiveness ratios, and overall radiation dose prior to diagnosis [24]. When it comes to per-scan radiation exposure, some studies have described a higher dose exposure per WBCT scan than traditional scanning [23, 25]. Another limitation lies in the challenge of capturing precise weight-bearing positions for certain patients, such as those with mobility limitations or severe pain, compromising the accuracy, reliability, and reproducibility of the obtained images [26]. Despite all its promises, WBCT still faces obstacles that need to be addressed to maximize its effectiveness and widen its clinical utility.

21.2.2 Portable Handheld X-Ray and Ultrasound Devices

Portable X-ray devices have revolutionized the field of musculoskeletal (MSK) diagnostics by providing the capability to capture high-quality radiographic images directly at the point of care with lower radiation doses compared to conventional X-rays and fluoroscopy devices (Fig. 21.2). This use of portable X-rays in the foot and ankle has not been widely investigated; however, it was shown to be effective in assessing ankle syndesmosis in a prospective clinical study on healthy individuals [27]. This technology offers numerous advantages, including increased accessibility, reduced radiation, real-time imaging, and improved patient comfort. By integrating portable X-rays into the diagnostic workflow, healthcare professionals can enhance the access, accuracy, and efficiency of diagnosing foot and ankle pathologies.

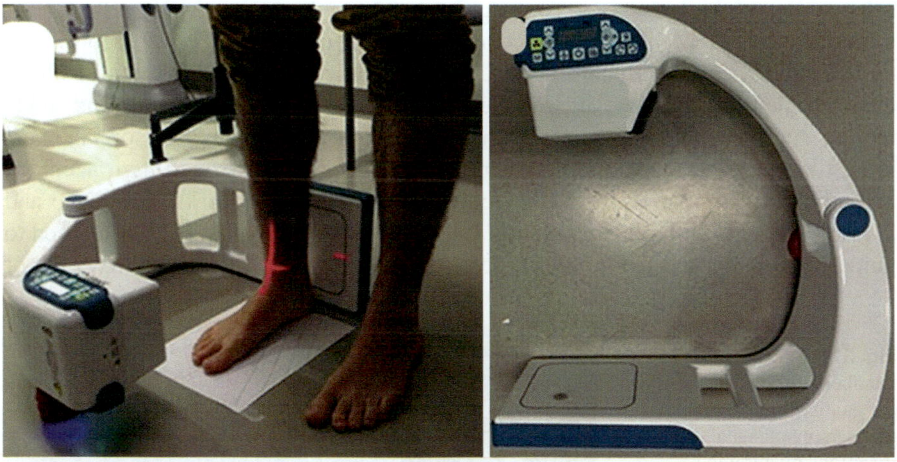

Fig. 21.2 Portable weight-bearing X-ray device and its acquisition of an ankle radiograph under (physiologic) loading

On the other hand, portable handheld ultrasonography (P-US) devices have emerged as a valuable tool in the diagnosis and management of foot and ankle pathologies. This emerging technology enables clinicians to perform real-time imaging at the point of care, providing immediate visual information and a reliably accurate evaluation [28]. The portability, ease of use, and lack of radiation make it a convenient option for foot and ankle orthopedic surgeons, allowing them to assess patients in various clinical settings. The use of portable ultrasonography has been investigated for diagnosing lateral ankle instability [29]. Using a cadaveric ligament transection model of ankle ligament injury, the results showed a strong correlation between the measurement values obtained via P-US and fluoroscopy. Moreover, P-US-based measurements were found to be reliable for measurements of the talar translation and the lateral clear space. P-US was also compared with X-rays for suspected ankle and fifth metatarsal bone fractures [30]. The results of the study indicated that P-US exhibited comparable diagnostic accuracy to radiographic imaging in the detection of ankle fractures. Another study assessed whether P-US could potentially reduce the need for radiographs in children with ankle injuries and fractures. The authors found that P-US exhibited high sensitivity (78%), specificity (71%), and positive likelihood ratio (2.7) in the detection of ankle fractures among children. Moreover, when combined with Ottawa Ankle Rules (OAR) upon presentation, P-US reduced the need for X-ray evaluation by 17% with no missed fractures. These findings suggest the potential of P-US as a reliable diagnostic tool for foot and ankle traumas in clinical settings as well as in battlefields or in austere settings where access to more advanced imaging tools is limited [31]. P-US allows for dynamic imaging of the Achilles tendon and detection of tendon ruptures, assessing tendon thickness, integrity, and vascularity [32]. Real-time assessment of the plantar fascia, reliably visualizing thickening of the heel fat pad, hypoechogenicity, and other characteristic changes associated with plantar fasciitis is achievable with high accuracy compared to gold standards (i.e., MRI), a high level of interobserver agreement, and acceptable learning curve using P-US according to the literature [33, 34]. Regarding P-US-guided procedures, multiple studies have demonstrated the efficacy of ultrasound-guided injections in the foot and ankle region, such as the first metatarsophalangeal joint, second metatarsophalangeal joint, tibiotalar joint, and Achilles peri tendinous space, and found high accuracy and reliability compared to blind injections or anatomic landmarks (Figs. 21.3 and 21.4). In patients with plantar fasciitis and Achilles tendinopathy, P-US was used to guide fasciotomy and tenotomy, respectively [33, 35–37]. The portability, ease of

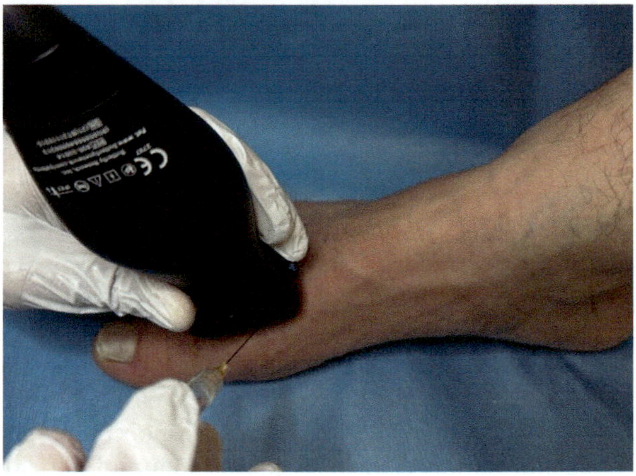

Fig. 21.3 Portable handheld ultrasound used for guiding metatarsophalangeal joint injections by a trained clinician at the clinic

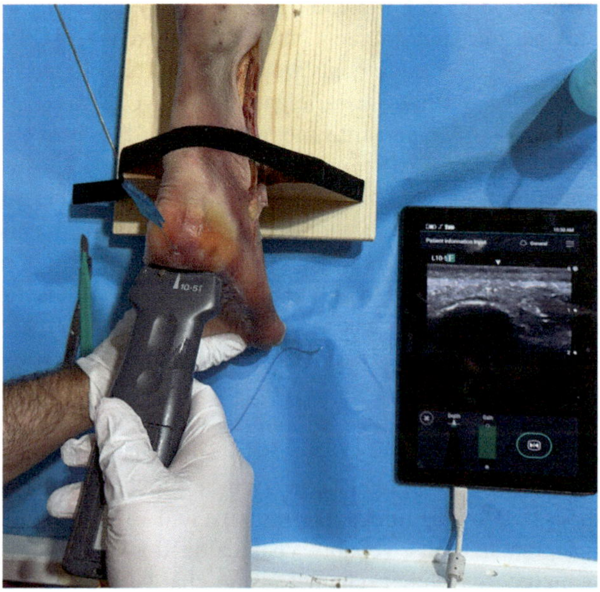

Fig. 21.4 Portable handheld ultrasound used for a cadaveric study on minimal invasive plantar fascia release

access, and ease of use—that only needs a smartphone or tablet—of modern portable X-ray and P-US machines have facilitated the integration of this technology into clinical practice, bridging the gaps in care caused by barriers such as time, cost, physical distance, and expertise [38].

21.2.3 Needle Arthroscopy

Minimally invasive arthroscopy, also known as needle arthroscopy, has been introduced to the orthopedic world in recent years, with numerous applications yet to be investigated [39]. The use of these new scopes has been demonstrated in the diagnosis of various foot and ankle conditions, including but not limited to ligament injuries, ankle sprains, ligament ruptures, ankle ligament impingements, ankle osteochondral lesions, intra-articular injections, and complications management such as bacterial

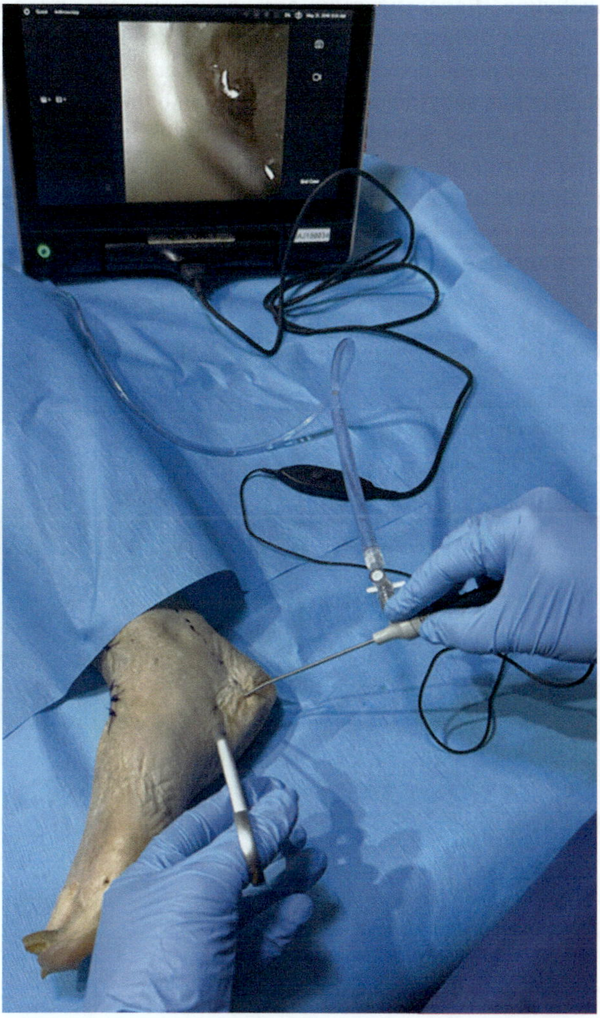

Fig. 21.5 Needle scope used during endoscopic plantar fascia release

arthritis following ankle fractures [40–42]. However, most of these implications need to be validated through future studies with larger populations. Needle scope is a promising tool that can provide a direct view of the joint in a less invasive manner; however, providing tools that can add intervention abilities to surgeons while using this novel diagnostic tool seems to be the next aim for developers of this device and should be further investigated and validated before being introduced for wider usage (Fig. 21.5).

21.3 Virtual and Augmented Reality Technologies

In simple words, virtual reality (VR) tools create a virtual environment for the assessment and diagnosis of various anatomical regions of the body, while augmented reality (AR) methods can superimpose a 3D real environment onto the real-time surgical field for a live experience of 3D simulation models during the procedure. These technologies have shown promise in various medical disciplines, including diagnostic radiology, neurosurgery, and orthopedic surgery [43–47]. There are different applications for VR and AR, both for diagnostic purposes, pre-operative planning, and post-operative care. VR and AR can be used to visualize and manipulate anatomical structures beneath the surface anatomy in a 3D manner, aiding in surgical planning as well as virtual education of the residents and fellows (Fig. 21.6) [44]. They can also be used to create virtual environments that simulate real-world scenarios, allowing for more realistic training experiences for healthcare professionals [48]. Furthermore, VR technologies can help establish better communication between surgeons and patients [49]. Developing solutions to obtain

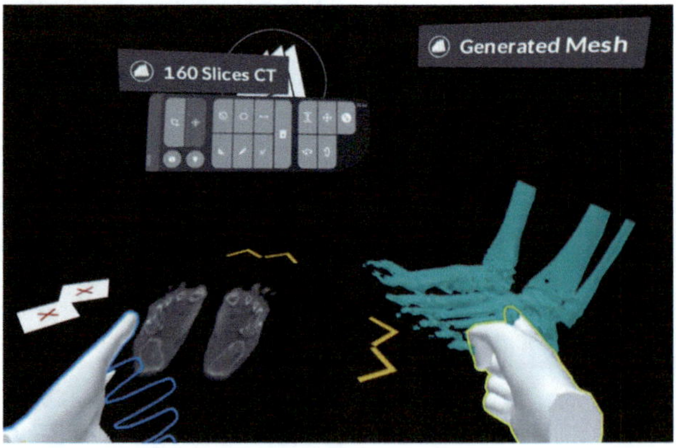

Fig. 21.6 Using a virtual reality platform (Medicalholodeck) via a head-mounted device for weight-bearing CT evaluation both in 2D and 3D by a clinician

patient-reported outcomes through VR applications, providing a platform through which surgeons can communicate with the patients, go over different steps of the treatment, the operation, and even post-operative care, guiding the patients regarding rehabilitation methods, assess the patients remotely, and examine the patients regarding the outcomes of the surgery such as range of motion, movements, pain points, skin changes, and other examinations that can be done without the presence of the physician and are only a few of the advantages that VR and AR can add to orthopedics, particularly to the foot and ankle realm.

21.4 Wearable Devices and Sensor Technologies

Advances in wearable technologies and sensors, for predictive, diagnostic, and monitoring purposes, have been significant in the past decade in foot and ankle surgery, from wearable accessories such as watches or earbuds to sensors incorporated in shoes, socks, bands, and clothing. Wearable devices and sensor technologies have shown promise in monitoring gait and biomechanics and providing objective and quantitative measurements of motion and gait parameters, allowing for a more detailed understanding of the underlying causes, disease progression, and treatment outcomes. For example, wearable sensors attached to shoes can objectively assess spatiotemporal gait parameters, such as stride length, stance phase time, and toe clearance in patients with Parkinson's disease, obviating the need for a specialized gait analysis facility [50]. Gait analysis using wearable sensors has shown great prospects in various health-related applications, including rehabilitation, clinical diagnosis, and sports [51]. These wearable sensors, such as inertial measurement units, provide valid and reliable data on motion and can be securely attached to different body segments for comprehensive biomechanical analysis that has been shown to be as reliable and valid as analysis performed by optical tracking systems [52]. Additionally, wearable sensors have been used to monitor activities and detect abnormalities in gait patterns, providing valuable information for differentiating between pathological and healthy controls [53] (Fig. 21.7). Monitoring patients with joint osteoarthritis, neuropathy, foot drop, Achilles tendinopathy and ruptures, and stress fractures are some of the applications of wearables and sensors in the field of foot and ankle surgery [54–59]. The combination of data obtained from wearables with AI algorithms and AI-based data

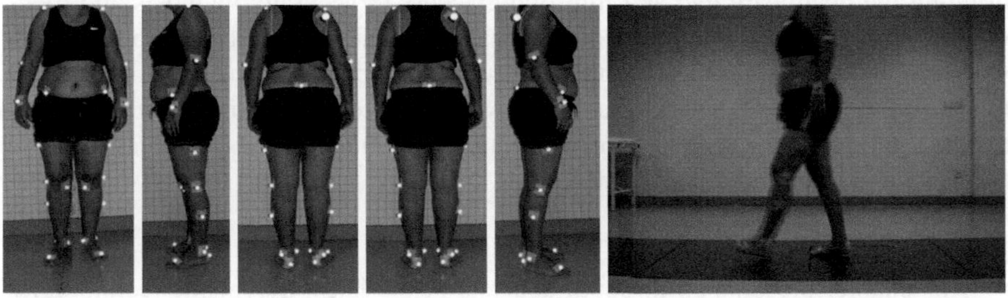

Fig. 21.7 Gait analysis performed with the help of wearable sensors and motion-tracking cameras

analysis methods has led to even more accurate and reliable outcomes regarding the prediction of pathologies, monitoring, and prediction of the outcomes in foot and ankle patients [60]. However, to develop the most efficient wearables, clinicians and engineers should work closely in order to combine clinical knowledge and needs with engineering science and sensor technologies. Developing sensor-based wearables for specific pathologies or clinical purposes needs a comprehensive knowledge on the capabilities of the sensors, materials, data-obtaining methods, data analysis, correlating the data and analysis outcomes with the clinical and anatomical conditions, and interpreting the outcomes and validating the product accurately.

21.5 Conclusion

The emerging diagnostic tools for foot and ankle pathologies present several challenges and opportunities for future development. One of the key challenges is the ethical considerations in adopting these emerging technologies. Ethical issues may arise in terms of patient privacy, data security, and informed consent. It is crucial to ensure that patient information is protected and that individuals are fully informed about the risks and benefits of using these technologies [61]. Additionally, regulatory hurdles and standards for diagnostic technologies need to be established to ensure the safety and efficacy of these tools. It is important to validate the diagnostic accuracy and reliability of these tools through rigorous clinical studies and comparisons with established and gold standard diagnostic methods [62]. Additionally, technical issues and malfunctions may occur with these technologies, requiring troubleshooting and maintenance to ensure their proper functioning. Continuous improvement of the databases, along with updates and improvements to the software and hardware components of these tools, is necessary to address any limitations and enhance their diagnostic capabilities. Education is also a crucial aspect of the adoption and utilization of emerging diagnostic tools. Healthcare professionals require routine training on the proper use and interpretation of these technologies to ensure accurate diagnosis and appropriate consequent treatment planning [62]. Additionally, patient education is important to ensure that individuals understand the benefits and limitations of these technologies and can make informed decisions about their healthcare. Moreover, efforts should be made to ensure that these technologies are accessible and affordable in underprivileged and underserved settings, bridging the healthcare gap and improving access to accurate diagnosis and treatment [63, 64]. With the continued development of new technologies, particularly with recent advance in Artificial Intelligence and computer vision, we should expect to see even greater advances in the accuracy and efficiency of diagnosis in the years to come [64]. This will lead to improved patient outcomes and a better quality of life for those with foot and ankle conditions.

Key Points

- This chapter presents an in-depth analysis of the latest diagnostic technologies for foot and ankle pathologies, emphasizing their precision and transformative impact on surgical procedures and patient outcomes.

- The integration of advanced diagnostic tools like portable ultrasound devices and weight-bearing CT scans has revolutionized the accuracy and efficiency of diagnosing foot and ankle issues by incorporating the advent of physiological loading.
- Minimally invasive surgical techniques are becoming more prevalent among foot and ankle surgeons. These techniques use novel tools and equipment, and their use has enhanced recovery times and reduced post-operative complications.
- The future of diagnostic techniques in foot and ankle surgery is ample, with opportunities for research and development. However, certain limitations, such as ethical considerations, patient education, and provider training regarding these advanced diagnostic modalities, need to be addressed.

References

1. Ghandour S, Ashkani-Esfahani S, Kwon JY. The emerging role of automation, measurement standardization, and artificial intelligence in foot and ankle imaging: an update. Foot Ankle Clin. 2023;28:667–80. https://doi.org/10.1016/j.fcl.2023.04.006.
2. Ashkani-Esfahani S, Nassour N, Hendriks JR, et al. Unmonitored patient demographic data changes can lead to bias in reported outcomes and data registry development. J Foot Ankle Surg. 2022;10:216–7. https://doi.org/10.5005/jp-journals-10040-1273.
3. Tariq A, Hussain HK. Evaluating the potential of artificial intelligence in orthopedic surgery for value-based healthcare. Int J Multidiscip Sci Arts. 2023;2:27–35. https://doi.org/10.47709/ijmdsa.v2i1.2394.
4. Guss D, Lucchese O, Ashkani-Esfahani S, et al. Automated volume measurement of the syndesmosis using 3D weightbearing CT. Foot Ankle Orthop. 2022;7:2473011421S00031. https://doi.org/10.1177/2473011421S00031.
5. Peiffer M, Van Den Borre I, Segers T, et al. Implementing automated 3D measurements to quantify reference values and side-to-side differences in the ankle syndesmosis. Sci Rep. 2023;13:13774. https://doi.org/10.1038/s41598-023-40599-3.
6. Lu K, Ma T, Yang C, et al. Risk prediction model for deep surgical site infection (DSSI) following open reduction and internal fixation of displaced intra-articular calcaneal fracture. Int Wound J. 2022;19:656–65. https://doi.org/10.1111/iwj.13663.
7. Nassour N, Tatara AM, Jones SV, et al. Prediction of surgical site infection following ankle fracture: a machine learning approach. Open Forum Infect Dis. 2023;10:1149. https://doi.org/10.1093/ofid/ofad500.1149.
8. Nassour N, Akhbari B, Ranganathan N, et al. Using machine learning in the prediction of symptomatic venous thromboembolism following ankle fracture. Foot Ankle Surg. 2023;30(2):110–6. https://doi.org/10.1016/j.fas.2023.10.003.
9. Ashkani-Esfahani S, Mojahed Yazdi R, Bhimani R, et al. Detection of ankle fractures using deep learning algorithms. Foot Ankle Surg. 2022;28:1259–65. https://doi.org/10.1016/j.fas.2022.05.005.
10. Ashkani-Esfahani S, Mojahed-Yazdi R, Bhimani R, et al. Deep learning algorithms improve the detection of subtle lisfranc malalignments on weightbearing radiographs. Foot Ankle Int. 2022;43:1118–26. https://doi.org/10.1177/10711007221093574.
11. Borjali A, Ashkani-Esfahani S, Bhimani R, et al. The use of deep learning enables high diagnostic accuracy in detecting syndesmotic instability on weight-bearing CT scanning. Knee Surg Sports Traumatol Arthrosc. 2023;31(12):6039–45. https://doi.org/10.1007/s00167-023-07565-y.
12. Lawlor MC, Kluczynski MA, Marzo JM. Weight-bearing cone-beam CT scan assessment of stability of supination external rotation ankle fractures in a cadaver model. Foot Ankle Int. 2018;39:850–7. https://doi.org/10.1177/1071100718761035.

13. Bhimani R, Ashkani-Esfahani S, Lubberts B, et al. Utility of volumetric measurement via weight-bearing computed tomography scan to diagnose syndesmotic instability. Foot Ankle Int. 2020;41:859–65. https://doi.org/10.1177/1071100720917682.
14. Bhimani R, Ashkani-Esfahani S, Lubberts B, et al. Utility of WBCT to diagnose syndesmotic instability in patients with Weber B lateral malleolar fractures. J Am Acad Orthop Surg. 2022;30:e423–33. https://doi.org/10.5435/JAAOS-D-21-00566.
15. Bhimani R, Sornsakrin P, Ashkani-Esfahani S, et al. Using area and volume measurement via weightbearing CT to detect Lisfranc instability. J Orthop Res. 2021;39:2497–505. https://doi.org/10.1002/jor.24970.
16. Sangoi D, Ranjit S, Bernasconi A, et al. 2D manual vs 3D automated assessment of alignment in normal and charcot-marie-tooth cavovarus feet using weightbearing CT. Foot Ankle Int. 2022;43:973–82. https://doi.org/10.1177/10711007221084308.
17. Bernasconi A, Cooper L, Lyle S, et al. Pes cavovarus in Charcot-Marie-Tooth compared to the idiopathic cavovarus foot: a preliminary weightbearing CT analysis. Foot Ankle Surg. 2021;27:186–95. https://doi.org/10.1016/j.fas.2020.04.004.
18. de Cesar Netto C, Shakoor D, Roberts L, et al. Hindfoot alignment of adult acquired flatfoot deformity: a comparison of clinical assessment and weightbearing cone beam CT examinations. Foot Ankle Surg. 2019;25:790–7. https://doi.org/10.1016/j.fas.2018.10.008.
19. Kwolek K, Gądek A, Kwolek K, et al. Automated decision support for Hallux Valgus treatment options using anteroposterior foot radiographs. World J Orthop. 2023;14:800–12. https://doi.org/10.5312/wjo.v14.i11.800.
20. Ortolani M, Leardini A, Pavani C, et al. Angular and linear measurements of adult flexible flatfoot via weight-bearing CT scans and 3D bone reconstruction tools. Sci Rep. 2021;11:16139. https://doi.org/10.1038/s41598-021-95708-x.
21. Alexander NB, Bettin CC, Sarfani S, et al. Cost analysis and utilization of weight bearing CT. Foot Ankle Orthop. 2022;7:2473011421S00075. https://doi.org/10.1177/2473011421S00075.
22. Alexander NB, Sarfani S, Strickland CD, et al. Cost analysis and reimbursement of weightbearing computed tomography. Foot Ankle Orthop. 2023;8:24730114231164143. https://doi.org/10.1177/24730114231164143.
23. Ashkani-Esfahani S, Lucchese O, Bhimani R, et al. Automation improves the efficiency of weightbearing CT scan 3D volumetric assessments of the syndesmosis. Foot Ankle Surg. 2024 May 22:S1268-7731(24)00112-7. https://doi.org/10.1016/j.fas.2024.05.010.
24. Bhimani R, Searle OF, Ashkani-Esfahani S, et al. Cost-effectiveness of weight-bearing computed tomography in diagnosing syndesmotic instability. Foot Ankle Orthop. 2022;7:2473011421S00115. https://doi.org/10.1177/2473011421S00115.
25. Mettler FA, Huda W, Yoshizumi TT, Mahesh M. Effective doses in radiology and diagnostic nuclear medicine: a catalog. Radiology. 2008;248:254–63. https://doi.org/10.1148/radiol.2481071451.
26. Sensiba PR, Coffey MJ, Williams NE, et al. Inter- and intraobserver reliability in the radiographic evaluation of adult flatfoot deformity. Foot Ankle Int. 2010;31:141–5. https://doi.org/10.3113/FAI.2010.0141.
27. Schermann H, Ashkani-Esfahani S, Zhao J, et al. Use of a lightweight portable fluoroscopy device for obtaining weightbearing ankle images. J Foot Ankle Surg. 2023;62:102–6. https://doi.org/10.1053/j.jfas.2022.05.006.
28. Reach JS, Easley ME, Chuckpaiwong B, Nunley JA. Accuracy of ultrasound guided injections in the foot and ankle. Foot Ankle Int. 2009;30:239–42. https://doi.org/10.3113/FAI.2009.0239.
29. Saengsin J, Bhimani R, Sato G, et al. Use of portable ultrasonography for the diagnosis of lateral ankle instability. J Orthop Res. 2022;40:2421–9. https://doi.org/10.1002/jor.25256.
30. Crombach A, Azizi N, Lameijer H, et al. Point-of-care bedside ultrasound examination for the exclusion of clinically significant ankle and fifth metatarsal bone fractures; a single blinded prospective diagnostic cohort study. J Foot Ankle Res. 2020;13:19. https://doi.org/10.1186/s13047-020-00387-y.

31. Nadler A, Tsung JW, Rabiner JE. Point-of-care ultrasonography for ankle injuries in children. Pediatr Emerg Care. 2022;38:e17–22. https://doi.org/10.1097/PEC.0000000000002594.
32. Atta M, Jafari S, Moore K. Analyzing the use of ultrasound: achilles tendon rupture. Open J Emerg Med. 2019;7:41–7. https://doi.org/10.4236/ojem.2019.73005.
33. Langer PR. Two emerging technologies for Achilles tendinopathy and plantar fasciopathy. Clin Podiatr Med Surg. 2015;32:183–93. https://doi.org/10.1016/j.cpm.2014.11.002.
34. Sabir N, Demirlenk S, Yagci B, et al. Clinical utility of sonography in diagnosing plantar fasciitis. J Ultrasound Med. 2005;24:1041–8. https://doi.org/10.7863/jum.2005.24.8.1041.
35. Sanchez PJ, Grady JF, Saxena A. Percutaneous ultrasonic tenotomy for achilles tendinopathy is a surgical procedure with similar complications. J Foot Ankle Surg. 2017;56:982–4. https://doi.org/10.1053/j.jfas.2017.06.015.
36. Smith J, Finnoff JT. Diagnostic and interventional musculoskeletal ultrasound: part 1. Fundamentals. PM R. 2009;1:64–75. https://doi.org/10.1016/j.pmrj.2008.09.001.
37. Smith WB, Melton W, Davies J. Midsubstance tendinopathy, percutaneous techniques (platelet-rich plasma, extracorporeal shock wave therapy, prolotherapy, radiofrequency ablation). Clin Podiatr Med Surg. 2017;34:161–74. https://doi.org/10.1016/j.cpm.2016.10.005.
38. Shen-Wagner J, Deutchman M. Point-of-care ultrasound: a practical guide for primary care. Fam Pract Manag. 2020;27:33–40.
39. Burt J, Smith V, Gee CW, et al. The role of outpatient needle arthroscopy in the diagnosis and management of musculoskeletal complaints: a systematic review of the Arthrex NanoScope. Knee. 2023;42:246–57. https://doi.org/10.1016/j.knee.2023.04.003.
40. Butler JJ, Brash AI, Azam MT, et al. The role of needle arthroscopy in the assessment and treatment of ankle sprains. Foot Ankle Clin. 2023;28:345–54. https://doi.org/10.1016/j.fcl.2023.01.005.
41. Duenes ML, Azam MT, Butler JJ, et al. In-office needle arthroscopy for the foot and ankle. Arthroscopy. 2023;39:1129–30. https://doi.org/10.1016/j.arthro.2023.01.005.
42. Walinga AB, Dahmen J, Stornebrink T, Kerkhoffs GM. Needle arthroscopic inspection and treatment of (osteo)chondral lesions of the ankle in unstable syndesmotic injuries treated with suture button fixation: a standardized approach. Arthrosc Tech. 2023;12:e1121–6. https://doi.org/10.1016/j.eats.2023.02.050.
43. Bonmarin M, Läuchli S, Navarini A. Augmented and virtual reality in dermatology—where do we stand and what comes next? Dermato. 2022;2:1–7. https://doi.org/10.3390/dermato2010001.
44. Chidambaram S, Palumbo MC, Stifano V, et al. The potential for using extended reality technology in interdisciplinary case discussions and case planning in stereotactic radiosurgery: proof-of-concept usability study. JMIR Neurotechnol. 2022;1:e36960. https://doi.org/10.2196/36960.
45. Douglas DB, Venets D, Wilke C, et al. Augmented reality and virtual reality: initial successes in diagnostic radiology. In: State of the art virtual reality and augmented reality knowhow. London: IntechOpen; 2018.
46. Ong CW, Tan MCJ, Lam M, Koh VTC. Applications of extended reality in ophthalmology: systematic review. J Med Internet Res. 2021;23:e24152. https://doi.org/10.2196/24152.
47. Verhey JT, Haglin JM, Verhey EM, Hartigan DE. Virtual, augmented, and mixed reality applications in orthopedic surgery. Int J Med Robot. 2020;16:e2067. https://doi.org/10.1002/rcs.2067.
48. Atilla OD, Yesilaras M, Kilic TY, et al. The accuracy of bedside ultrasonography as a diagnostic tool for fractures in the ankle and foot. Acad Emerg Med. 2014;21:1058–61. https://doi.org/10.1111/acem.12467.
49. McKnight RR, Pean CA, Buck JS, et al. Virtual reality and augmented reality-translating surgical training into surgical technique. Curr Rev Musculoskelet Med. 2020;13:663–74. https://doi.org/10.1007/s12178-020-09667-3.
50. Schlachetzki JCM, Barth J, Marxreiter F, et al. Wearable sensors objectively measure gait parameters in Parkinson's disease. PLoS One. 2017;12:e0183989. https://doi.org/10.1371/journal.pone.0183989.

51. Tao W, Liu T, Zheng R, Feng H. Gait analysis using wearable sensors. Sensors. 2012;12:2255–83. https://doi.org/10.3390/s120202255.
52. Kobsar D, Masood Z, Khan H, et al. Wearable inertial sensors for gait analysis in adults with osteoarthritis-a scoping review. Sensors. 2020;20:7143. https://doi.org/10.3390/s20247143.
53. Rivera B, Cano C, Luis I, Elias DA. A 3D-printed knee wearable goniometer with a mobile-app interface for measuring range of motion and monitoring activities. Sensors. 2022;22:763. https://doi.org/10.3390/s22030763.
54. Faragó P, Grama L, Farago M-A, Hintea S. A novel wearable foot and ankle monitoring system for the assessment of gait biomechanics. Appl Sci. 2021;11:268. https://doi.org/10.3390/app11010268.
55. Giraldo-Pedroza A, Lee WC-C, Lam W-K, et al. Effects of wearable devices with biofeedback on biomechanical performance of running-a systematic review. Sensors. 2020;20:6637. https://doi.org/10.3390/s20226637.
56. Martinez-Hernandez U, Rubio-Solis A, Cedeno-Campos V, Dehghani-Sanij AA. Towards an intelligent wearable ankle robot for assistance to foot drop. In: 2019 IEEE international conference on systems, man and cybernetics (SMC). IEEE: Piscataway; 2019. p. 3410–5.
57. Singh S, Lee T, Giles J. Development of a biofeedback wearable sensor system to improve rehabilitation following surgical repair of achilles tendon tears. In: EPiC Series in Health Sciences. Stockport: EasyChair; 2019. p. 334–7.
58. Tavakkoli Oskouei S, Malliaras P, Hill KD, et al. Evaluating daily physical activity and biomechanical measures using wearable technology in people with Achilles tendinopathy: a descriptive exploratory study. Musculoskelet Sci Pract. 2022;58:102534. https://doi.org/10.1016/j.msksp.2022.102534.
59. Wang X, Qiu J, Fong D. The applications of wearable devices in the rehabilitation of ankle injuries: a systematic review and meta-analysis. Med Novel Technol Devices. 2023;17:100210. https://doi.org/10.1016/j.medntd.2023.100210.
60. Davarzani S, Saucier D, Talegaonkar P, et al. Closing the wearable gap: foot–ankle kinematic modeling via deep learning models based on a smart sock wearable. Wearable Technol. 2023;4:e4. https://doi.org/10.1017/wtc.2023.3.
61. Huda M. Empowering application strategy in the technology adoption: insights from professional and ethical engagement. J Sci Technol Policy Manage. 2018;10:172–92. https://doi.org/10.1108/JSTPM-09-2017-0044.
62. Stauffer TP, Kim BI, Grant C, et al. Robotic technology in foot and ankle surgery: a comprehensive review. Sensors. 2023;23:686. https://doi.org/10.3390/s23020686.
63. Stojmanovski Mercieca LA, Formosa C, Chockalingam N. A scoping review of foot and ankle telemedicine guidelines. Health Sci Rep. 2023;6:e1076. https://doi.org/10.1002/hsr2.1076.
64. Foot and Ankle Surgery – FIXUS AI. 2023. https://fixus.mgh.harvard.edu/footandankle/. Accessed 4 December 2023.

GPSR Compliance

The European Union's (EU) General Product Safety Regulation (GPSR) is a set of rules that requires consumer products to be safe and our obligations to ensure this.

If you have any concerns about our products, you can contact us on ProductSafety@springernature.com

In case Publisher is established outside the EU, the EU authorized representative is:

Springer Nature Customer Service Center GmbH
Europaplatz 3
69115 Heidelberg, Germany

Batch number: 10726133

Printed by Printforce, the Netherlands